W9-AQL-837

WIND POWER IN AMERICA'S FUTURE

20% WIND ENERGY BY 2030

WIND POWER IN AMERICA'S FUTURE

20% WIND ENERGY BY 2030

U.S. DEPARTMENT OF ENERGY

DOVER PUBLICATIONS, INC.
MINEOLA, NEW YORK

Planet Friendly Publishing
✓ Made in the United States
✓ Printed on Recycled Paper
Text: 10% Cover: 10%
Learn more: www.greenedition.org

At Dover Publications we're committed to producing books in an earth-friendly manner and to helping our customers make greener choices.

Manufacturing books in the United States ensures compliance with strict environmental laws and eliminates the need for international freight shipping, a major contributor to global air pollution.

And printing on recycled paper helps minimize our consumption of trees, water and fossil fuels. The text of *Wind Power in America's Future: 20% Wind Energy by 2030* was printed on paper made with 10% post-consumer waste, and the cover was printed on paper made with 10% post-consumer waste. According to Environmental Defense's Paper Calculator, by using this innovative paper instead of conventional papers, we achieved the following environmental benefits:

Trees Saved: 15 • Air Emissions Eliminated: 1,318 pounds
Water Saved: 6,349 gallons • Solid Waste Eliminated: 385 pounds

For more information on our environmental practices, please visit us online at www.doverpublications.com/green

Bibliographical Note

This Dover edition, first published in 2010, is an unabridged republication of *20% Wind Energy by 2030: Increasing Wind Energy's Contribution to U.S. Electricity Supply*, a report issued by the U.S. Department of Energy, Washington, D.C., in July 2008.

The report is available electronically at http://www.osti.gov/bridge.

Library of Congress Cataloging-in-Publication Data

Wind power in America's future : 20% wind energy by 2030 / U.S. Department of Energy.
p. cm.
"Unabridged republication of 20% Wind Energy by 2030: Increasing Wind Energy's Contribution to U.S. Electricity Supply, a report issued by the U.S. Department of Energy, Washington, D.C., in July 2008."
Includes bibliographical references and index.
ISBN-13: 978-0-486-47500-4
ISBN-10: 0-486-47500-X
1. Wind power. 2. Electric power-plants. I. United States. Dept. of Energy.

TK1541.W57 2010
333.9'20973—dc22

2009029323

Manufactured in the United States by Courier Corporation
47500X01
www.doverpublications.com

GRATEFUL APPRECIATION TO PARTNERS

The U.S. Department of Energy would like to acknowledge the in-depth analysis and extensive research conducted by the National Renewable Energy Laboratory and the major contributions and manuscript reviews by the American Wind Energy Association and many wind industry organizations that contributed to the production of this report. The costs curves for energy supply options and the WinDS modeling assumptions were developed in cooperation with Black & Veatch. The preparation of this technical report was coordinated by Energetics Incorporated of Washington, DC and Renewable Energy Consulting Services, Inc. of Palo Alto, CA. All authors and reviewers who contributed to the preparation of the report are listed in Appendix D.

Table of Contents

List of Figures

List of Tables

Abbreviations and Acronyms

ACE	area control error
AEO	*Annual Energy Outlook*
AEP	American Electric Power
AGATE	Advanced General Aviation Transport Experiments
AGC	automatic generation control
ALA	American Lung Association
AMA	American Medical Association
API	American Petroleum Institute
APPA	American Public Power Association
ATTU	*Annual Turbine Technology Update*
AWEA	American Wind Energy Association
AWST	AWS Truewind
BACI	before-and-after-control impact
Berkeley Lab	Lawrence Berkeley National Laboratory
BLM	Bureau of Land Management
BPA	Bonneville Power Administration
BSH	Bundesamt für Seeschiffahrt und Hydrographie
BTM	BTM Consult ApS
Btu	British thermal unit
BWEC	Bat and Wind Energy Cooperative
CAA	Clean Air Act
CAIR	Clean Air Interstate Rule
CAISO	California Independent System Operator
CAMR	Clean Air Mercury Rule
CapX 2020	Capacity Expansion Plan for 2020
CBO	Congressional Budget Office
CDEAC	Clean and Diversified Energy Advisory Committee
CEC	California Energy Commission
CEQA	California Environmental Quality Act
CESA	Clean Energy States Alliance
CF	capacity factor
CFRP	carbon filament-reinforced plastic
CNV	California/Nevada
CO_2	carbon dioxide
Coal-IGCC	integrated gasification combined cycle coal plants
Coal-new	new pulverized coal plants
COD	commercial operation date
COE	cost of energy
CREZ	Competitive Renewable Energy Zones
CT	combustion turbine
dB	decibels
DEA	Danish Energy Authority
DEIS	draft environmental impact statement
DOD	U.S. Department of Defense
DOE	U.S. Department of Energy
DOI	U.S. Department of Interior
DWT	distributed wind technology

ECAR	East Central Area Reliability Coordinating Agreement
EEI	Edison Electric Institute
EERE	Office of Energy Efficiency and Renewable Energy
EFTA	European Free Trade Agreement
EIA	Energy Information Administration
EIR	environmental impact review
EIS	environmental impact statement
ELCC	effective load-carrying capability
EPA	U.S. Environmental Protection Agency
EPAct	Energy Policy Act
EPC	engineering, procurement, and construction
EPRI	Electric Power Research Institute
ERCOT	Electric Reliability Council of Texas
ERO	Electric Reliability Organization
EU	European Union
EUI	Energy Unlimited Inc.
EWEA	European Wind Energy Association
FAA	Federal Aviation Administration
FACTS	flexible AC transmission system
FEIR	final environmental impact report
FERC	Federal Energy Regulatory Commission
FL	Florida
FRCC	Florida Reliability Coordinating Council
FTE	full-time equivalent
GaAs	gallium arsenide
Gas-CC	combined cycle natural gas plants
Gas-CT	gas combustion turbine
GE	General Electric International
GHG	greenhouse gas
GIS	geographic information system
GRP	glass fiber-reinforced-plastic
GS3C	Grassland/Shrub-Steppe Species Collaborative
GVW	gross vehicle weight
GW	gigawatt
GWh	gigawatt-hour
Hg	mercury
HSIL	high-surge impedance-loading (transmission line)
HVDC	high-voltage direct current
Hz	hertz
IEA	International Energy Agency
IEC	International Electrotechnical Commission
IEEE	Institute of Electrical and Electronics Engineers
IGCC	integrated gasification combined cycle
IOU	investor-owned utility
IPCC	Intergovernmental Panel on Climate Change
IRP	integrated resource planning
ISET	Institute for Solar Energy Technology (Institut für Solare Energieversorgungstechnik)
ISO	independent system operator

ISO-NE	ISO New England
ITC	investment tax credit
JEDI	Jobs and Economic Development Impact (model)
kg	kilogram
km^2	square kilometers
kV	kilovolt
kW	kilowatt
kWh	kilowatt-hour
lb	pound
LC	levelized cost
LDC	load duration curve
LIDAR	light detection and ranging
LLC	Limited Liability Company
LNG	liquefied natural gas
LOLP	loss of load probability
m	meter
m^2	square meter
MAAC	Mid-Atlantic Area Council
MACRS	Modified Accelerated Cost Recovery System
MAIN	Mid-American Interconnected Network
MAPP	Mid-Continent Area Power Pool
Midwest ISO	Midwest Independent System Operator
MMBtu	million British thermal units
MMS	Minerals Management Service
MMTCE	million metric tons of carbon equivalent
MNDOC	Minnesota Department of Commerce
MOU	Memorandum of Understanding
MRO	Midwest Reliability Organization
MTEP	MISO Transmission Expansion Plan
MVA	megavolt amperes
MW	megawatt
MWh	megawatt-hour
MW-mile	megawatt-mile
NAICS	North American Industrial Classification System
NAS	National Academy of Sciences
NCAR	National Center for Atmospheric Research
NCEP	National Commission on Energy Policy
NE	New England
NEMS	National Energy Modeling System
NEPA	National Environmental Policy Act
NERC	North American Electric Reliability Corporation
NESCAUM	Northeast States for Coordinated Air Use Management
NGOs	nongovernmental organizations
nm	nautical mile
NOAA	National Oceanic and Atmospheric Administration
NOI	notice of intent
NOx	nitrogen oxides
NPCC	Northeast Power Coordinating Council
NPV	net present value

NRC	National Research Council
NRECA	National Rural Electric Cooperative Association
NREL	National Renewable Energy Laboratory
NSTC	National Science and Technology Council
NWCC	National Wind Coordinating Collaborative
NWF	National Wildlife Federation
NWS	National Weather Service
NY	New York
NYISO	New York Independent System Operator
NYSERDA	New York State Energy Research and Development Authority
O_3	ozone
O&M	operations and maintenance
OE	Office of Electricity Delivery and Energy Reliability
OCS	Outer Continental Shelf
OMB	Office of Management and Budget
PBF	Public Benefits Fund
PGE	Portland General Electric
PJM	Pennsylvania-New Jersey-Maryland Interconnection
PMA	Power Marketing Administration
PNM	Public Service Company of New Mexico
POI	point of interconnection
PPA	power purchase agreement
PSE	Puget Sound Energy
PTC	production tax credit
PUC	Public Utility Commission
PURPA	Public Utility Regulatory Policies Act
QF	qualifying or qualified facility
R&D	research and development
RMA	Rocky Mountain Area
RD&D	research, development & demonstration
REC	renewable energy credit
REPI	Renewable Energy Production Incentive
REPP	Renewable Energy Policy Project
RFC	ReliabilityFirst Corporation
RGGI	Regional Greenhouse Gas Initiative
RMATS	Rocky Mountain Area Transmission Study
RPS	Renewable Portfolio Standards
RTO	Regional Transmission Organization
s	second
Sandia	Sandia National Laboratories
SCADA	supervisory control and data acquisition
SEAC	Strategic Energy Analysis Center
SEPA	Southeastern Power Administration
SERC	Southeastern Electric Reliability Council
SF_6	sulfur hexafluoride (one of six greenhouse gases identified in the Kyoto Protocol)
SiC	silicon carbide
SO_2	sulfur dioxide
SODAR	sonic detection and ranging

SO_x	sulfur oxides
SPP	Southwest Power Pool
ST	steam turbine
Std. Dev.	standard deviation
SWPA	Southwestern Power Administration
TRE	Texas Regional Entity
TVA	Tennessee Valley Authority
TWh	terawatt-hours
UCTE	Union for the Co-ordination of Transmission of Electricity
UKERC	UK Energy Research Centre
USACE	U.S. Army Corps of Engineers
USCAP	U.S. Climate Action Partnership
USDA	U.S. Department of Agriculture
USFS	U.S. Department of Agriculture Forest Service
USFWS	U.S. Fish & Wildlife Service
USGS	U.S. Geological Survey
UWIG	Utility Wind Integration Group
V	volt
VAR	volt-ampere-reactive
W	watt
WEST	Western EcoSystems Technology
Western	Western Area Power Administration (formerly WAPA)
WCI	Western Climate Initiative
WECC	Western Electricity Coordinating Council
WGA	Western Governors' Association
Wh	watt-hour
WinDS	Wind Energy Deployment System Model
WindPACT	Wind Partnerships for Advanced Component Technology
WPA	Wind Powering America
WRA	Western Resource Advocates
WRCAI	Western Regional Climate Action Initiative
WWG	Wildlife Workgroup

1

Chapter 1. Executive Summary & Overview

1.1 Introduction and Collaborative Approach

Energy prices, supply uncertainties, and environmental concerns are driving the United States to rethink its energy mix and develop diverse sources of clean, renewable energy. The nation is working toward generating more energy from domestic resources—energy that can be cost-effective and replaced or "renewed" without contributing to climate change or major adverse environmental impacts.

In 2006, President Bush emphasized the nation's need for greater energy efficiency and a more diversified energy portfolio. This led to a collaborative effort to explore a modeled energy scenario in which wind provides 20% of U.S. electricity by 2030. Members of this 20% Wind collaborative (see 20% Wind Scenario sidebar) produced this report to start the discussion about issues, costs, and potential outcomes associated with the 20% Wind Scenario. A 20% Wind Scenario in 2030, while ambitious, could be feasible if the significant challenges identified in this report are overcome.

20% Wind Scenario: Wind Energy Provides 20% of U.S. Electricity Needs by 2030

Key Issues to Examine:

- Does the nation have sufficient wind energy resources?
- What are the wind technology requirements?
- Does sufficient manufacturing capability exist?
- What are some of the key impacts?
- Can the electric network accommodate 20% wind?
- What are the environmental impacts?
- Is the scenario feasible?

Assessment Participants:

- U.S. Department of Energy (DOE)
  - Office of Energy Efficiency and Renewable Energy (EERE), Office of Electricity Delivery and Energy Reliability (OE), and Power Marketing Administrations (PMAs)
  - National Renewable Energy Laboratory (NREL)
  - Lawrence Berkeley National Laboratory (Berkeley Lab)
  - Sandia National Laboratories (SNL)
- Black & Veatch engineering and consulting firm
- American Wind Energy Association (AWEA)
  - Leading wind manufacturers and suppliers
  - Developers and electric utilities
  - Others in the wind industry

This report was prepared by DOE in a joint effort with industry, government, and the nation's national laboratories (primarily the National Renewable Energy Laboratory and Lawrence Berkeley National Laboratory). The report considers some associated challenges, estimates the impacts, and discusses specific needs and outcomes in the areas of technology, manufacturing and employment, transmission and grid integration, markets, siting strategies, and potential environmental effects associated with a 20% Wind Scenario.

In its Annual Energy Outlook 2007, the U.S. Energy Information Administration (EIA) estimates that U.S. electricity demand will grow by 39% from 2005 to 2030, reaching 5.8 billion megawatt-hours (MWh) by 2030. To meet 20% of that demand, U.S. wind power capacity would have to reach more than 300 gigawatts (GW) or more than 300,000 megawatts (MW). This growth represents an increase of more than 290 GW within 23 years.[1]

The data analysis and model runs for this report were concluded in mid-2007. All data and information in the report are based on wind data available through the end of 2006. At that time, the U.S. wind power fleet numbered 11.6 GW and spanned 34 states. In 2007, 5,244 MW of new wind generation were installed.[2] With these additions, American wind plants are expected to generate an estimated 48 billion kilowatt-hours (kWh) of wind energy in 2008, more than 1% of U.S. electricity supply. This capacity addition of 5,244 MW in 2007 exceeds the more conservative growth trajectory developed for the 20% Wind Scenario of about 4,000 MW/year in 2007 and 2008. The wind industry is on track to grow to a size capable of installing 16,000 MW/year, consistent with the latter years in the 20% Wind Scenario, more quickly than the trajectory used for this analysis.

1.1.1 SCOPE

This report examines some of the costs, challenges, and key impacts of generating 20% of the nation's electricity from wind energy in 2030. Specifically, it investigates requirements and outcomes in the areas of technology, manufacturing, transmission and integration, markets, environment, and siting.

The modeling done for this report estimates that wind power installations with capacities of more than 300 gigawatts (GW) would be needed for the 20% Wind Scenario. Increasing U.S. wind power to this level from 11.6 GW in 2006 would require significant changes in transmission, manufacturing, and markets. This report presents an analysis of one specific scenario for reaching the 20% level and contrasts it to a scenario of no wind growth beyond the level reached in 2006. Major assumptions in the analysis have been highlighted throughout the document and have been summarized in the appendices. These assumptions may be considered optimistic. In this report, no sensitivity analyses have been done to estimate the impact that changes in the assumptions would have on the information presented here. As summarized at the end of this chapter, the analysis provides an overview of some potential impacts of these two scenarios by 2030. This report does not compare the Wind Scenario to other energy portfolio options, nor does it outline an action plan.

To successfully address energy security and environmental issues, the nation needs to pursue a portfolio of energy options. None of these options by itself can fully address these issues; there is no "silver bullet." This technical report examines one potential scenario in which wind power serves as a significant element in the portfolio. However, the 20% Wind Scenario is not a prediction of the future. Instead, it paints a picture of what a particular 20% Wind Scenario could mean for the nation.

---

[1] AEO data from 2007 were used in this report. AEO released new data in March of 2008, which were not incorporated into this report. While the new EIA data could change specific numbers in the report, it would not change the overall message of the report.

[2] According to AWEA's 2007 Market Report of January 2008, the U.S. wind energy industry installed 5,244 MW in 2007, expanding the nation's total wind power generating capacity by 45% in a single calendar year and more than doubling the 2006 installation of 2,454 MW. Government sources for validation of 2007 installations were not available at the time this report was written.

1.1.2 CONTRIBUTORS

Report contributors include a broad cross section of key stakeholders, including leaders from the nation's utility sector, environmental communities, wildlife advocacy groups, energy industries, the government and policy sectors, investors, and public and private businesses. In all, the report reflects input from more than 50 key energy stakeholder organizations and corporations. Appendix D contains a list of contributors. Research and modeling was conducted by experts within the electric industry, government, and other organizations.

This report is not an authoritative expression of policy perspectives or opinions held by representatives of DOE.

1.1.3 ASSUMPTIONS AND PROCESS

To establish the groundwork for this report, the engineering company Black & Veatch (Overland Park, Kansas) analyzed the market potential for significant wind energy growth, quantified the potential U.S. wind supply, and developed cost supply curves for the wind resource. In consultation with DOE, NREL, AWEA, and wind industry partners, future wind energy cost and performance projections were developed. Similar projections for conventional generation technologies were developed based on Black & Veatch experience with power plant design and construction (Black & Veatch 2007).

To identify a range of challenges, possible solutions, and key impacts of providing 20% of the nation's electricity from wind, the stakeholders in the 20% Wind Scenario effort convened expert task forces to examine specific areas

Wind Energy Deployment System Model Assumptions (See Appendices A and B)

- The assumptions used for the WinDS model were obtained from a number of sources, including technical experts (see Appendix D), the WinDS base case (Denholm and Short 2006), AEO 2007 (EIA 2007), and a study performed by Black & Veatch (2007). These assumptions include projections of future costs and performance for all generation technologies, transmission system expansion costs, wind resources as a function of geographic location within the continental United States, and projected growth rates for wind generation.
- Wind energy generation is prescribed annually on a national level in order to reach 20% wind energy by 2030:
  - A stable policy environment supports accelerated wind deployment.
  - Balance of generation is economically optimized with no policy changes from those in place today (e.g., no production tax credit [PTC] beyond 12/31/08).
  - Technology cost and performance assumptions as well as electric grid expansion and operation assumptions that affect the direct electric system cost.
- Land-based and offshore wind energy technology cost reductions and performance improvements are expected by 2030 (see tables A-1, B-10, and B-11). Assumes that capital costs would be reduced by 10% over the next two decades and capacity factors would be increased by about 15% (corresponding to a 15% increase in annual energy generation by a wind plant)
- Future environmental study and permit requirements do not add significant costs to wind technology.
- Fossil fuel technology costs and performance are generally flat between 2005 and 2030 (see tables A-1 and B-13).
- Nuclear technology cost reductions are expected by 2030 (see tables A-1 and B-13).
- Reserve and capacity margins are calculated at the North American Electric Reliability Corporation (NERC) region level, and new transmission capacity is added as needed (see sections A.2.2 and B.3).
- Wind resource as a function of geographic location from various sources (see Table B-8).
- Projected electricity demand, financing assumptions, and fuel prices are based on *Annual Energy Outlook* (EIA 2007; see sections B.1, B.2, and B.4.2).
- Cost of new transmission is generally split between the originating project, be it wind or conventional generation, and the ratepayers within the region.
- Ten percent of existing grid capacity is available for wind energy.
- Existing long-term power purchase agreements are not implemented in WinDS. The model assumes that local load is met by the generation technologies in a given region.
- Assumes that the contributions to U.S. electricity supplies from other renewable sources of energy would remain at 2006 levels in both scenarios.

critical to this endeavor: Technology and Applications, Manufacturing and Materials, Environmental and Siting Impacts, Electricity Markets, Transmission and Integration, and Supporting Analysis. These teams conducted in-depth analyses of potential impacts, using related studies and various analytic tools to examine the benefits and costs. (See Appendix D for the task force participants.)

NREL's Wind Deployment System (WinDS) model[3] was employed to create a scenario that paints a "picture" of this level of wind energy generation and evaluates some impacts associated with wind. Assumptions about the future of the U.S. electric generation and transmission sector were developed in consultation with the task forces and other parties. Some assumptions in this analysis could be considered optimistic. Examples of assumptions used in this analysis are listed in the "Wind Energy Deployment System Model Assumptions" text box and are presented in detail in Appendices A and B. For comparison, the modeling team contrasted the 20% Wind Scenario impacts to a reference case characterized by no growth in U.S. wind capacity or other renewable energy sources after 2006.

In the course of the 20% Wind Scenario process, two workshops were held to define and refine the work plan, present and discuss preliminary results, and obtain relevant input from key stakeholders external to the report preparation effort.

1.1.4 Report Structure

The 20% Wind Scenario in 2030 would require improved turbine technology to generate wind power, significant changes in transmission systems to deliver it through the electric grid, and large expanded markets to purchase and use it. In turn, these essential changes in the power generation and delivery process would involve supporting changes and capabilities in manufacturing, policy development, and environmental regulation. As shown in Figure 1-1, the chapters of this report address some of the requirements and impacts in each of these areas. Detailed discussions of the modeling process, assumptions, and results can be found in Appendices A through C.

Figure 1-1. Report chapters

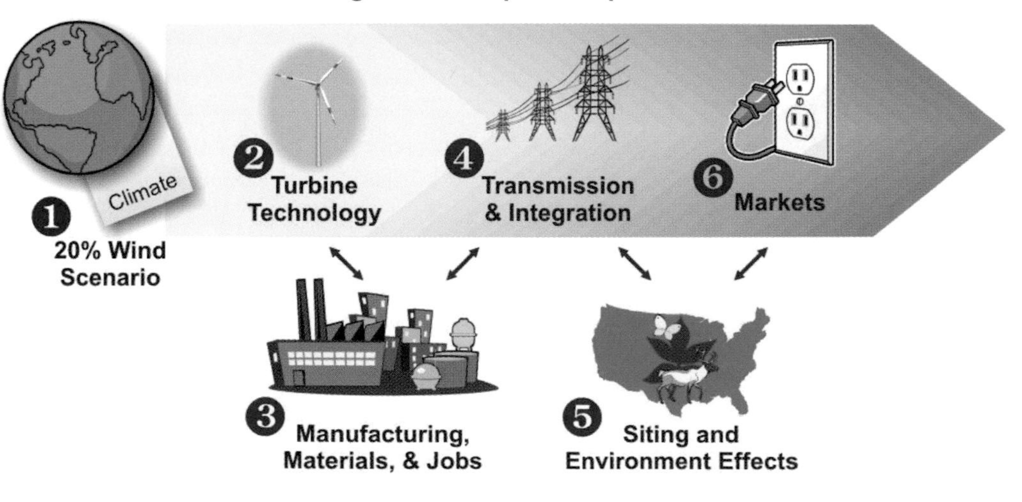

[3] The model, developed by NREL's Strategic Energy Analysis Center (SEAC), is designed to address the principal market issues related to the penetration of wind energy technologies into the electric sector. For additional information and documentation, see text box entitled "Wind Energy Deployment System Model Assumptions," Appendices A and B, and http://www.nrel.gov/analysis/winds/.

1.1.5 SETTING THE CONTEXT: TODAY'S U.S. WIND INDUSTRY

After experiencing strong growth in the mid-1980s, the U.S. wind industry hit a plateau during the electricity restructuring period in the 1990s and then regained momentum in 1999. Industry growth has since responded positively to policy incentives when they are in effect (see Figure 1-2). Today, the U.S. wind industry is growing rapidly, driven by sustained production tax credits (PTCs), rising concerns about climate change, and renewable portfolio standards (RPS) or goals in roughly 50% of the states.

Figure 1-2. Cumulative U.S. wind capacity, by year (in megawatts [MW])

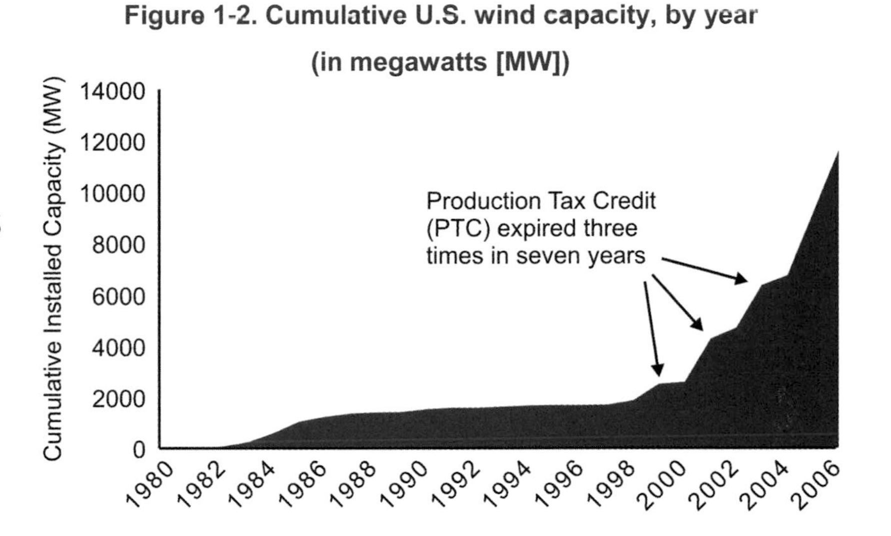

U.S. turbine technology has advanced steadily to offer improved performance, and these efforts are expected to continue (see "Initiatives to Improve Wind Turbine Performance" sidebar). In 2006 alone, average turbine size increased by more than 11% over the 2005 level to an average size of 1.6 MW. In addition, average capacity factors have improved 11% over the past two years. To meet the growing demand for wind energy, U.S. manufacturers have expanded their capacity to produce and assemble the essential components. Despite this growth, U.S. components continue to represent a relatively small share of total turbine and tower materials, and U.S. manufacturers are struggling to keep pace with rising demand (Wiser & Bolinger 2007).

Initiatives to Improve Wind Turbine Performance

Avoid problems before installation

- Improve reliability of turbines and components
- Full-scale testing prior to commercial introduction
- Development of appropriate design criteria, specifications, and standards
- Validation of design tools

Monitor performance

- Monitor and evaluate turbine and wind-plant performance
- Performance tracking by independent parties
- Early identification of problems

Rapid deployment of problem resolution

- Develop and communicate problem solutions
- Focused activities with stakeholders to address critical issues (e.g., Gearbox Reliability Collaborative)

In 2005 and 2006, the United States led the world in new wind installations. By early 2007, global wind power capacity exceeded 74 GW, and U.S. wind power capacity totaled 11.6 GW. This domestic wind power has been installed across 35 states and delivers roughly 0.8% of the electricity consumed in the nation (Wiser and Bolinger 2007).

A Brief History of the U.S. Wind Industry

The U.S. wind industry got its start in California during the 1970s, when the oil shortage increased the price of electricity generated from oil. The California wind industry benefited from federal and state ITCs as well as state-mandated standard utility contracts that guaranteed a satisfactory market price for wind power. By 1986, California had installed more than 1.2 GW of wind power, representing nearly 90% of global installations at that time.

Expiration of the federal ITC in 1985 and the California incentive in 1986 brought the growth of the U.S. wind energy industry to an abrupt halt in the mid-1980s. Europe took the lead in wind energy, propelled by aggressive renewable energy policies enacted between 1974 and 1985. As the global industry continued to grow into the 1990s, technological advances led to significant increases in turbine power and productivity. Turbines installed in 1998 had an average capacity 7 to 10 times greater than that of the 1980s turbines, and the price of wind-generated electricity dropped by nearly 80% (AWEA 2007). By 2000, Europe had more than 12,000 MW of installed wind power, versus only 2,500 MW in the United States, and Germany became the new international leader.

With low natural gas prices and U.S. utilities preoccupied by industry restructuring during the 1990s, the federal production tax credit (PTC) enacted in 1992 (as part of the Energy Policy Act [EPAct]) did little to foster new wind installations until just before its expiration in June 1999. Nearly 700 MW of new wind generation were installed in the last year before the credit expired—more than in any previous 12-month period since 1985. After the PTC expired in 1999, it was extended for two brief periods, ending in 2003. It was then reinstated in late 2004. Although this intermittent policy support led to sporadic growth, business inefficiencies inherent in serving this choppy market inhibited investment and restrained market growth.

Energy Policy Act of 1992

The PTC gave power producers 1.5 cents (increased annually with inflation) for every kilowatt-hour (kWh) of electricity produced from wind during the first 10 years of operation.

To promote renewable energy systems, many states began requiring electricity suppliers to obtain a small percentage of their supply from renewable energy sources, with percentages typically increasing over time. With Iowa and Texas leading the way, more than 20 states have followed suit with RPSs, creating an environment for stable growth.

After a decade of trailing Germany and Spain, the United States reestablished itself as the world leader in new wind energy in 2005. This resurgence is attributed to increasingly supportive policies, growing interest in renewable energy, and continued improvements in wind technology and performance. The United States retained its leadership of wind development in 2006 and, because of its very large wind resources, is likely to remain a major force in the highly competitive wind markets of the future.

1.2 SCENARIO DESCRIPTION

The 20% Wind Scenario presented here would require U.S. wind power capacity to grow from 11.6 GW in 2006 to more than 300 GW over the next 23 years (see Figure 1-3). This ambitious growth could be achieved in many different ways, with varying challenges, impacts, and levels of success. The 20% Wind Scenario would require an installation rate of 16 GW per year after 2018 (see Figure 1-4). This report examines one particular scenario for achieving this dramatic growth and contrasts it to another scenario that—for analytic simplicity—assumes no wind growth after 2006. The authors recognize that U.S. wind capacity is currently growing rapidly (although from a very small base) and that wind energy technology will be a part of any future electricity generation scenario for the United States. At the same time, a great deal of uncertainty remains about the level of contribution that wind could or is likely to make. In the 2007 *Annual Energy Outlook* (EIA 2007), an additional 7 GW beyond the 2006 installed capacity of 11.6 GW is forecast by 2030.[4] Other organizations are projecting higher capacity additions, and it would be difficult to develop a "most likely" forecast given today's uncertainties. The analysis presented here sidesteps these uncertainties and contrasts some of the challenges and impacts of producing 20% of the nation's electricity from wind with a scenario in which no additional wind is added after 2006. This results in an estimate, expressed in terms of parameters, of the impacts associated with increased reliance on wind energy generation under given assumptions. The analysis was also simplified by assuming that the contributions to U.S. electricity supplies from other renewable sources of energy would remain at 2006 levels in both scenarios (see Figure A-6 for resource mix).

Figure 1-3. Required growth in U.S. capacity (GW) to implement the 20% Wind Scenario

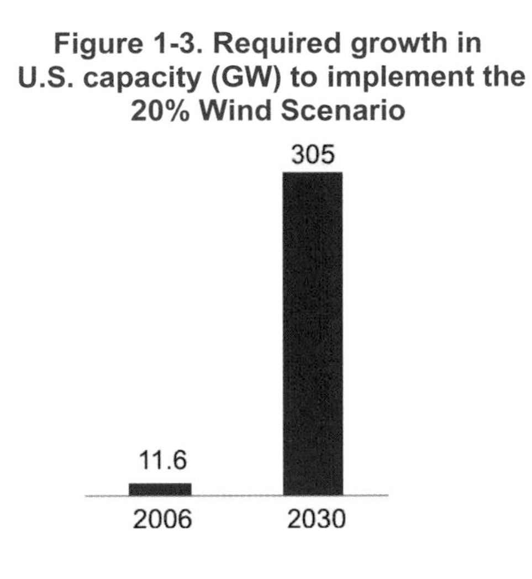

Figure 1-4. Annual and cumulative wind installations by 2030

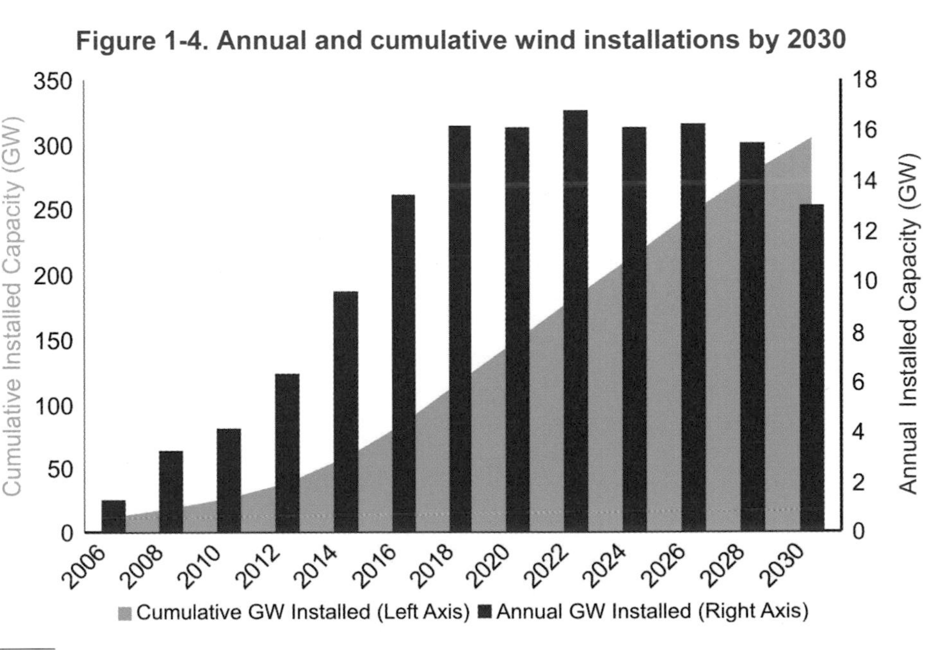

The 20% Wind Scenario has been carefully defined to provide a base of

[4] AEO data from 2007 were used in this report. AEO released new data in March 2008, which were not incorporated into this report. While new EIA data could change specific numbers in this report, it would not change the overall message of the report.

common assumptions for detailed analysis of all impact areas. Broadly stated, this 20% scenario is designed to consider incremental costs while recognizing realistic constraints and considerations (see the "Considerations in the 20% Wind Scenario" sidebar in Appendix A). Specifically, the scenario describes the mix of wind resources that would need to be captured, the geographic distribution of wind power installations, estimated land needs, the required utility and transmission infrastructure, manufacturing requirements, and the pace of growth that would be necessary.

1.2.1 Wind Geography

The United States possesses abundant wind resources. As shown in Figure 1-5, current "bus-bar" energy costs for wind (based on costs of the wind plant only, excluding transmission and integration costs and the PTC) vary by type of location (land-based or offshore) and by class of wind power density (higher classes offer greater productivity). Transmission and integration will add additional costs, which are discussed in Chapter 4. The nation has more than 8,000 GW of available land-based wind resources (Black & Veatch 2007) that industry estimates can be captured economically. NREL periodically classifies wind resources by wind speed, which forms the basis of the Black & Veatch study. See Appendix B for further details.

Electricity must be transmitted from where it is generated to areas of high electricity demand, using the existing transmission system or new transmission lines where necessary. As shown in Figure 1-6, the delivered cost of wind power increases when costs associated with connecting to the existing electric grid are included. The assumptions used in this report are different than EIA's assumptions and are documented in Appendices A and B. The cost and performance assumptions of the 20% Wind Scenario are based on real market data from 2007. Cost and performance for all technologies either decrease or remain flat over time. The data suggest that as

Figure 1-5. Supply curve for wind energy—current bus-bar energy costs

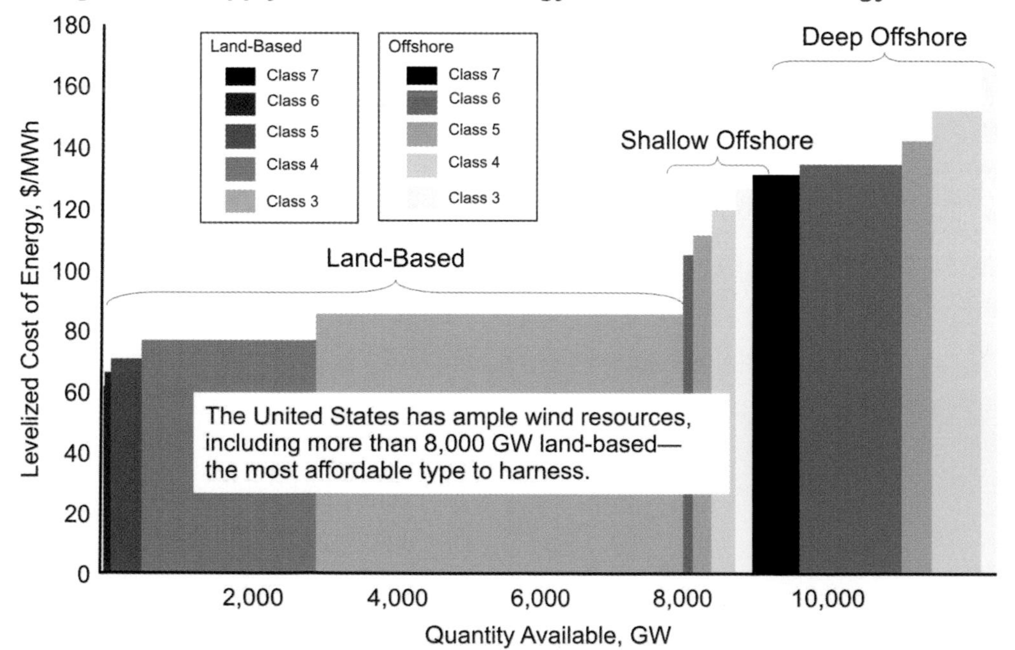

Note: See Appendix B for wind technology cost and performance projections; PTC and transmission and integration costs are excluded.

Figure 1-6. Supply curve for wind energy—energy costs including connection to 10% of existing transmission grid capacity

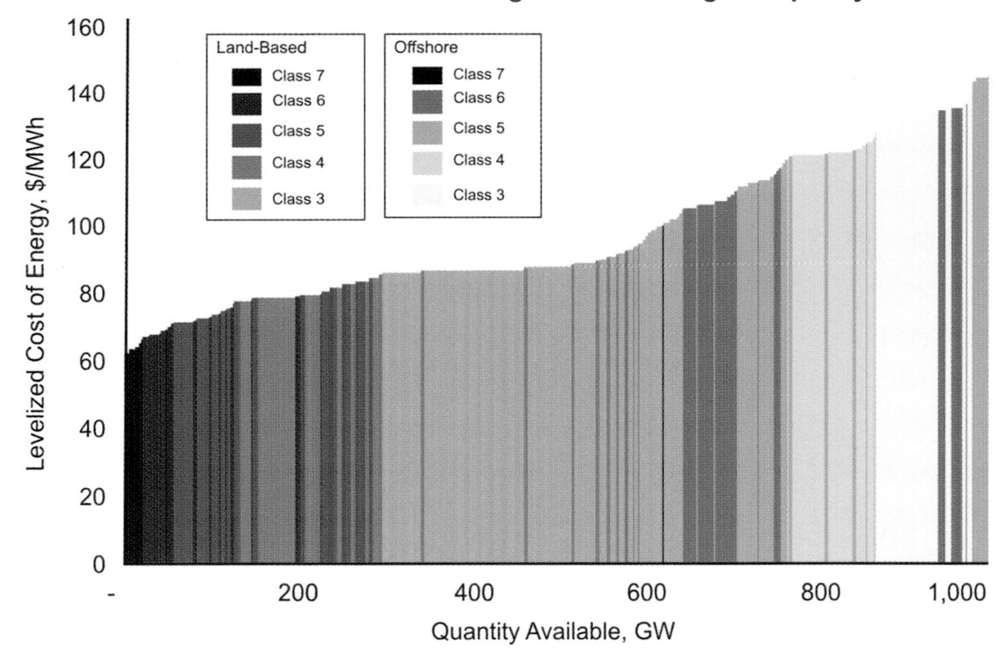

Note: See Appendix B for wind technology cost and performance projections. Excludes PTC, includes transmission costs to access existing electric transmission within 500 miles of wind resource.

much as 600 GW of wind resources could be available for $60 to $100 per megawatt-hour (MWh), including the cost of connecting to the existing transmission system. Including the PTC reduces the cost by about $20/MWh, and costs are further reduced if technology improvements in cost and performance are projected. In some cases, new transmission lines connecting high-wind resource areas to load centers could be cost-effective, and in other cases, high transmission costs could offset the advantage of land-based generation, as in the case of large demand centers along wind-rich coastlines.

NREL's WinDS model estimated the overall U.S. generation capacity expansion that is required to meet projected electricity demand growth through 2030. Both wind technology and conventional generation technology (i.e., coal, nuclear) were included in the modeling, but other renewables were not included. Readers should refer to Appendices A and B to see a more complete list of the modeling assumptions. Wind energy development for the 20% Wind Scenario optimized the total delivered costs, including future reductions in cost per kilowatt-hour for wind sites both near to and remote from demand sites from 2000 through 2030.[5] Chapter 2 presents additional discussion of wind technology potential. Of the 293 GW that would be added, the model specifies more than 50 GW of offshore wind energy (see Figure 1-7), mostly along the northeastern and southeastern seaboards.

[5] The modeling assumptions prescribed annual wind energy generation levels that reached 20% of projected demand by 2030 so as to demonstrate technical feasibility and quantify costs and impacts. Policy options that would help induce this growth trajectory were not included. It is assumed that a stable policy environment that recognizes wind's benefits could lead to growth rates that would result in the 20% Wind Scenario.

Figure 1-7. 20% cumulative installed wind power capacity required to produce 20% of projected electricity by 2030

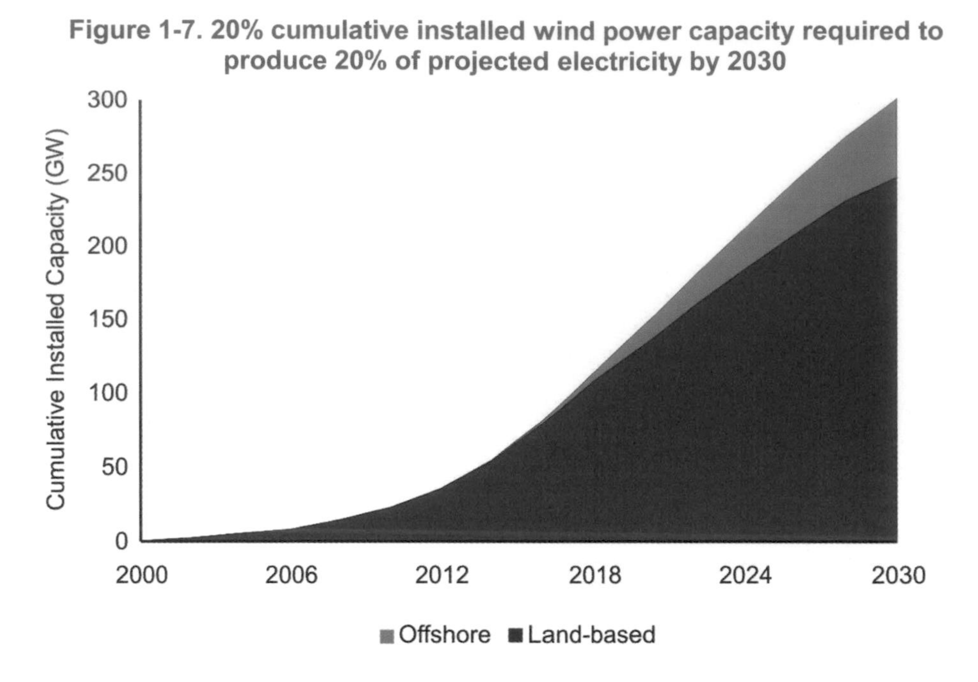

Based on this least-cost optimization algorithm (which incorporates future cost per kilowatt-hour of wind and cost of transmission), the WinDS model estimated the wind capacity needed by state by 2030. As shown in Figure 1-8, most states would have the opportunity to develop their wind resources. Total land requirements are extensive, but only about 2% to 5% of the total would be dedicated entirely to the wind installation. In addition, the visual impacts and other siting concerns of wind energy projects must be taken into account in assessing land requirements. Chapter 5 contains additional discussion of land use and visual impacts. Again, the 20% Wind Scenario presented here is not a prediction. Figure 1-8 simply shows one way in which a 20% wind future could evolve.

Figure 1-8. 46 states would have substantial wind development by 2030

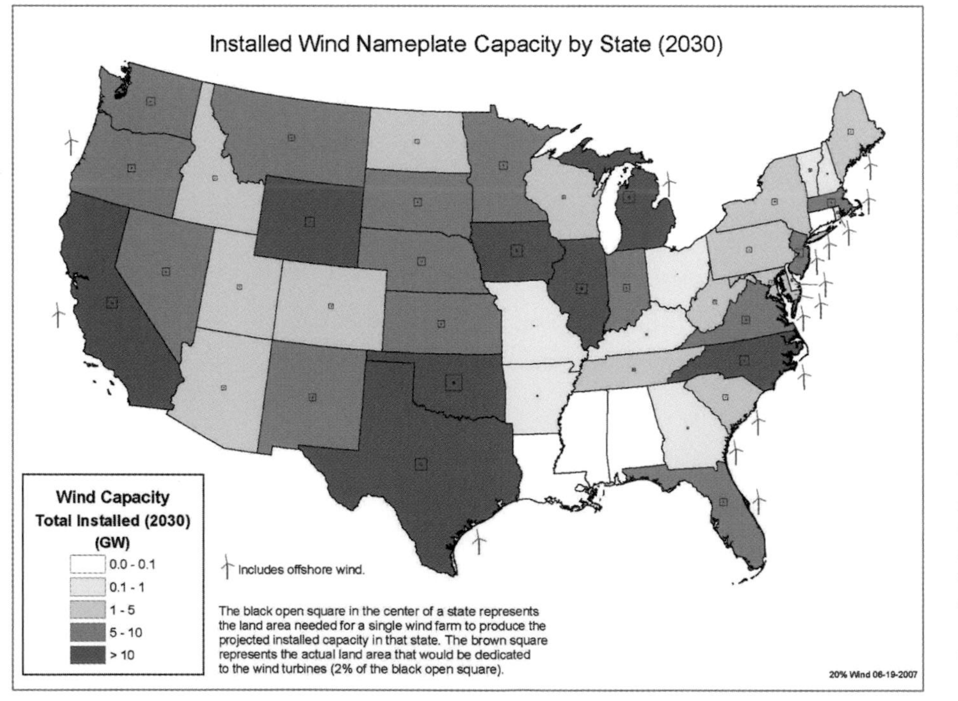

Land Requirements

Altogether, new land-based installations would require approximately 50,000 square kilometers (km^2) of land, yet the actual footprint of land-based turbines and related infrastructure would require only about 1,000 to 2,500 km^2 of dedicated land—slightly less than the area of Rhode Island.

The 20% Wind Scenario envisions 251 GW of land-based and 54 GW of shallow offshore wind capacity to optimize delivered costs, which include both generation and transmission.

Wind capacity levels in each state depend on a variety of assumptions and the national optimization of electricity generation expansion. Based on the perspectives of industry experts and near-term wind development plans, wind capacity in Ohio was modified and offshore wind development in Texas was included. In reality, each state's wind capacity level will vary significantly as electricity markets evolve and state policies promote or restrict the energy production of electricity from wind and other renewable and conventional energy sources.

1.2.2 Wind Power Transmission and Integration

Development of 293 GW of new wind capacity would require expanding the U.S. transmission grid in a manner that not only accesses the best wind resource regions of the country but also relieves current congestion on the grid, including new transmission lines to deliver wind power to electricity consumers. Figure 1-9 conceptually illustrates the optimized use of wind resources within the local areas as well as the transmission of wind-generated electricity from high-resource areas to high-demand centers. This data was generated by the WinDS model (given prescribed constraints). The figure does not represent proposals for specific transmission lines.

Figure 1-9. All new electricity generation including wind energy would require expansion of U.S. transmission by 2030

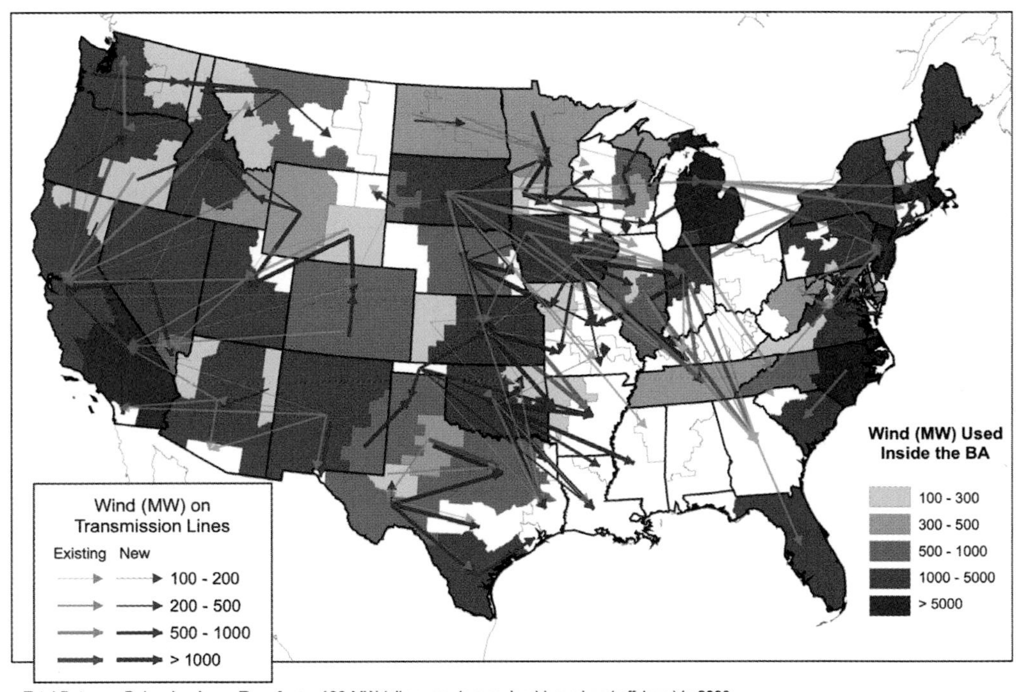

Total Between Balancing Areas Transfer >= 100 MW (all power classes, land-based and offshore) in 2030.
Wind power can be used locally within a Balancing Area (BA), represented by purple shading, or transferred out of the area on new or existing transmission lines, represented by red or blue arrows. Arrows originate and terminate at the centroid of the BA for visualization purposes; they do not represent physical locations of transmission lines.

Figure 1-10 displays transmission needs in the form of one technically feasible transmission grid as a 765 kV overlay. A complete discussion of transmission issues can be found in Chapter 4.

Until recently, concerns had been prevalent in the electric utility sector about the difficulty and cost of dealing with the variability and uncertainty of energy production from wind plants and other weather-driven renewable technologies. But utility engineers in some parts of the United States now have extensive experience with wind plant impacts, and their analyses of these impacts have helped to reduce these concerns. As discussed in detail in Chapter 4, wind's variability is being accommodated, and given optimistic assumptions, studies suggest the cost impact could be as little as the current level—10% or less of the value of the wind energy generated.

Figure 1-10. Conceptual transmission plan to accommodate 400 GW of wind energy (AEP 2007)

Wind Power Classification

Wind Power Class	Resource Potential	Wind Power Density at 50 m W/m²	Wind Speed* at 50 m m/s	Wind Speed* at 50 m mph
3	Fair	300 - 400	6.4 - 7.0	14.3 - 15.7
4	Good	400 - 500	7.0 - 7.5	15.7 - 16.8
5	Excellent	500 - 600	7.5 - 8.0	16.8 - 17.9
6	Outstanding	600 - 800	8.0 - 8.8	17.9 - 19.7
7	Superb	800 - 1600	8.8 - 11.1	19.7 - 24.8

* Wind speeds are based on a Weibull k value of 2.0

This map shows the wind resource data used by the WinDS model for the 20% Wind Scenario. It is a combination of high resolution and low resolution datasets produced by NREL and other organizations. The data was screened to eliminate areas unlikely to be developed onshore due to land use or environmental issues. In many states, the wind resource on this map is visually enhanced to better show the distribution on ridge crests and other features.

Transmission Lines
Voltage (kV)
234 - 499
500 - 699
700 - 799
1000 (DC)
Source: POWERmap, powermap.platts.com, ©2007 Platts, a division of the McGraw-Hill Companies

Conceptual 765 kV Network
Existing 765 kV
New 765 kV
AC-DC-AC Link
Source: American Electric Power (AEP)

1.2.3 ELECTRICAL ENERGY MIX

The U.S. Energy Information Administration (EIA) estimates that U.S. electricity demand will grow by 39% from 2005 to 2030, reaching 5.8 billion MWh by 2030. The 20% Wind Scenario would require delivery of nearly 1.16 billion MWh of wind energy in 2030, altering U.S. electricity generation as shown in Figure 1-11. In this scenario, wind would supply enough energy to displace about 50% of electric utility natural gas consumption and 18% of coal consumption by 2030. This amounts to an 11% reduction in natural gas across all industries. (Gas-fired generation would probably be displaced first, because it typically has a higher cost.)

Figure 1-11. U.S. electrical energy mix

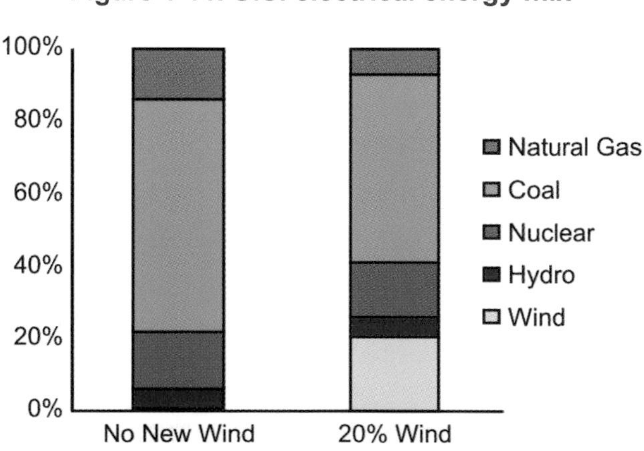

The increased wind development in this scenario could reduce the need for new coal and combined cycle natural gas capacity, but would increase the need for additional combustion turbine natural gas capacity to maintain electric system reliability. These units, though, would be run only as needed.[6]

1.2.4 PACE OF NEW WIND INSTALLATIONS

Manufacturing capacity would require time to ramp up enough to support rapid growth in new U.S. wind installations. The 20% Wind Scenario estimates that the installation rate would need to

[6] Appendix A presents a full analysis of changes in the capacity mix and energy generation under the 20% Wind Scenario.

increase from installing 3 GW per year in 2006 to more than 16 GW per year by 2018 and to continue at roughly that rate through 2030, as seen in Figure 1-4. This increase in installation rate, although quite large, is comparable to the recent annual installation rate of natural gas units, which totaled more than 16 GW in 2005 alone (EIA 2005).

The assumptions of the 20% Wind Scenario form the foundation for the technical analyses presented in the remaining chapters. This overview is provided as context for the potential impacts and technical challenges discussed in the next sections.

Wind vs. Traditional Electricity Generation

Wind power avoids several of the negative effects of traditional electricity generation from fossil fuels:

- Emissions of mercury or other heavy metals into the air
- Emissions associated with extracting and transporting fuels
- Lake and streambed acidification from acid rain or mining
- Water consumption associated with mining or electricity generation
- Production of toxic solid wastes, ash, or slurry
- Greenhouse gas (GHG) emissions

1.3 IMPACTS

The 20% Wind Scenario presented here offers potentially positive impacts in terms of greenhouse gas (GHG) reductions, water conservation, and energy security, as compared to the base case of no wind growth in this analysis. However, tapping this resource at this level would entail large front-end capital investments to install wind capacity and expanded transmission systems. The impacts described in this section are based largely on the analytical tools and methodology discussed in detail in Appendices A, B, and C.

Wind power would be a critical part of a broad and near-term strategy to substantially reduce air pollution, water pollution, and global climate change associated with traditional generation technologies (see “Wind vs. Traditional Electricity Generation” sidebar). As a domestic energy resource, wind power would also stabilize and diversify national energy supplies.

20% Wind Scenario: Projected Impacts

- **Environment:** Avoids air pollution and reduces GHG emissions; reduces electric sector CO_2 emissions by 825 million metric tons annually
- **Water savings:** Reduces cumulative water use in the electric sector by 8% (4 trillion gallons)
- **U.S. energy security:** Diversifies electricity portfolio and represents an indigenous energy source with stable prices not subject to fuel volatility
- **Energy consumers:** Potentially reduces demand for fossil fuels, in turn reducing fuel prices and stabilizing electricity rates
- **Local economics:** Creates new income source for rural landowners and tax revenues for local communities in wind development areas
- **American workers:** Generates well-paying jobs in sectors that support wind development, such as manufacturing, engineering, construction, transportation, and financial services; new manufacturing will cause significant growth in wind industry supply chain (see Appendix C)

1.3.1 GREENHOUSE GAS REDUCTIONS

Supplying 20% of U.S. electricity from wind could reduce annual electric sector carbon dioxide (CO_2) emissions by 825 million metric tons by 2030.

1

20% Wind Scenario: Major Challenges

- Investment in the nation's transmission system, so that the power generated is delivered to urban centers that need the increased supply;
- Larger electric load balancing areas, in tandem with better regional planning, so that regions can depend on a diversity of generation sources, including wind power;
- Continued reduction in wind capital costs and improvement in turbine performance through technology advancement and improved manufacturing capabilities; and
- Addressing potential concerns about local siting, wildlife, and environmental issues within the context of generating electricity.

The threat of climate change and the growing attention paid to it are helping to position wind power as an increasingly attractive option for new power generation. U.S. electricity demand is growing rapidly, and cleaner power sources (e.g., renewable energy) and energy-saving practices (i.e., energy efficiency) could help meet much of the new demand while reducing GHG emissions. Today, wind energy represents approximately 35% of new capacity additions (AWEA 2008). Greater use of wind energy, therefore, presents an opportunity for reducing emissions today as the nation develops additional clean power options for tomorrow.

Concerns about climate change have spurred many industries, policy makers, environmentalists, and utilities to call for reductions in GHG emissions. Although the cost of reducing emissions is uncertain, the most affordable near-term strategy likely involves wider deployment of currently available energy efficiency and clean energy technologies. Wind power is one of the potential supply-side solutions to the climate change problem (Socolow and Pacala 2006).

GHG Reduction

Under the 20% Wind Scenario, a cumulative total of 7,600 million metric tons of CO_2 emissions would be avoided by 2030, and more than 15,000 million metric tons of CO_2 emissions would be avoided through 2050.

Governments at many levels have enacted policies to actively support clean electricity generation, including the renewable energy PTC and state RPS. A growing number of energy and environmental organizations are calling for expanded wind and other renewable power deployment to try to reduce society's carbon footprint.

According to EIA, The United States annually emits approximately 6,000 million metric tons of CO_2. These emissions are expected to increase to nearly 7,900 million metric tons by 2030, with the electric power sector accounting for approximately 40% of the total (EIA 2007). As shown in Figure 1-12, based on the analysis completed for this report, generating 20% of U.S. electricity from wind could avoid approximately 825 million metric tons of CO_2 emissions in the electric sector in 2030. The 20% Wind Scenario would also reduce *cumulative* emissions from the electric sector through that same year by more than 7,600 million metric tons of CO_2 (2,100 million metric tons of carbon equivalent).[7] See Figures 1-12 and 1-13 . In general, CO_2 emission reductions are not only a wind energy benefit but could be achieved under other energy-mix scenarios.

The Fourth Assessment Report of the United Nations Environment Program and World Meteorological Organization's Intergovernmental Panel on Climate Change (IPCC) notes that "Renewable energy generally has a positive effect on energy

[7] CO_2 can be converted to carbon equivalent by multiplying by 12/44. Appendix A presents results in carbon equivalent, not CO_2. Because it assumes a higher share of coal-fired generation, the WinDS model projects higher CO_2 emissions than the EIA model.

Figure 1-12. Annual CO_2 emissions avoided (vertical bars) would reach 825 million metric tons by 2030

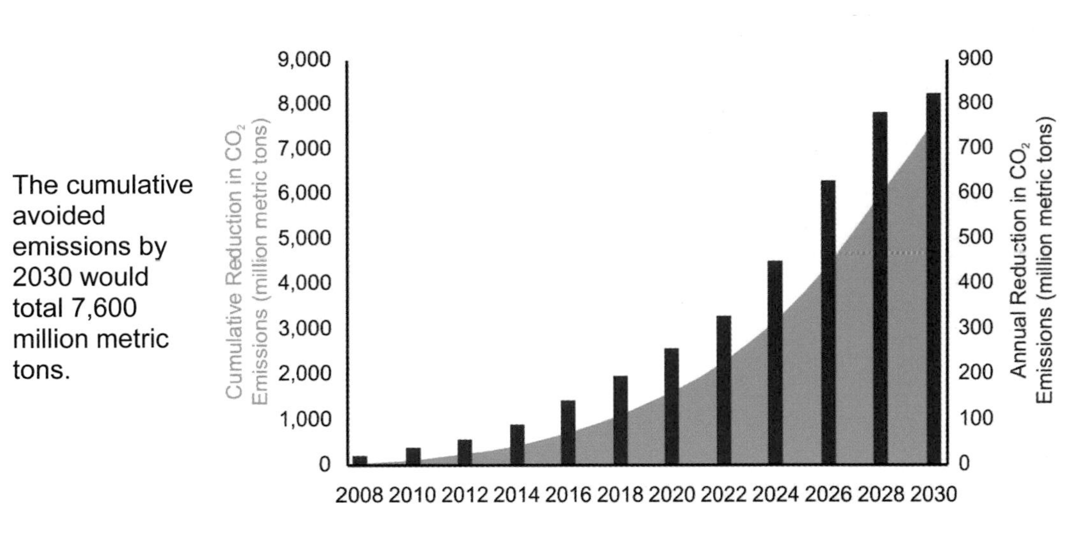

Figure 1-13. CO_2 emissions from the electricity sector

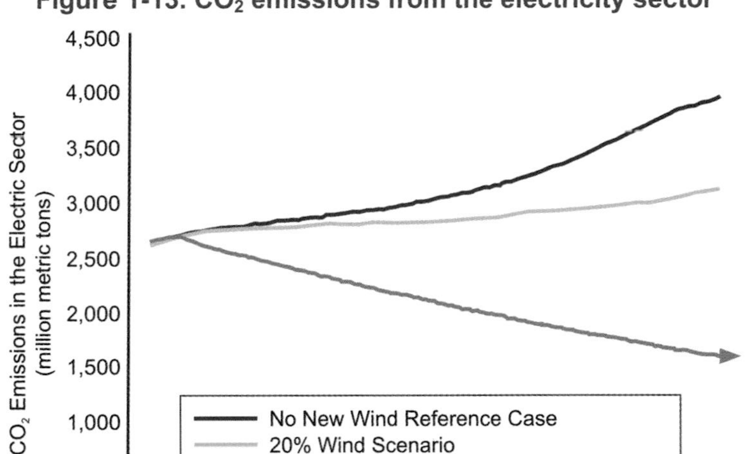

security, employment, and air quality. Given costs relative to other supply options, renewable electricity can have a 30% to 35% share of the total electricity supply in 2030. Deployment of low-GHG (greenhouse gas) emission technologies would be required for achieving stabilization and cost reductions" (IPCC 2007).

More than 30 U.S. states have created climate action plans. In addition, the Regional Greenhouse Gas Initiative (RGGI) is a 10-state collaborative in the Northeast to address CO_2 emissions. All of these state and regional efforts include wind energy as part of a portfolio strategy to reduce overall emissions from energy production (RGGI 2006).

Because wind turbines typically have a service life of at least 20 years and transmission lines can last more than 50 years, investments in achieving 20% wind power by 2030 could continue to supply clean energy through at least 2050. As a result, the cumulative climate change impact of achieving 20% wind power could grow to more than 15,000 million metric tons of CO_2 emissions avoided by mid-century (4,182 million metric tons of carbon equivalent).

The 20% Wind Scenario constructed here would displace a significant amount of fossil fuel generation. According to the WinDS model, by 2030, wind generation is projected to displace 50% of electricity generated from natural gas and 18% of that generated from coal. The displacement of coal is of particular interest because it provides a comparatively higher carbon emissions reduction opportunity. Recognizing that coal power will continue to play a major role in future electricity generation, a large increase in total wind capacity could potentially defer the need to build some new coal capacity, avoiding or postponing the associated increases in carbon emissions. Current DOE projections anticipate construction of approximately 140 GW of new coal plant capacity by 2030 (EIA 2007); the 20% Wind Scenario could avoid construction of more than 80 GW of new coal capacity.[8]

Wind energy that displaces fossil fuel generation can also help meet existing regulations for emissions of conventional pollutants, including sulfur dioxide, nitrogen oxides, and mercury.

1.3.2 Water Conservation

The 20% scenario would potentially reduce cumulative water consumption in the electric sector by 8% (or 4 trillion gallons) from 2007 through 2030—significantly reducing water consumption in the arid states of the interior West. In 2030, annual water consumption in the electric sector would be reduced by 17%.

Wind Reduces Vulnerability

Continued reliance on natural gas for new power generation is likely to put the United States in growing competition in world markets for liquefied natural gas (LNG)—some of which will come from Russia, Qatar, Iran, and other nations in less-than-stable regions.

Water scarcity is a significant problem in many parts of the United States. Even so, few U.S. citizens realize that electricity generation accounts for nearly 50% of all water withdrawals in the nation, with irrigation withdrawals coming in second at 34% (USGS 2005). Water is used for the cooling of natural gas, coal, and nuclear power plants and is an increasing part of the challenge in developing those resources.

Although a significant portion of the water withdrawn for electricity production is recycled back through the system, approximately 2% to 3% of the water withdrawn is consumed through evaporative losses. Even this small fraction adds up to approximately 1.6 to 1.7 trillion gallons of water consumed for power generation each year.

As additional wind generation displaces fossil fuel generation, each megawatt-hour generated by wind could save as much as 600 gallons of water that would otherwise

[8] Carbon mitigation policies were not modeled in either the 20% Wind or No New Wind Scenarios, which results in conventional generation mixes typical of current generation capacity. Under carbon mitigation scenarios, additional technologies could be implemented to reduce the need for conventional generation technology (see Appendix A).

Figure 1-14. National water savings from the 20% Wind Scenario

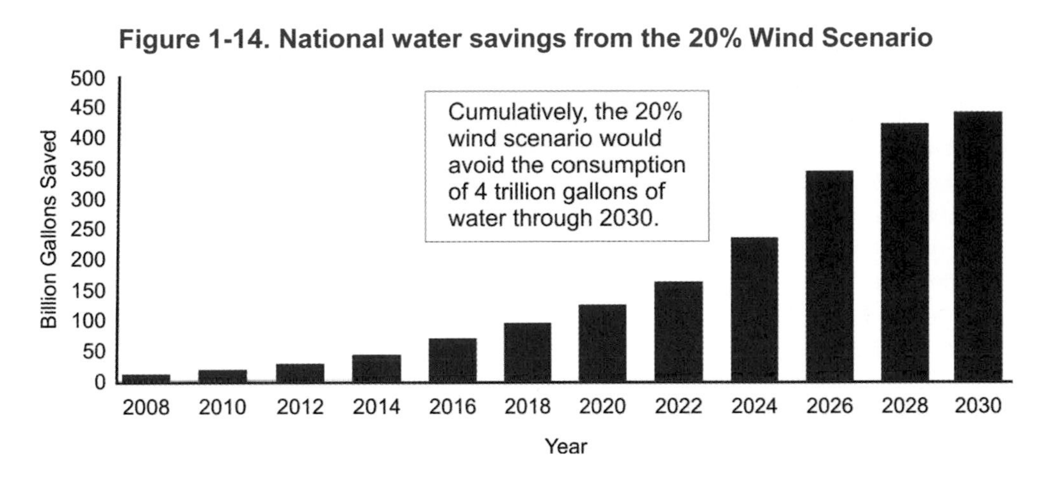

be lost to fossil plant cooling.[9] Because wind energy generation uses a negligible amount of water, the 20% Wind Scenario would avoid the consumption of 4 trillion gallons of water through 2030, a cumulative reduction of 8%, with annual reductions through 2030 shown in Figure 1-14. The annual savings in 2030 is approximately 450 billion gallons. This savings would reduce the expected annual water consumption for electricity generation in 2030 by 17%. The projected water savings are dependent on a future generation mix, which is discussed further in Appendix A.

Based on the WinDS modeling results, nearly 30% of the projected water savings from the 20% Wind Scenario would occur in western states, where water resources are particularly scarce. The Western Governors Association (WGA) highlights this concern in its Clean and Diversified Energy Initiative, which recognizes increased water consumption as a key challenge in accommodating rapid growth in electricity demand. In its 2006 report on water needs, the WGA states that "difficult political choices will be necessary regarding future economic and environmental uses of water and the best way to encourage the orderly transition to a new equilibrium" (WGA 2006).

1.3.3 Energy Security and Stability

There is broad and growing recognition that the nation should diversify its energy portfolio so that a supply disruption affecting a single energy source will not significantly disrupt the national economy. Developing domestic energy sources with known and stable costs would significantly improve U.S. energy stability and security.

When electric utilities have a Power Purchase Agreement or own wind turbines, the price of energy is expected to remain relatively flat and predictable for the life of the wind project, given that there are no fuel costs and assuming that the machines are well maintained. In contrast, a large part of the cost of coal- and gas-fired electricity is in the fuel, for which prices are often volatile and unpredictable. Fuel price risks reduce security and stability for U.S. manufacturers and consumers, as well as for the U.S. economy as a whole. Even small reductions in the amount of energy available or changes in the price of fuel can cause large economic disruptions across the nation. This capacity to disrupt was clearly illustrated by the 1973 embargo imposed by the Organization of Arab Petroleum Exporting Countries (the "Arab oil embargo"); the 2000–2001 California electricity market problems; and the gasoline

[9] See Appendix A for specific assumptions.

and natural gas shortages and price spikes that followed the 2005 hurricane damage to oil refinery and natural gas processing facilities along the Gulf Coast.

Using wind energy increases security and stability by diversifying the national electricity portfolio. Just as those investing for retirement are advised to diversify investments across companies, sectors, and stocks and bonds, diversification of electricity supplies helps distribute the risks and stabilize rates for electricity consumers.

Wind energy reduces reliance on foreign energy sources from politically unstable regions. As a domestic energy source, wind requires no imported fuel, and the turbine components can be either produced on U.S. soil or imported from any friendly nation with production capabilities.

Energy security concerns for the electric industry will likely increase in the foreseeable future as natural gas continues to be a leading source of new generation supply. With declining domestic natural gas sources, future natural gas supplies are expected to come in the form of liquefied natural gas (LNG) imported on tanker ships. U.S. imports of LNG could quadruple by 2030 (EIA 2007). Almost 60% of uncommitted natural gas reserves are in Iran, Qatar, and Russia. These countries, along with others in the Middle East, are expected to be major suppliers to the global LNG market. Actions by those sources can disrupt international energy markets and thus have indirect adverse effects on our economy. Additional risks arise from competition for these resources caused by the growing energy demands of China, India, and other developing nations. According to the WinDS model results, under the 20% Wind Scenario, wind energy could displace approximately 11% of natural gas consumption, which is equivalent to 60% of expected LNG imports in 2030.[10] This displacement would reduce the nation's energy vulnerability to uncertain natural gas supplies. See Appendix A for gas demand reduction assumptions and calculations.

Continued reliance on fossil energy sources exposes the nation to price risks and supply uncertainties. Although the electric sector does not rely heavily on petroleum, which represents one of the nation's biggest energy security threats, diversifying the electric generation mix with increased domestic renewable energy would still enhance national energy security by increasing energy diversity and price stability.

1.3.4 COST OF THE 20% WIND SCENARIO

The overall economic cost of the 20% Wind Scenario accrues mainly from the incremental costs of wind energy relative to other generation sources. This is impacted by the assumptions behind the scenario, listed in Table A-1. Also, some incremental transmission would be required to connect wind to the electric power system. This transmission investment would be in addition to the significant investment in the electric grid that will be needed to serve continuing load growth, whatever the mix of new generation. The market cost of wind energy remains higher than that of conventional energy sources in many areas across the country. In addition, the transmission grid would have to be expanded and upgraded in wind-rich areas and across the existing system to deliver wind energy to many demand centers. An integrated approach to expanding the transmission system would need to include furnishing access to wind resources as well as meeting other system needs.

[10] Compared to consumption of the high price scenario of EIA (2007), used in this report.

Compared to other generation sources, the 20% Wind Scenario entails higher initial capital costs (to install wind capacity and associated transmission infrastructure) in many areas, yet offers lower ongoing energy costs for operations, maintenance, and fuel. Given the optimistic cost and performance assumptions of wind and

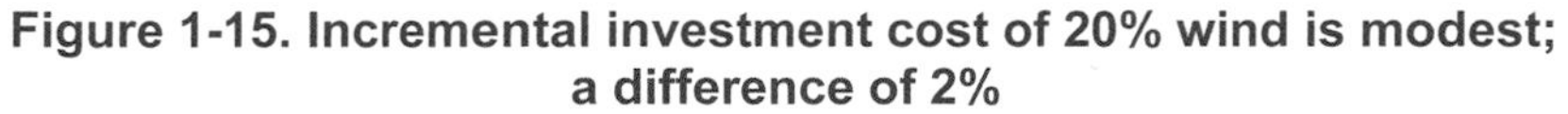

Figure 1-15. Incremental investment cost of 20% wind is modest; a difference of 2%

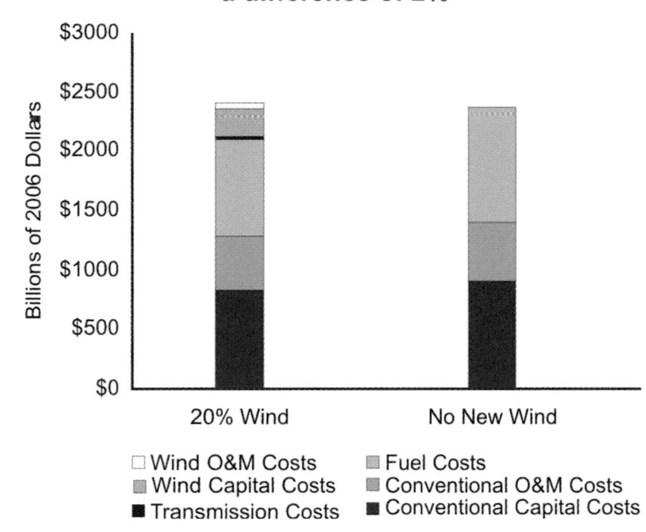

conventional energy sources (detailed in Appendix B), the 20% Wind Scenario could require an incremental investment of as little as $43 billion net present value (NPV) more than the base-case scenario involving no new wind power generation (No New Wind Scenario). This would represent less than 0.06 cents (6 one-hundredths of 1 cent) per kilowatt-hour of total generation by 2030, or roughly 50 cents per month per household. Figure 1-15 shows this cost comparison. The base-case costs are calculated under the assumption of no major changes in fuel availability or environmental restrictions. In this scenario, the cost differential would be about 2% of a total NPV expenditure exceeding $2 trillion.

This analysis is intended to identify the incremental cost of pursuing the 20% Wind Scenario. In regions where the capital costs of the 20% Wind Scenario exceed those of building little or no additional wind capacity, the differential could be offset by the operating costs and benefits discussed earlier. For example, even though Figure 1-15 shows that under optimistic assumptions, the 20% Wind Scenario could increase total capital costs by nearly $197 billion, most of those costs would be offset by the nearly $155 billion in decreased fuel expenditures, resulting in a net incremental cost of approximately $43 billion in NPV. These monetary costs do not reflect other potential offsetting positive impacts.

As estimated by the NREL WinDS model, given optimistic assumptions, the specific cost of the proposed transmission expansion for the 20% Wind Scenario is $20 billion in NPV. The actual required grid investment could also involve significant costs for permitting delays, construction of grid extensions to remote areas with wind resources, and investments in advanced grid controls, integration, and training to enable regional load balancing of wind resources.

The total installed costs for wind plants include costs associated with siting and permitting of these plants. It has become clear that wind power expansion would require careful, logical, and fact-based consideration of local and environmental concerns, allowing siting issues to be addressed within a broad risk framework. Experience in many regions has shown that this can be done, but efficient, streamlined procedures will likely be needed to enable installation rates in the range of 16 GW per year. Chapter 5 covers these issues in more detail.

1.4 Conclusion

There are significant costs, challenges, and impacts associated with the 20% Wind Scenario presented in this report. There are also substantial positive impacts from wind power expansion on the scale and pace described in this chapter that are not likely to be realized in a business-as-usual future. Achieving the 20% Wind Scenario would involve a major national commitment to clean, domestic energy sources with minimal emissions of GHGs and other environmental pollutants.

1.5 References and Other Suggested Reading

AEP 2007. *Interstate Transmission Vision for Wind Integration.* American Electric Power Transmission. http://www.aep.com/about/i765project/technicalpapers.asp.

AWEA 2007. American Wind Energy Association Web site, Oct.1, 2007: http://www.awea.org/faq/cost.html.

AWEA 2008. 2007 Market Report. January 2008 http://www.awea.org/projects/pdf/Market_Report_Jan08.pdf.

Black & Veatch 2007. Twenty Percent *Wind Energy Penetration in the United State: A Technical Analysis of the Energy Resource*. Walnut Creek, CA.

BTM Consult. 2007. *International Wind Energy Development, World Market Update 2006.* Ringkøbing, Denmark: BTM.

Denholm, P., and W. Short. 2006. *Documentation of WinDS Base Case.* Version AEO 2006 (1). Golden, CO: National Renewable Energy Laboratory (NREL). http://www.nrel.gov/analysis/winds/pdfs/winds_data.pdf.

Edmonds, J.A., M.A. Wise, J.J. Dooley, S.H. Kim, S.J. Smith, P.J. Runci, L.E. Clarke, E.L. Malone, and G.M. Stokes. 2007. *Global Energy Technology Strategy: Addressing Climate Change*. Richland, WA: Global Energy Strategy Technology Project. http://www.pnl.gov/gtsp/docs/gtsp_2007_final.pdf

EIA (Energy Information Administration). 2005. *Electric Power Annual.* Washington, DC: EIA. Table 2.6. http://www.eia.doe.gov/cneaf/electricity/epa/epat2p6.html.

EIA. 2007. *Annual Energy Outlook*. Washington, DC: EIA. http://www.eia.doe.gov/oiaf/aeo/index.html.

IPCC (Intergovernmental Panel on Climate Change). 2007. *Climate Change 2007: Impacts, Adaptation and Vulnerability: Working Group II Contribution to the Fourth Assessment Report of the Intergovernmental Panel on Climate change. PCC* Report presented at 8th session of Working Group II of the IPCC, April 2007, Brussels, Belgium. http://www.ipcc.ch/ipccreports/ar4-wg2.htm

Johnston, L., E. Hausman, A. Sommer, B. Biewald, T. Woolf, D. Schlissel, A. Roschelle, and D. White. 2006. *Climate Change and Power: Carbon Dioxide Emissions Costs and Electricity Resource Planning*. Cambridge, MA: Synapse Energy Economics, Inc.

RGGI (Regional Greenhouse Gas Initiative). 2006. "About RGGI." http://www.rggi.org/about.htm.

Socolow, R.H., and S.W. Pacala. 2006. "A Plan to Keep Carbon in Check," *Scientific American*, September.

Teske, S., A. Zervos, and O. Schafer. 2007. *Energy [R]evolution: A Blueprint for Solving Global Warming*, USA National Energy Scenario. Amsterdam: Greenpeace International. http://www.greenpeace.org/raw/content/usa/press-center/reports4/energy-r-evolution-introduc.pdf

USCAP (U.S. Climate Action Partnership). 2007. A Call for Action. http://www.us-cap.org/USCAPCallForAction.pdf

USGS (U.S. Geological Survey). 2005. *Estimated Use of Water in the United States in 2000.* http://pubs.usgs.gov/circ/2004/circ1268/htdocs/figure01.html

WGA (Western Governors' Association). 2006. *Water Needs and Strategies for a Sustainable Future,* p. 4. http://www.westgov.org/wga/publicat/Water06.pdf

Wiser, R. and M. Bolinger. 2007. *Annual Report on U.S. Wind Power Installation, Cost, and Performance Trends: 2006.* DOE/GO - 102007-2433. Golden, CO: NREL.
http://www.osti.gov/bridge/product.biblio.jsp?query_id=0&page=0&osti_id=908214

Wiser, R., M. Bolinger, and M. St. Clair. 2005. *Easing the Natural Gas Crisis: Reducing Natural Gas Prices through Increased Deployment of Renewable Energy and Energy Efficiency*. Berkeley, CA: Berkeley Lab. Report No. LBNL-56756. http://eetd.lbl.gov/EA/EMP/reports/56756.pdf

Wood Mackenzie. 2007. *Impact of a Federal Renewable Portfolio Standard.* Edinburgh, Scotland: Wood Mackenzie.

Chapter 2. Wind Turbine Technology

2

Today's wind technology has enabled wind to enter the electric power mainstream. Continued technological advancement would be required under the 20% Wind Scenario.

2.1 INTRODUCTION

Current turbine technology has enabled wind energy to become a viable power source in today's energy market. Even so, wind energy provides approximately 1% of total U.S. electricity generation. Advancements in turbine technology that have the potential to increase wind energy's presence are currently being explored. These areas of study include reducing capital costs, increasing capacity factors, and mitigating risk through enhanced system reliability. With sufficient research, development, and demonstration (RD&D), these new advances could potentially have a significant impact on commercial product lines in the next 10 years.

A good parallel to wind energy evolution can be derived from the history of the automotive industry in the United States. The large-scale production of cars began with the first Model T production run in 1910. By 1940, after 30 years of making cars and trucks in large numbers, manufacturers had produced vehicles that could reliably move people and goods across the country. Not only had the technology of the vehicle improved, but the infrastructure investment in roads and service stations made their use practical. Yet 30 years later, in 1970, one would hardly recognize the vehicles or infrastructure as the same as those in 1940. Looking at the changes in automobiles produced over that 30-year span, we see how RD&D led to the continuous infusion of modern electronics; improved combustion and manufacturing processes; and ultimately, safer, more reliable cars with higher fuel efficiency. In a functional sense, wind turbines now stand roughly where the U.S. automotive fleet stood in 1940. Gradual improvements have been made in the past 30 years over several generations of wind energy products. These technology advances enable today's turbines to reliably deliver electricity to the grid at a reasonable cost.

Through continued RD&D and infrastructure development, great strides will be made to produce even more advanced machines supporting future deployment of wind power technology. This chapter describes the status of wind technology today and provides a brief history of technology development over the past three decades. Prospective improvements to utility-scale land-based wind turbines as well as offshore wind technology are discussed. Distributed wind technology [100 kilowatts (kW) or less] is also addressed in this chapter.

2.2 TODAY'S COMMERCIAL WIND TECHNOLOGY

Beginning with the birth of modern wind-driven electricity generators in the late 1970s, wind energy technology has improved dramatically up to the present. Capital costs have decreased, efficiency has increased, and reliability has improved. High-quality products are now routinely delivered by major suppliers of turbines around the world, and complete wind generation plants are being engineered into the grid infrastructure to meet utility needs. In the 20% Wind Scenario outlined in this report, it is assumed that capital costs would be reduced by 10% over the next two decades, and capacity factors would be increased by about 15% (corresponding to a 15% increase in annual energy generation by a wind plant).

2.2.1 WIND RESOURCES

Wind technology is driven by the nature of the resource to be harvested. The United States, particularly the Midwestern region from Texas to North Dakota, is rich in wind energy resources as shown in Figure 2-1, which illustrates the wind resources measured at a 50-meter (m) elevation. Measuring potential wind energy generation at a 100-m elevation (the projected operating hub height of the next generation of modern turbines) greatly increases the U.S. land area that could be used for wind deployment, as shown in Figure 2-2 for the state of Indiana. Taking these measurements into account, current U.S. land-based and offshore wind resources are estimated to be sufficient to supply the electrical energy needs of the entire country several times over. For a description of U.S. wind resources, see Appendix B.

Figure 2-1. The wind resource potential at 50 m above ground on land and offshore

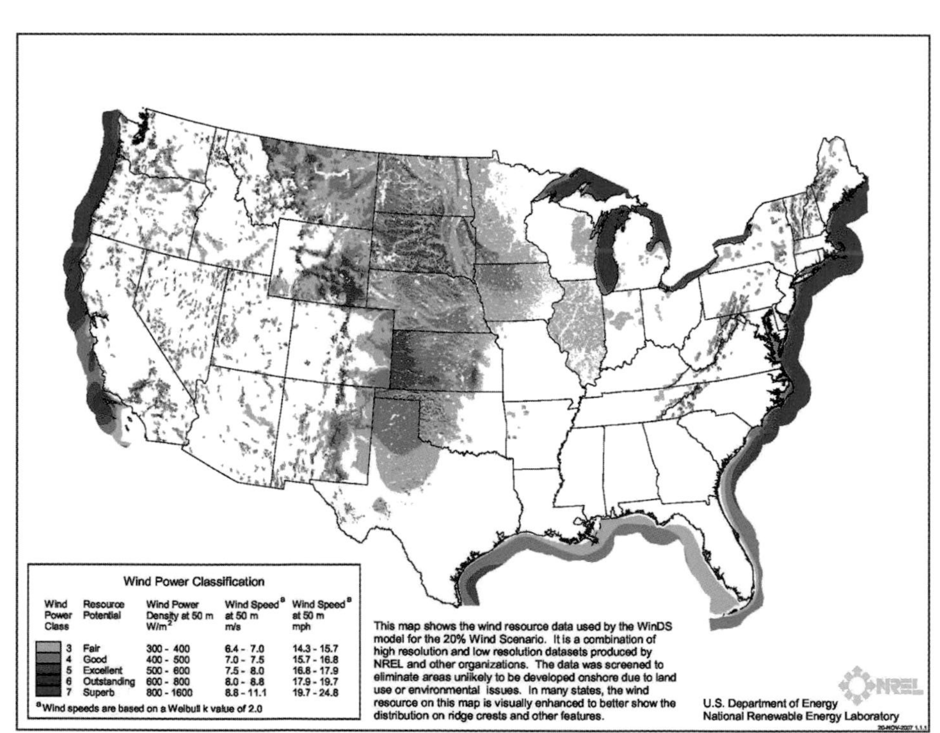

Identifying the good wind potential at high elevations in states such as Indiana and off the shore of both coasts is important because it drives developers to find ways to harvest this energy. Many of the opportunities being pursued through advanced

Figure 2-2. Comparison of the wind energy resource at 50 m, 70 m, and 100 m for Indiana

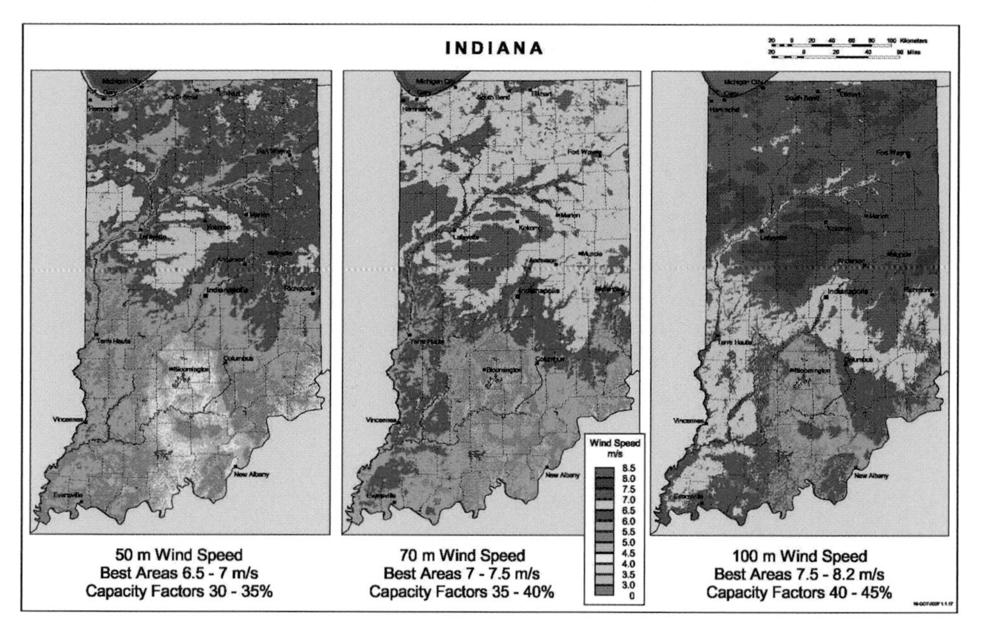

technology are intended to achieve higher elevations, where the resource is much greater, or to access extensive offshore wind resources.

2.2.2 Today's Modern Wind Turbine

Modern wind turbines, which are currently being deployed around the world, have three-bladed rotors with diameters of 70 m to 80 m mounted atop 60-m to 80-m towers, as illustrated in Figure 2-3. Typically installed in arrays of 30 to 150 machines, the average turbine installed in the United States in 2006 can produce approximately 1.6 megawatts (MW) of electrical power. Turbine power output is controlled by rotating the blades around their long axis to change the angle of attack with respect to the relative wind as the blades spin around the rotor hub. This is called controlling the blade pitch. The turbine is pointed into the wind by rotating the nacelle around the tower. This is called controlling the yaw. Wind sensors on the nacelle tell the yaw controller where to point the turbine. These wind sensors, along with sensors on the generator and drivetrain, also tell the blade pitch controller how to regulate the power output and rotor speed to prevent overloading the structural components. Generally, a turbine will start producing power in winds of about 5.36 m/s and reach maximum power output at about 12.52 m/s–13.41 m/s. The turbine will pitch or feather the blades to stop power production and rotation at about 22.35 m/s. Most utility-scale turbines are upwind machines, meaning that they operate with the blades upwind of the tower to avoid the blockage created by the tower.

The amount of energy in the wind available for extraction by the turbine increases with the cube (the third power) of wind speed; thus, a 10% increase in wind speed creates a 33% increase in available energy. A turbine can capture only a portion of this cubic increase in energy, though, because power above the level for which the electrical system has been designed, referred to as the rated power, is allowed to pass through the rotor.

Figure 2-3. A modern 1.5-MW wind turbine installed in a wind power plant

In general, the speed of the wind increases with the height above the ground, which is why engineers have found ways to increase the height and the size of wind turbines while minimizing the costs of materials. But land-based turbine size is not expected to grow as dramatically in the future as it has in the past. Larger sizes are physically possible; however, the logistical constraints of transporting the components via highways and of obtaining cranes large enough to lift the components present a major economic barrier that is difficult to overcome. Many turbine designers do not expect the rotors of land-based turbines to become much larger than about 100 m in diameter, with corresponding power outputs of about 3 MW to 5 MW.

2.2.3 Wind Plant Performance and Price

The performance of commercial turbines has improved over time, and as a result, their capacity factors have slowly increased. Figure 2-4 shows the capacity factors at commercial operation dates (CODs) ranging from 1998 to 2005. The data show that turbines in the Lawrence Berkeley National Laboratory (Berkeley Lab) database (Wiser and Bolinger 2007) that began operating commercially before 1998 have an average capacity factor of about 22%. The turbines that began commercial operation after 1998, however, show an increasing capacity factor trend, reaching 36% in 2004 and 2005.

The cost of wind-generated electricity has dropped dramatically since 1980, when the first commercial wind plants began operating in California. Since 2003, however, wind energy prices have increased. Figure 2-5 (Wiser and Bolinger 2007)

Figure 2-4. Turbine capacity factor by commercial operation date (COD) using 2006 data

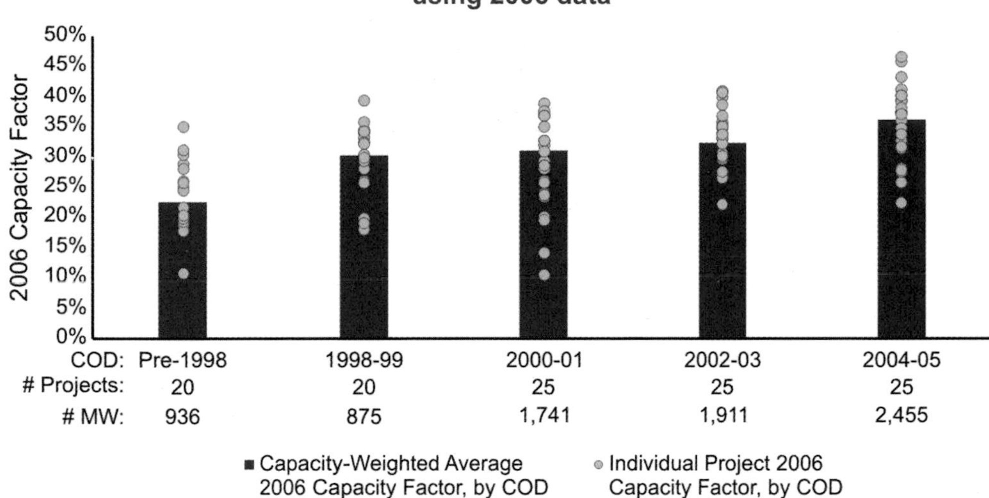

Figure 2-5. Wind energy price by commercial operation date (COD) using 2006 data

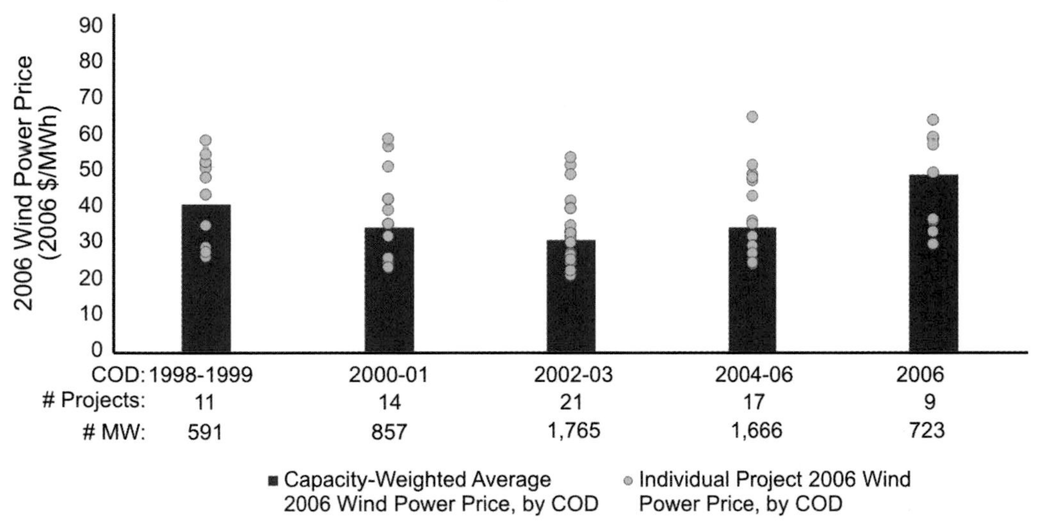

shows that in 2006 the price paid for electricity generated in large wind farms was between 3.0 and 6.5 cents/kilowatt-hour (kWh), with an average near 5 cents/kWh (1 cent/kWh = $10/megawatt-hour [MWh]). This price includes the benefit of the federal production tax credit (PTC), state incentives, and revenue from the sale of any renewable energy credits.

Wind energy prices have increased since 2002 for the following reasons (Wiser and Bolinger 2007):

- Shortages of turbines and components, resulting from the dramatic recent growth of the wind industry in the United States and Europe
- The weakening U.S. dollar relative to the euro (many major turbine components are imported from Europe, and there are relatively few wind turbine component manufacturers in the United States)

2

- A significant rise in material costs, such as steel and copper, as well as transportation fuels over the last three years
- The on-again, off-again cycle of the wind energy PTC (uncertainty hinders investment in new turbine production facilities and encourages hurried and expensive production, transportation, and installation of projects when the tax credit is available).

Expected future reductions in wind energy costs would come partly from expected investment in the expansion of manufacturing volume in the wind industry. In addition, a stable U.S. policy for renewable energy and a heightened RD&D effort could also lower costs.

2.2.4 Wind Technology Development

Until the early 1970s, wind energy filled a small niche market, supplying mechanical power for grinding grain and pumping water, as well as electricity for rural battery charging. With the exception of battery chargers and rare experiments with larger electricity-producing machines, the windmills of 1850 and even 1950 differed very little from the primitive devices from which they were derived. Increased RD&D in the latter half of the twentieth century, however, greatly improved the technology.

In the 1980s, the practical approach of using low-cost parts from agricultural and boat-building industries produced machinery that usually worked, but was heavy, high-maintenance, and grid-unfriendly. Little was known about structural loads caused by turbulence, which led to the frequent and early failure of critical parts, such as yaw drives. Additionally, the small-diameter machines were deployed in the California wind corridors, mostly in densely packed arrays that were not aesthetically pleasing in such a rural setting. These densely packed arrays also often blocked the wind from neighboring turbines, producing a great deal of turbulence for the downwind machines. Reliability and availability suffered as a result.

Recognizing these issues, wind operators and manufacturers have worked to develop better machines with each new generation of designs. Drag-based devices and simple lift-based designs gave way to experimentally designed and tested high-lift rotors, many with full-span pitch control. Blades that had once been made of sail or sheet metal progressed through wood to advanced fiberglass composites. The direct current (DC) alternator gave way to the grid-synchronized induction generator, which has now been replaced by variable-speed designs employing high-speed solid-state switches of advanced power electronics. Designs moved from mechanical cams and linkages that feathered or furled a machine to high-speed digital controls. A 50 kW machine, considered large in 1980, is now dwarfed by the 1.5 MW to 2.5 MW machines being routinely installed today.

Many RD&D advances have contributed to these changes. Airfoils, which are now tested in wind tunnels, are designed for insensitivity to surface roughness and dirt. Increased understanding of aeroelastic loads and the ability to incorporate this knowledge into finite element models and structural dynamics codes make the machines of today more robust but also more flexible and lighter on a relative basis than those of a decade ago.

As with any maturing technology, however, many of the simpler and easier improvements have already been incorporated into today's turbines. Increased

RD&D efforts and innovation will be required to continue to expand the wind energy industry.

2.2.5 Current Turbine Size

Throughout the past 20 years, average wind turbine ratings have grown almost linearly, as illustrated by Figure 2-6. Each group of wind turbine designers has predicted that its latest machine is the largest that a wind turbine will ever be. But with each new generation of wind turbines (roughly every five years), the size has grown along the linear curve and has achieved reductions in life-cycle cost of energy (COE).

Figure 2-6. The development path and growth of wind turbines

As discussed in Section 2.2.2, this long-term drive to develop larger turbines is a direct result of the desire to improve energy capture by accessing the stronger winds at higher elevations. (The increase in wind speed with elevation is referred to as wind shear.) Although the increase in turbine height is a major reason for the increase in capacity factor over time, there are economic and logistical constraints to this continued growth to larger sizes.

The primary argument for limiting the size of wind turbines is based on the square-cube law. This law roughly states that as a wind turbine rotor grows in size, its energy output increases as the rotor swept area (the diameter squared), while the volume of material, and therefore its mass and cost, increases as the cube of the diameter. In other words, at some size, the cost for a larger turbine will grow faster than the resulting energy output revenue, making scaling a losing economic game.

Engineers have successfully skirted this law by either removing material or using it more efficiently as they increase size. Turbine performance has clearly improved, and cost per unit of output has been reduced, as illustrated in Figures 2-4 and 2-5. A Wind Partnerships for Advanced Component Technology (WindPACT) study has also shown that in recent years, blade mass has been scaling at an exponent of about 2.3 as opposed to the expected 3.0 (Ashwill 2004), demonstrating how successive

Figure 2-7. Growth in blade weight

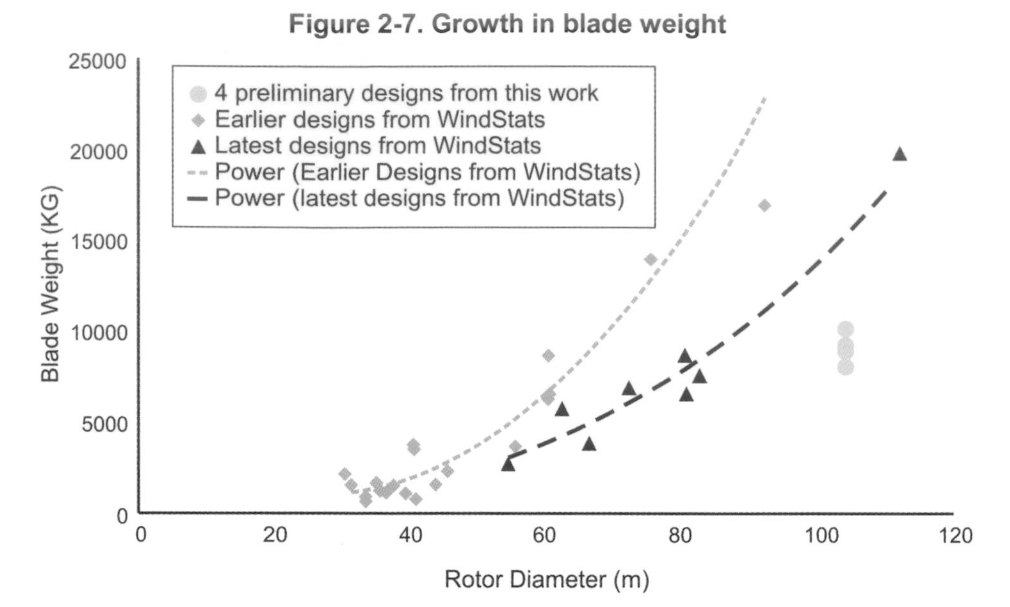

generations of blade design have moved off the cubic weight growth curve to keep weight down (see Figure 2-7). The latest designs continue to fall below the cubic line of the previous generation, indicating the continued infusion of new technology into blade design. If advanced RD&D were to result in even better design methods, as well as new materials and manufacturing methods that allow the entire turbine to scale as the diameter squared, continuing to innovate around this size limit would be possible.

Land transportation constraints can also limit wind turbine growth for turbines installed on land. Cost-effective road transportation is achieved by remaining within standard over-the-road trailer dimensions of 4.1 m high by 2.6 m wide and a gross vehicle weight (GVW) under 80,000 pounds (lb.; which translates to a cargo weight of about 42,000 lb.). Loads that exceed 4.83 m in height trigger expensive rerouting (to avoid obstructions) and often require utility and law enforcement assistance along the roadways. These dimension limits have the most impact on the base diameter of wind turbine towers. Rail transportation is even more dimensionally limited by tunnel and overpass widths and heights. Overall widths should remain within 3.4 m, and heights are limited to 4.0 m. Transportation weights are less of an issue in rail transportation, with GVW limits of up to 360,000 lb. (Ashwill 2004).

Once turbines arrive at their destination, their physical installation poses other practical constraints that limit their size. Typically, 1.5 MW turbines are installed on 80-m towers to maximize energy capture. Crane requirements are quite stringent because of the large nacelle mass in combination with the height of the lift and the required boom extension. As the height of the lift to install the rotor and nacelle on the tower increases, the number of available cranes with the capability to make this lift is fairly limited. In addition, cranes with large lifting capacities are difficult to transport and require large crews, leading to high operation, mobilization, and demobilization costs. Operating large cranes in rough or complex, hilly terrain can also require repeated disassembly to travel between turbine sites (NREL 2002).

2.2.6 Current Status of Turbine Components

The Rotor

Typically, a modern turbine will cut in and begin to produce power at a wind speed of about 5 m/s (see Figure 2-8). It will reach its rated power at about 12 m/s to 14

Figure 2-8. Typical power output versus wind speed curve

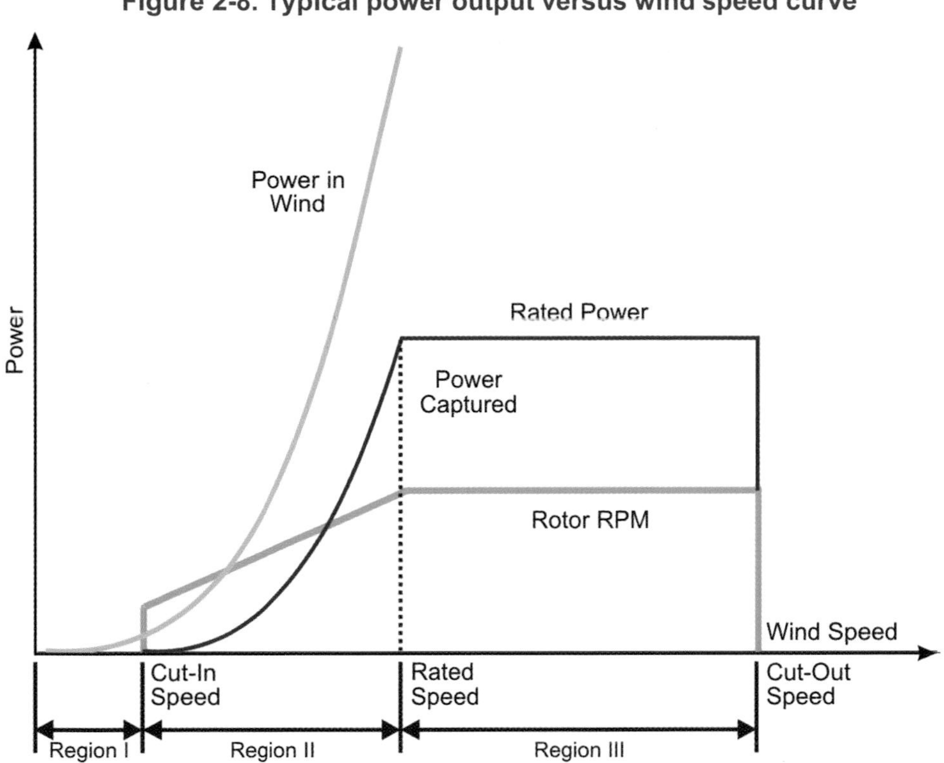

m/s, where the pitch control system begins to limit power output and prevent generator and drivetrain overload. At around 22 m/s to 25 m/s, the control system pitches the blades to stop rotation, feathering the blades to prevent overloads and damage to the turbine's components. The job of the rotor is to operate at the absolute highest efficiency possible between cut-in and rated wind speeds, to hold the power transmitted to the drivetrain at the rated power when the winds go higher, and to stop the machine in extreme winds. Modern utility-scale wind turbines generally extract about 50% of the energy in this stream below the rated wind speed, compared to the maximum energy that a device can theoretically extract, which is 59% of the energy stream (see "The Betz Limit" sidebar).

Most of the rotors on today's large-scale machines have an individual mechanism for pitch control; that is, the mechanism rotates the blade around its long axis to control the power in high winds. This device is a significant improvement over the first generation of fixed-pitch or collective-pitch linkages, because the blades can now be rotated in high winds to feather them out of the wind. This reduces the maximum loads on the system when the machine is parked. Pitching the blades out of high winds also reduces operating loads, and the combination of pitchable blades with a variable-speed generator allows the turbine to maintain generation at a constant rated-power output. The older generation of constant-speed rotors sometimes had instantaneous

The Betz Limit

Not all of the energy present in a stream of moving air can be extracted; some air must remain in motion after extraction. Otherwise, no new, more energetic air can enter the device. Building a wall would stop the air at the wall, but the free stream of energetic air would just flow around the wall. On the other end of the spectrum, a device that does not slow the air is not extracting any energy, either. The maximum energy that can be extracted from a fluid stream by a device with the same working area as the stream cross section is 59% of the energy in the stream. Because it was first derived by wind turbine pioneer Albert Betz, this maximum is known as the Betz Limit.

power spikes up to twice the rated power. Additionally, this pitch system operates as the primary safety system because any one of the three independent actuators is capable of stopping the machine in an emergency.

Blades

As wind turbines grow in size, so do their blades—from about 8 m long in 1980 to more than 40 m for many land-based commercial systems and more than 60 m for offshore applications today. Rigorous evaluation using the latest computer analysis tools has improved blade designs, enabling weight growth to be kept to a much lower rate than simple geometric scaling (see Figure 2-7). Designers are also starting to work with lighter and stronger carbon fiber in highly stressed locations to stiffen blades and improve fatigue resistance while reducing weight. (Carbon fiber, however, costs about 10 times as much as fiberglass.) Using lighter blades reduces the load-carrying requirements for the entire supporting structure and saves total costs far beyond the material savings of the blades alone.

By designing custom airfoils for wind turbines, developers have improved blades over the past 20 years. Although these airfoils were primarily developed to help optimize low-speed wind aerodynamics to maximize energy production while limiting loads, they also help prevent sensitivity to blade fouling that is caused by dirt and bug accumulation on the leading edge. This sensitivity reduction greatly improves blade efficiency (Cohen et al. 2008).

Current turbine blade designs are also being customized for specific wind classes. In lower energy sites, the winds are lighter, so design loads can be relaxed and longer blades can be used to harvest more energy in lower winds. Even though blade design methods have improved significantly, there is still much room for improvement, particularly in the area of dynamic load control and cost reduction.

Controls

Today's controllers integrate signals from dozens of sensors to control rotor speed, blade pitch angle, generator torque, and power conversion voltage and phase. The controller is also responsible for critical safety decisions, such as shutting down the turbine when extreme conditions are encountered. Most turbines currently operate in variable-speed mode, and the control system regulates the rotor speed to obtain peak efficiency in fluctuating winds. It does this by continuously updating the rotor speed and generator loading to maximize power and reduce drivetrain transient torque loads. Operating in variable-speed mode requires the use of power converters, which offer additional benefits (which are discussed in the next subsection). Research into the use of advanced control methods to reduce turbulence-induced loads and increase energy capture is an active area of work.

Electrical controls with power electronics enable machines to deliver fault-ride-through control, voltage control, and volt-ampere-reactive (VAR) support to the grid. In the early days of grid-connected wind generators, the grid rules required that wind turbines go offline when any grid event was in progress. Now, with penetration of wind energy approaching 10% in some regions of the United States, more than 8% nationally in Germany, and more than 20% of the average generation in Denmark, the rules are being changed (Wiser and Bolinger 2007). Grid rules on both continents are requiring more support and fault-ride-through protection from the wind generation component. Current electrical control systems are filling this need with wind plants carefully engineered for local grid conditions

The Drivetrain (Gearbox, Generator, and Power Converter)

Generating electricity from the wind places an unusual set of requirements on electrical systems. Most applications for electrical drives are aimed at using electricity to produce torque, instead of using torque to produce electricity. The applications that generate electricity from torque usually operate at a constant rated power. Wind turbines, on the other hand, must generate at all power levels and spend a substantial amount of time at low power levels. Unlike most electrical machines, wind generators must operate at the highest possible aerodynamic and electrical efficiencies in the low-power/low-wind region to squeeze every kilowatt-hour out of the available energy. For wind systems, it is simply not critical for the generation system to be efficient in above-rated winds in which the rotor is letting energy flow through to keep the power down to the rated level. Therefore, wind systems can afford inefficiencies at high power, but they require maximum efficiency at low power—just the opposite of almost all other electrical applications in existence.

Torque has historically been converted to electrical power by using a speed-increasing gearbox and an induction generator. Many current megawatt-scale turbines use a three-stage gearbox consisting of varying arrangements of planetary gears and parallel shafts. Generators are either squirrel-cage induction or wound-rotor induction, with some newer machines using the doubly fed induction design for variable speed, in which the rotor's variable frequency electrical output is fed into the collection system through a solid-state power converter. Full power conversion and synchronous machines are drawing interest because of their fault-ride-through and other grid support capacities.

As a result of fleet-wide gearbox maintenance issues and related failures with some designs in the past, it has become standard practice to perform extensive dynamometer testing of new gearbox configurations to prove durability and reliability before they are introduced into serial production. The long-term reliability of the current generation of megawatt-scale drivetrains has not yet been fully verified with long-term, real-world operating experience. There is a broad consensus that wind turbine drivetrain technology will evolve significantly in the next several years to reduce weight and cost and improve reliability.

The Tower

The tower configuration used almost exclusively in turbines today is a steel monopole on a concrete foundation that is custom designed for the local site conditions. The major tower variable is height. Depending on the wind characteristics at the site, the tower height is selected to optimize energy capture with respect to the cost of the tower. Generally, a turbine will be placed on a 60-m to 80-m tower, but 100-m towers are being used more frequently. Efforts to develop advanced tower configurations that are less costly and more easily transported and installed are ongoing.

Balance of Station

The balance of the wind farm station consists of turbine foundations, the electrical collection system, power-conditioning equipment, supervisory control and data acquisition (SCADA) systems, access and service roads, maintenance buildings, service equipment, and engineering permits. Balance-of-station components contribute about 20% to the installed cost of a wind plant.

Operations and Availability

Operation and maintenance (O&M) costs have also dropped significantly since the 1980s as a result of improved designs and increased quality. O&M data from the technology installed well before 2000 show relatively high annual costs that increase with the age of the equipment. Annual O&M costs are reported to be as high as \$30-\$50/MWh for wind power plants with 1980s technology, whereas the latest generation of turbines has reported annual O&M costs below \$10/MWh (Wiser and Bolinger 2007). Figure 2-9 shows annual O&M expenses by wind project age and equipment installation year. Relative to wind power prices shown in Figure 2-5, the O&M costs can be a significant portion of the price paid for wind-generated electricity. Since the late 1990s, modern equipment operation costs have been reduced for the initial operating years. Whether annual operation costs grow as these modern turbines age is yet to be determined and will depend greatly on the quality of these new machines.

Figure 2-9. Operation and maintenance costs for large-scale wind plants installed within the last 10 years for the early years of operation (Wiser and Bolinger 2007)

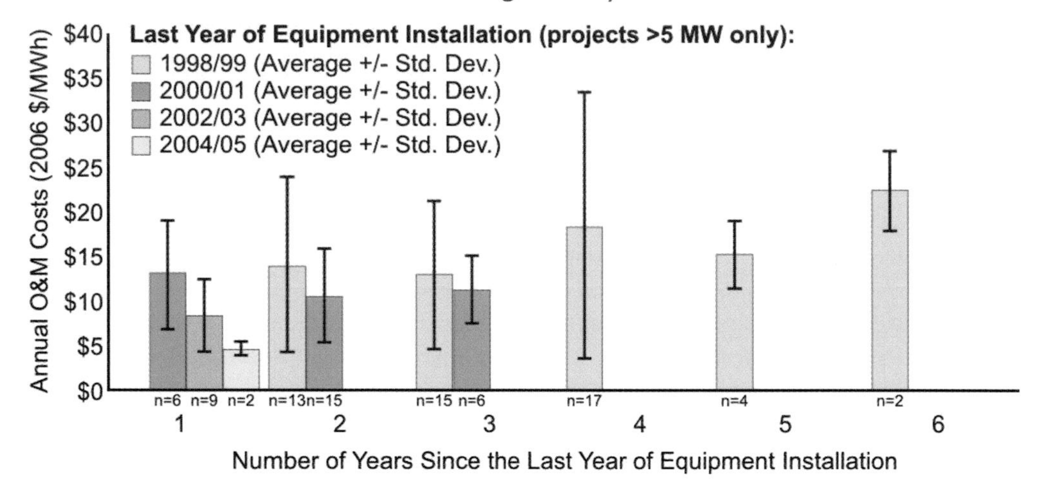

SCADA systems are being used to monitor very large wind farms and dispatch maintenance personnel rapidly and efficiently. This is one area where experience in managing large numbers of very large machines has paid off. Availability, defined as the fraction of time during which the equipment is ready to operate, is now more than 95% and often reported to exceed 98%. These data indicate the potential for improving reliability and reducing maintenance costs (Walford 2006).

2.3 TECHNOLOGY IMPROVEMENTS ON THE HORIZON

Technology improvements can help meet the cost and performance challenges embedded in this 20% Wind Scenario. The required technological improvements are relatively straightforward: taller towers, larger rotors, and continuing progress through the design and manufacturing learning curve. No single component or design innovation can fulfill the need for technology improvement. By combining a number of specific technological innovations, however, the industry can introduce new advanced architectures necessary for success. The 20% Wind Scenario does not require success in all areas; progress can be made even if only some of the technology innovations are achieved.

2.3.1 FUTURE IMPROVEMENTS TO TURBINE COMPONENTS

2

Many necessary technological advances are already in the active development stages. Substantial research progress has been documented, and individual companies are beginning the development process for these technologies. The risk of introducing new technology at the same time that manufacturing production is scaling up and accelerating to unprecedented levels is not trivial. Innovation always carries risk. Before turbine manufacturers can stake the next product on a new feature, the performance of that innovation needs to be firmly established and the durability needs to be characterized as well as possible. These risks are mitigated by RD&D investment, including extensive component and prototype testing before deployment.

The following are brief summaries of key wind energy technologies that are expected to increase productivity through better efficiency, enhanced energy capture, and improved reliability.

The Rotor

The number one target for advancement is the means by which the energy is initially captured—the rotor. No indicators currently suggest that rotor design novelties are on their way, but there are considerable incentives to use better materials and innovative controls to build enlarged rotors that sweep a greater area for the same or lower loads. Two approaches are being developed and tested to either reduce load levels or create load-resistant designs. The first approach is to use the blades themselves to attenuate both gravity- and turbulence-driven loads (see the following subsection). The second approach lies in an active control that senses rotor loads and actively suppresses the loads transferred from the rotor to the rest of the turbine structure. These improvements will allow the rotor to grow larger and capture more energy without changing the balance of the system. They will also improve energy capture for a given capacity, thereby increasing the capacity factor (Ashwill 2004).

Another innovation already being evaluated at a smaller scale by Energy Unlimited Inc. (EUI; Boise, Idaho) is a variable-diameter rotor that could significantly increase capacity factor. Such a rotor has a large area to capture more energy in low winds and a system to reduce the size of the rotor to protect the system in high winds. Although this is still considered a very high-risk option because of the difficulty of building such a blade without excessive weight, it does provide a completely different path to a very high capacity factor (EUI 2003).

Blades

Larger rotors with longer blades sweep a greater area, increasing energy capture. Simply lengthening a blade without changing the fundamental design, however, would make the blade much heavier. In addition, the blade would incur greater structural loads because of its weight and longer moment arm. Blade weight and resultant gravity-induced loads can be controlled by using advanced materials with higher strength-to-weight ratios. Because high-performance materials such as carbon fibers are more expensive, they would be included in the design only when the payoff is maximized. These innovative airfoil shapes hold the promise of maintaining excellent power performance, but have yet to be demonstrated in full-scale operation.

Figure 2-10. Curvature-based twist coupling

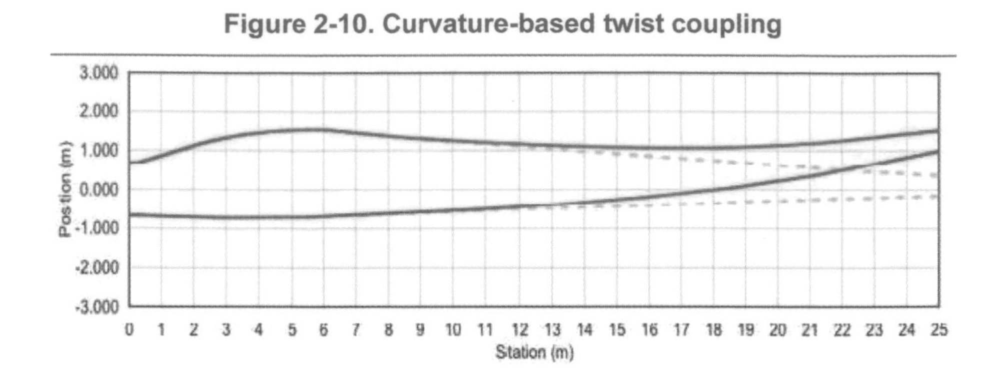

One elegant concept is to build directly into the blade structure a passive means of reducing loads. By carefully tailoring the structural properties of the blade using the unique attributes of composite materials, the internal structure of the blade can be built in a way that allows the outer portion of the blade to twist as it bends (Griffin 2001). "Flap-pitch" or "bend-twist" coupling, illustrated in Figure 2-10, is accomplished by orienting the fiberglass and carbon plies within the composite layers of the blade. If properly designed, the resulting twisting changes the angle of attack over much of the blade, reducing the lift as wind gusts begin to load the blade and therefore passively reducing the fatigue loads. Yet another approach to achieving flap-pitch coupling is to build the blade in a curved shape (see Figure 2-11) so that the aerodynamic loads apply a twisting action to the blade, which varies the angle of attack as the aerodynamic loads fluctuate.

Figure 2-11. Twist-flap coupled blade design (material-based twist coupling)

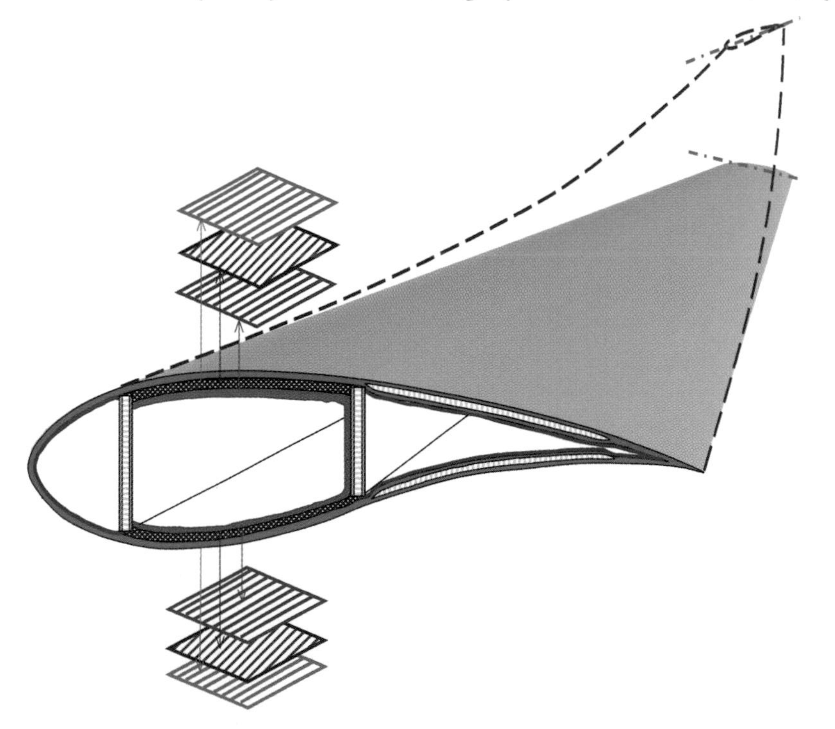

To reduce transportation costs, concepts such as on-site manufacturing and segmented blades are also being explored. It might also be possible to segment molds and move them into temporary buildings close to the site of a major wind installation so that the blades can be made close to, or actually at, the wind site.

Active Controls

Active controls using independent blade pitch and generator torque can be used to reduce tower-top motion, power fluctuations, asymmetric rotor loads, and even individual blade loads. Actuators and controllers already exist that can achieve most of the promised load reductions to enable larger rotors and taller towers. In addition, some researchers have published control algorithms that could achieve the load reductions (Bossanyi 2003). Sensors capable of acting as the eyes and ears of the control system will need to have sufficient longevity to monitor a high-reliability, low-maintenance system. There is also concern that the increased control activity will accelerate wear on the pitch mechanism. Thus, the technical innovation that is essential to enabling some of the most dramatic improvements in performance is not a matter of exploring the unknown, but rather of doing the hard work of mitigating the innovation risk by demonstrating reliable application through prototype testing and demonstration.

Towers

To date, there has been little innovation in the tower, which is one of the more mundane components of a wind installation. But because placing the rotor at a higher elevation is beneficial and because the cost of steel continues to rise rapidly, it is highly likely that this component will be examined more closely in the future, especially for regions of higher than average wind shear.

Because power is related to the cube (the third power) of wind speed, mining upward into these rich veins of higher wind speed potentially has a high payoff—for example, a 10% increase in wind speed produces about a 33% increase in available power. Turbines could sit on even taller towers than those in current use if engineers can figure out how to make them with less steel. Options for using materials other than steel (e.g., carbon fiber) in the tower are being investigated. Such investigations could bear fruit if there are significant adjustments in material costs. Active controls that damp out tower motion might be another enabling technology. Some tower motion controls are already in the research pipeline. New tower erection technologies might play a role in O&M that could also help drive down the system cost of energy (COE) (NREL 2002).

Tower diameters greater than approximately 4 m would incur severe overland transportation cost penalties. Unfortunately, tower diameter and material requirements conflict directly with tower design goals—a larger diameter is beneficial because it spreads out the load and actually requires less material because its walls are thinner. On-site assembly allows for larger diameters but also increases the number of joints and fasteners, raising labor costs as well as concerns about fastener reliability and corrosion. Additionally, tower wall thickness cannot be decreased without limit; engineers must adhere to certain minima to avoid buckling. New tower wall topologies, such as corrugation, can be employed to alleviate the buckling constraint, but taller towers will inevitably cost more.

The main design impact of taller towers is not on the tower itself, but on the dynamics of a system with the bulk of its mass atop a longer, more slender structure. Reducing tower-top weight improves the dynamics of such a flexible system. The tall tower dilemma can be further mitigated with smarter controls that attenuate tower motion by using blade pitch and generator torque control. Although both approaches have been demonstrated, they are still rarely seen in commercial applications.

2

The Drivetrain (Gearbox, Generator, and Power Conversion)

Parasitic losses in generator windings, power electronics, gears and bearings, and other electrical devices are individually quite small. When summed over the entire system, however, these losses add up to significant numbers. Improvements that remove or reduce the fixed losses during low power generation are likely to have an important impact on raising the capacity factor and reducing cost. These improvements could include innovative power-electronic architectures and large-scale use of permanent-magnet generators. Direct-drive systems also meet this goal by eliminating gear losses. Modular (transportable) versions of these large generation systems that are easier to maintain will go a long way toward increasing the productivity of the low-wind portion of the power curve.

Currently, gearbox reliability is a major issue, and gearbox replacement is quite expensive. One solution is a direct-drive power train that entirely eliminates the gearbox. This approach, which was successfully adopted in the 1990s by Enercon-GmbH (Aurich, Germany), is being examined by other turbine manufacturers. A less radical alternative reduces the number of stages in the gearbox from three to two or even one, which enhances reliability by reducing the parts count. The fundamental gearbox topology can also be improved, as Clipper Windpower (Carpinteria, California) did with its highly innovative multiple-drive-path gearbox, which divides mechanical power among four generators (see Figure 2-12). The multiple-drive-path design radically decreases individual gearbox component loads, which reduces gearbox weight and size, eases erection and maintenance demands, and improves reliability by employing inherent redundancies.

The use of rare-earth permanent magnets in generator rotors instead of wound rotors also has several advantages. High energy density eliminates much of the weight associated with copper windings, eliminates problems associated with insulation degradation and shorting, and reduces electrical losses. Rare-earth magnets cannot be subjected to elevated temperatures, however, without permanently degrading magnetic field strength, which imposes corresponding demands on generator cooling reliability. The availability of rare-earth permanent magnets is a potential concern because key raw materials are not available in significant quantities within the United States (see Chapter 3).

Power electronics have already achieved elevated performance and reliability levels, but opportunities for significant improvement remain. New silicon carbide (SiC) devices entering the market could allow operation at higher temperature and higher frequency, while improving reliability, lowering cost, or both. New circuit topologies could furnish better control of power quality, enable higher voltages to be used, and increase overall converter efficiency.

Distributed Energy Systems (Wallingford, Connecticut; formerly Northern Power Systems) has built an advanced prototype power electronics system that will deliver lower losses and conversion costs for permanent-magnet generators (Northern Power Systems 2006). Peregrine Power (Wilsonville, Oregon) has concluded that using SiC devices would reduce power losses, improve reliability, and shrink components by orders of magnitude (Peregrine Power 2006). A study completed by BEW Engineering (San Ramon, California; Behnke, Erdman, and Whitaker Engineering 2006) shows that using medium-voltage power systems for multimegawatt turbines could reduce the cost, weight, and volume of turbine electrical components as well as reduce electrical losses.

Figure 2-12. Clipper Windpower multiple-drive-path gearbox

The most dramatic change in the long-term application of wind generation may come from the grid support provided by the wind plant. Future plants will not only support the grid by delivering fault-ride-through capability as well as frequency, voltage, and VAR control, but will also carry a share of power control capability for the grid. Plants can be designed so that they furnish a measure of dispatch capability, carrying out some of the traditional duties of conventional power plants. These plants would be operated below their maximum power rating most of the time and would trade some energy capture for grid ancillary services. Paying for this trade-off will require either a lower capital cost for the hardware, contractual arrangements that will pay for grid services at a high enough rate to offset the energy loss, or optimally, a combination of the two. Wind plants might transition, then, from a simple energy source to a power plant that delivers significant grid support.

2.3.2 Learning-Curve Effect

Progressing along the design and manufacturing learning curve allows engineers to develop technology improvements (such as those listed in Section 2.3.1) and reduce capital costs. The more engineers and manufacturers learn by conducting effective RD&D and producing greater volumes of wind energy equipment, the more proficient and efficient the industry becomes. The learning curve is often measured by calculating the progress ratio, defined as the ratio of the cost after doubling cumulative production to the cost before doubling.

The progress ratio for wind energy from 1984 to 2000 was calculated for the high volume of machines installed in several European countries that experienced a healthy combination of steadily growing manufacturing output, external factors, and research investment during that time. Results show that progress ratio estimates were approximately the same for Denmark (91%), Germany (94%), and Spain (91%) (ISET 2003). At the time this report was written, there was not enough reliable data on U.S.-based manufacturing of wind turbines to determine a U.S. progress ratio. Figure 2-13 shows the data for Spain.

Figure 2-13. Cost of wind turbines delivered from Spain between 1984 and 2000

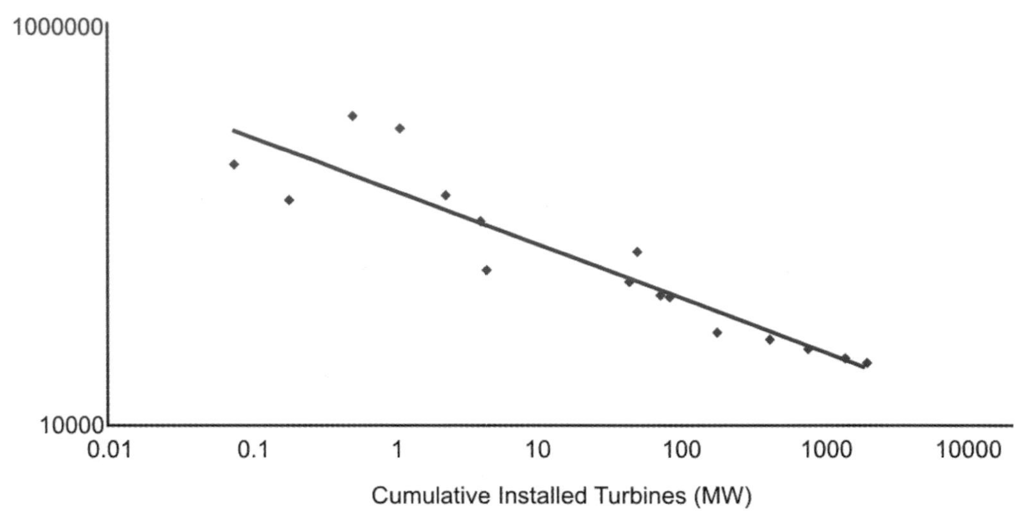

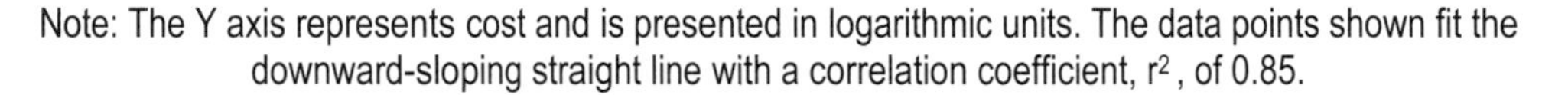
Note: The Y axis represents cost and is presented in logarithmic units. The data points shown fit the downward-sloping straight line with a correlation coefficient, r^2, of 0.85.

Moving from the current level of installed wind capacity of roughly 12 gigawatts (GW) to the 20% Wind Scenario total of 305 GW will require between four and five doublings of capacity. If the progress ratio of 91% shown in Figure 2-13 continues, prices could drop to about 65% of current costs, a 35% reduction. The low-hanging fruit of cost reduction, however, has already been harvested. The industry has progressed from machines based on designs created without any design tools and built almost entirely by hand to the current state of advanced engineering capability. The assumption in the 20% Wind Scenario is that a 10% reduction in capital cost could accelerate large-scale deployment. In order to achieve this reduction, a progress ratio of only 97.8% is required to produce a learning curve effect of 10% with 4.6 doublings of capacity. With sustained manufacturing growth and technological advancement, there is no technical barrier to achieving 10% capital cost reduction. See Appendix B for further discussion.

2.3.3 The System Benefits of Advanced Technology

A cost study conducted by the U.S. Department of Energy (DOE) Wind Program identified numerous opportunities for technology advancement to reduce the life-cycle COE (Cohen and Schweizer et al. 2008). Based on machine performance and cost, this study used advanced concepts to suggest pathways that integrate the individual contributions from component-level improvements into system-level estimates of the capital cost, annual energy production, reliability, O&M, and balance of station. The results, summarized in Table 2-1, indicate significant potential impacts on annual energy production and capital cost. Changes in annual energy production are equivalent to changes in capacity factor because the turbine rating was fixed. A range of values represents the best, most likely, and least beneficial outcomes.

The Table 2-1 capacity factor improvement of 11% that results from taller towers reflects the increase in wind resources at a hub height of 120 m, conservatively assuming the standard wind shear distribution meteorologists use for open country. Uncertainty in these capacity factor improvements are reflected in the table below. Depending on the success of new tower technology, the added costs could range from 8% to 20%, but there will definitely be an added cost if the tower is the only component in the system that is modified to take the rotor to higher elevations. An advantage would come from a system design in which the tower head mass is significantly reduced with the integration of a rotor and drivetrain that are significantly lighter.

Table 2-1. Areas of potential technology improvement

Technical Area	Potential Advances	Performance and Cost Increments (Best/Expected/Least Percentages)	
		Annual Energy Production	Turbine Capital Cost
Advanced Tower Concepts	• Taller towers in difficult locations • New materials and/or processes • Advanced structures/foundations • Self-erecting, initial, or for service	+11/+11/+11	+8/+12/+20
Advanced (Enlarged) Rotors	• Advanced materials • Improved structural-aero design • Active controls • Passive controls • Higher tip speed/lower acoustics	+35/+25/+10	-6/-3/+3
Reduced Energy Losses and Improved Availability	• Reduced blade soiling losses • Damage-tolerant sensors • Robust control systems • Prognostic maintenance	+7/+5/0	0/0/0
Drivetrain (Gearboxes and Generators and Power Electronics)	• Fewer gear stages or direct-drive • Medium/low speed generators • Distributed gearbox topologies • Permanent-magnet generators • Medium-voltage equipment • Advanced gear tooth profiles • New circuit topologies • New semiconductor devices • New materials (gallium arsenide [GaAs], SiC)	+8/+4/0	-11/-6/+1
Manufacturing and Learning Curve*	• Sustained, incremental design and process improvements • Large-scale manufacturing • Reduced design loads	0/0/0	-27/-13/-3
Totals		+61/+45/+21	-36/-10/+21

*The learning curve results from the NREL report (Cohen and Schweizer et al. 2008) are adjusted from 3.0 doublings in the reference to the 4.6 doublings in the 20% Wind Scenario.

2

The capital cost reduction shown for the drivetrain components is mainly attributed to the reduced requirements on the structure when lighter components are placed on the tower top. Performance increases as parasitic losses in mechanical and electrical components are reduced. Such components are designed specifically to optimize the performance for wind turbine characteristics. The improvements shown in Table 2-1 are in the single digits, but are not trivial.

Without changing the location of the rotor, energy capture can also be increased by using longer blades to sweep more area. A 10% to 35% increase in capacity factor is produced by 5% to 16% longer blades for the same rated power output. Building these longer blades at an equal or lower cost is a challenge, because blade weight must be capped while turbulence-driven loads remain no greater than what the smaller rotor can handle. With the potential of new structurally efficient airfoils, new materials, passive load attenuation, and active controls, it is estimated that this magnitude of blade growth can be achieved in combination with a modest system cost reduction.

Technology advances can also reduce energy losses in the field. Improved O&M techniques and monitoring capabilities can reduce downtime for repairs and scheduled maintenance. It is also possible to mitigate losses resulting from degradation of performance caused by wear and dirt over time. These improvements are expected to be in the single digits at best, with an approximate 5% improvement in lifetime energy capture.

Doubling the number of manufactured turbines several times over the years will produce a manufacturing learning-curve effect that can also help reduce costs. The learning-curve effects shown in Table 2-1 are limited to manufacturing-related technology improvements and do not reflect issues of component selection and design. As discussed in Section 2.3.2, the learning curve reflects efficiencies driven by volume production and manufacturing experience as well as the infusion of manufacturing technology and practices that encourage more manufacturing-friendly design in the future. Although these changes do not target any added energy capture, they are expected to result in continuous cost reductions. The only adjustment from the NREL reference (Cohen and Schweizer et al. 2008) is that the 20% Wind Scenario by 2030 requires 4.6 doublings of cumulative capacity rather than the 3.0 doublings used in the reference targeted at the year 2012. The most likely 13% cost reduction assumes a conservative progress ratio of 97% per doubling of capacity. However, there are a range of possible outcomes.

The potential technological advances outlined here support the technical feasibility of the 20% Wind Scenario by outlining several possible pathways to a substantial increase in capacity factor accompanied by a modest but double-digit reduction in capital cost.

2.3.4 Targeted RD&D

While there is an expected value to potential technology improvements, the risk of implementing them has not yet been reduced to the level that allows those improvements to be used in commercial hardware. The issues are well known and offer an opportunity for focused RD&D efforts. In the past, government and industry collaboration has been successful in moving high-risk, high-potential technologies into the marketplace.

One example of such collaboration is the advanced natural gas turbine, which improved the industry efficiency standard—which had been capped at 50%—to almost 60%. DOE invested $100 million in the H-system turbine and General Electric (GE) invested $500 million. Although it was known that higher operating temperatures would lead to higher efficiency, there were no materials for the turbine blades that could withstand the environment. The research program focused on advanced cooling techniques and new alloys to handle combustion that was nearly 300°F hotter. The project produced the world's largest single crystal turbine blades capable of resisting high-temperature cracking. The resulting "H system" gas turbine is 11.89 m long, 4.89 m in diameter, and weighs more than 811,000 lb. Each turbine is expected to save more than $200 million in operating costs over its lifetime (DOE 2000).

A similar example comes from the aviation world. The use of composite materials was known to provide excellent benefits for light-jet airframes, but the certification process to characterize the materials was onerous and expensive. NASA started a program to "reduce the cost of using composites and develop standardized procedures for certifying composite materials" (Brown 2007). The Advanced General Aviation Transport Experiments (AGATE), which began in 1994, solved those problems and opened the door for new composite material technology to be applied to the light-jet application. A technology that would have been too high-risk for the individual companies to develop was bridged into the marketplace through a cooperative RD&D effort by NASA, the Federal Aviation Administration (FAA), industry, and universities. The Adam aircraft A500 turboprop and the A700 very light jet are examples of new products based on this composite technology.

Some might claim that wind technology is a finished product that no longer needs additional RD&D, or that all possible improvements have already been made. The reality is that the technology is substantially less developed than fossil energy technology, which is still being improved after a century of generating electricity. A GE manager who spent a career in the gas turbine business and then transferred to manage the wind turbine business noted the complexity of wind energy technology: "Our respect for wind turbine technology has grown tremendously. The practical side is so complex and forces are so dramatic. We would never have imagined how complex turbines are" (Knight and Harrison 2005).

Already, there is a clear understanding of the materials, controls, and aerodynamics issues that must be resolved to make progress toward greater capacity factors. The combination of reduced capital cost and increased capacity factor will lead to reduced COE. Industry feels the risk of bringing new technology into the marketplace without a full-scale development program is too great and believes sustained RD&D would help reduce risk and help enable the transfer of new technology to the marketplace.

2.4 Addressing Technical and Financial Risks

Risks tend to lessen industry's desire to invest in wind technology. The wind plant performance track record, in terms of generated revenues and operating costs compared with the estimated revenues used in plant financing, will drive the risk level of future installations. The consequences of these risks directly affect the revenues of owners of wind manufacturing and operating capabilities.

2.4.1 Direct Impacts

When owners of wind manufacturing and operating capabilities directly bear the costs of failure, the impacts are said to be direct. This direct impact on revenue is often caused by:

- **Increasing O&M costs:** As discussed previously and illustrated in Figure 2-9, there is mounting evidence that O&M costs are increasing as wind farms age. Most of these costs are associated with unplanned maintenance or components wearing out before the end of their intended design lives. Some failures can be traced to poor manufacturing or installation quality. Others are caused by design errors, many of which are caused by weaknesses in the technology's state of the art, generally codified by the design process. Figures 2-14 and 2-15 both show steadily rising O&M costs for wind farms installed in the United States in the two decades before the turn of the century, and Figure 2-14 shows the components that have caused these increasing costs. The numbers and costs of component failures increase with time, and the risk to the operators grows accordingly. In Figure 2-14, the solid lines represent expected repairs that may not be completely avoidable, and the dashed lines show potential early failures that can significantly increase risk.
- **Poor availability driven by low reliability:** Energy is not generated while components are being repaired or replaced. Although a single failure of a critical component stops production from only one turbine, such losses can mount up to significant sums of lost revenue.
- **Poor wind plant array efficiency:** If turbines are placed too close together, their wakes interact, which can cause the downwind turbines to perform poorly. But if they are placed too far apart, land and plant maintenance costs increase.

Figure 2-14. Unplanned repair cost, likely sources, and risk of failure with wind plant age

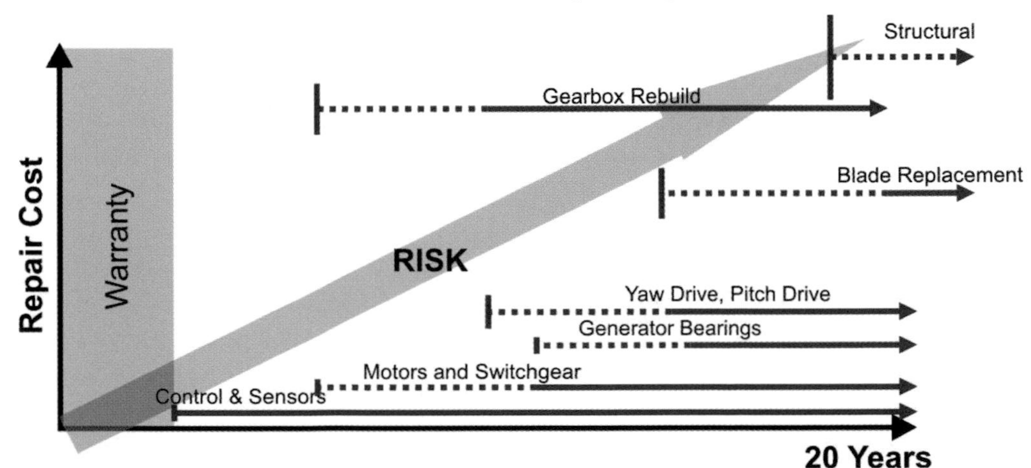

Figure 2-15. Average O&M costs of wind farms in the United States

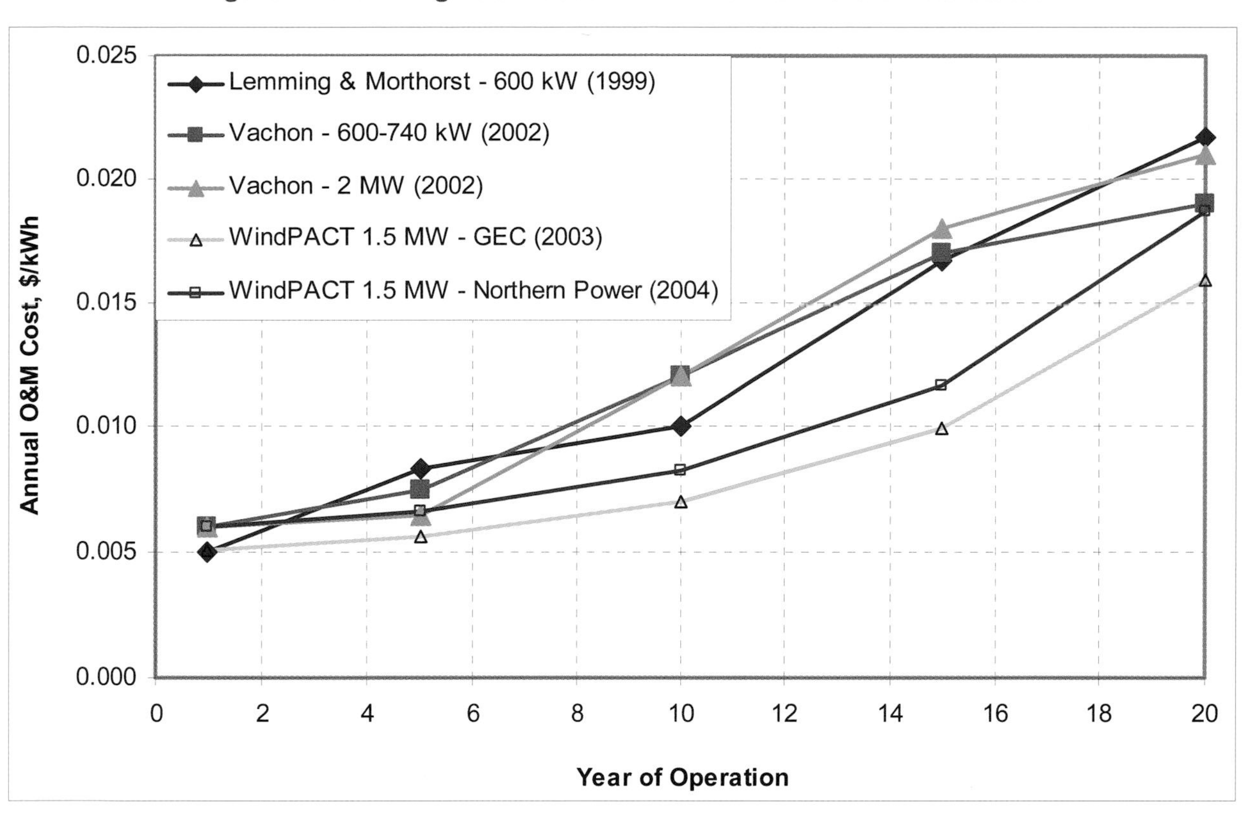

2.4.2 Indirect Impacts

Although the wind industry has achieved high levels of wind plant availability and reliability, unpredictable or unreliable performance would threaten the credibility of this emerging technology in the eyes of financial institutions. The consequences of real or perceived reliability problems would extend beyond the direct cost to the plant owners. These consequences on the continued growth of investment in wind could include:

- **Increased cost of insurance and financing:** Low interest rates and long-term loans are critical to financing power plants that are loaded with upfront capital costs. Each financial institution will assess the risk of investing in wind energy and charge according to those risks. If wind power loses credibility, these insurance and financing costs could increase.
- **Slowing or stopping development:** Lost confidence contributed to the halt of development in the United States in the late 1980s through the early 1990s. Development did not start again until the robust European market supported the technology improvements necessary to reestablish confidence in reliable European turbines. As a result, the current industry is dominated by European wind turbine companies. Active technical supporters of RD&D must anticipate and resolve problems before they threaten industry development.
- **Loss of public support:** If wind power installations do not operate continuously and reliably, the public might be easily convinced that

renewable energy is not a viable source of energy. The public's confidence in the technology is crucial. Without public support, partnerships working toward a new wind industry future cannot be successful.

2.4.3 Risk Mitigation Through Certification, Validation, and Performance Monitoring

To reduce risk, the wind industry requires turbines to adhere to international standards. These standards, which represent the collective experience of the industry's leading experts, imply a well-developed design process that relies on the most advanced design tools, testing for verification, and disciplined quality control.

Certification

Certification involves high-level, third-party technical audits of a manufacturer's design development. It includes a detailed review of design analyses, material selections, dynamic modeling, and component test results. The wind industry recognizes that analytical reviews are not sufficient to capture weaknesses in the design process. Therefore, consensus standard developers also require full-scale testing of blades, gearboxes, and the complete system prototype (see "Industry Standards" sidebar).

Actively complying with these standards encourages investment in wind energy by ensuring that turbines reliably achieve the maximum energy extraction needed to expand the industry.

Industry Standards

The American National Standards Institute (ANSI) has designated the American Wind Energy Association (AWEA) as the lead organization for the development and publication of industry consensus standards for wind energy equipment and services in the United States. AWEA also participates in the development of international wind energy standards through its representation on the International Electrotechnical Commission (IEC) TC-88 Subcommittee. Information on these standards can be accessed on AWEA's Web site (http://www.awea.org/standards).

Full-Scale Testing

Testing standards were drafted to ensure that accredited third-party laboratories are conducting tests consistently. These tests reveal many design and manufacturing deficiencies that are beyond detection by analytical tools. They also provide the final verification that the design process has worked and give the financial community the confidence needed to invest in a turbine model.

Full-scale test facilities and trained test engineers capable of conducting full-scale tests are rare. The facilities must have equipment capable of applying tremendous loads that mimic the turbulence loading that wind applies over the entire life of the blade or gearbox. Full-scale prototype tests are conducted in the field at locations with severe wind conditions. Extensive instrumentation is applied to the machine, according to a test plan prescribed by international standards, and comprehensive data are recorded over a specified range of operating conditions. These data give the certification agent a means for verifying the accuracy of the design's analytical basis. The industry and financial communities depend on these facilities and skilled test engineers to support all new turbine component development.

As turbines grow larger and more products come on the market, test facilities must also grow and become more efficient. New blades are reaching 50 m in length, and the United States has no facilities that can test blades longer than 50 m. Furthermore, domestic dynamometer facilities capable of testing gearboxes or new drivetrains are limited in capacity to 1.5 MW. The limited availability of facilities and qualified test engineers increases the deployment risk of new machines that are not subjected to the rigors of current performance validation in accredited facilities.

At full-scale facilities, it is also difficult to conduct tests accurately and capture the operating conditions that are important to verify the machine's reliability. These tests are expensive to conduct and accreditation is expensive to maintain for several reasons. First, the scale of the components is one of the largest of any commercial industry. Because blades are approaching sizes of half the length of a football field and can weigh more than a 12.2 m yacht, they are very difficult and expensive to transport on major highways. The magnitude of torque applied to the drivetrains for testing is among the largest of any piece of rotating equipment ever constructed. Figure 2-16 shows the largest blades being built and the approximate dates when U.S. blade test facilities were built to accommodate their testing.

Although it is very expensive for each manufacturer to develop and maintain

Figure 2-16. Blade growth and startup dates for U.S. blade test facilities

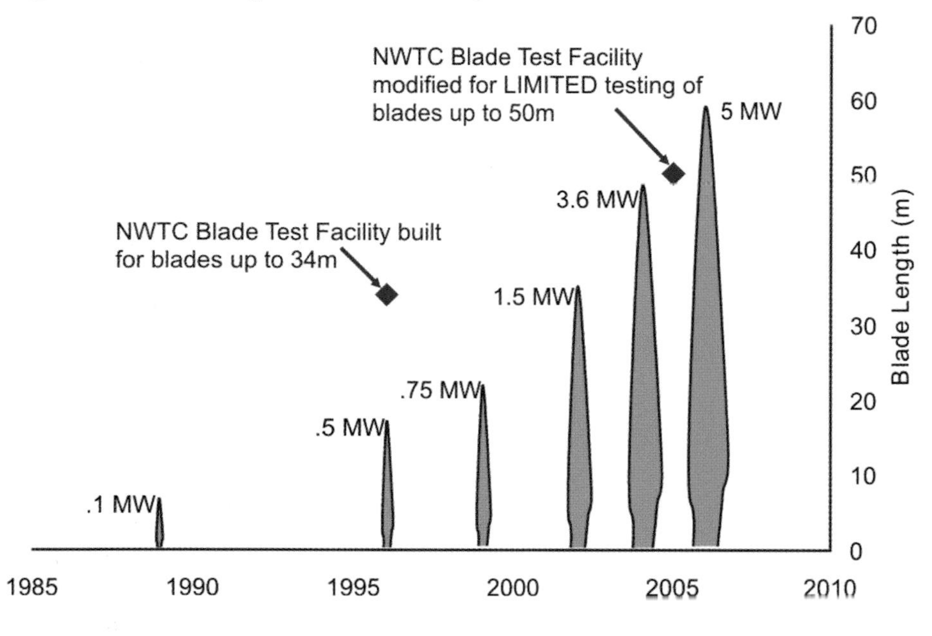

facilities of this scale for its own certification testing needs, without these facilities, rapid technological progress will be accompanied by high innovation risk. Wind energy history has proven that these kinds of tests are crucial for the industry's success and the financial community's confidence. These tests, then, are an essential element of any risk mitigation strategy.

Performance Monitoring and O&M

One of the main elements of power plant management is strategic monitoring of reliability. Other industries have established anonymous databases that serve to benchmark their reliability and performance, giving operators both the ability to recognize a drop in reliability and the data they need to determine the source of low reliability. The wind industry needs such a strategically designed database, which would give O&M managers the tools to recognize and pinpoint drops in reliability,

2

along with a way to collectively resolve technical problems. Reliability databases are an integral part of more sophisticated O&M management tools. Stiesdal and Madsen (2005) describe how databases can be used for managing O&M and improving future designs.

In mature industries, O&M management tools are available to help maximize maintenance efficiency. Achieving this efficiency is a key factor in minimizing the COE and maximizing the life of wind plants, thereby increasing investor confidence. Unlike central generation facilities, wind plants require maintenance strategies that minimize human attention and maximize remote health monitoring and automated fault data diagnosis. This requires intimate knowledge of healthy plant operating characteristics and an ability to recognize the characteristics of very complex faults that might be unique to a specific wind plant. Such tools do not currently exist for the wind industry, and their development will require RD&D to study wind plant systems interacting with complex atmospheric conditions and to model the interactions. The resultant deeper understanding will allow expert systems to be developed, systems that will aid operators in their quest to maximize plant performance and minimize operating costs through risk mitigation. These systems will also produce valuable data for improving the next generation of turbine designs.

2.5 Offshore Wind Technology

Offshore wind energy installations have a broadly dispersed, abundant resource and the economic potential for cost competitiveness that would allow them to make a large impact in meeting the future energy needs of the United States (Musial 2007). Of the contiguous 48 states, 28 have a coastal boundary. U.S. electric use data show that these same states use 78% of the nation's electricity (EIA 2006). Of these 28 states, only 6 have a sufficient land-based wind energy resource to meet more than 20% of their electric requirements through wind power. If shallow water offshore potential (less than 30 m in depth) is included in the wind resource mix, though, 26 of the 28 states would have the wind resources to meet at least 20% of their electric needs, with many states having sufficient offshore wind resources to meet 100% of their electric needs (Musial 2007). For most coastal states, offshore wind resources are the only indigenous energy source capable of making a significant energy contribution. In many congested energy-constrained regions, offshore wind plants might be necessary to supplement growing demand and dwindling fossil supplies.

Twenty-six offshore wind projects with an installed capacity of roughly 1,200 MW now operate in Europe. Most of these projects were installed in water less than 22 m deep. One demonstration project in Scotland is installed in water at a depth of 45 m. Although some projects have been hampered by construction overruns and higher-than-expected maintenance requirements, projections show strong growth in many European Union (EU) markets. For example, it is estimated that offshore wind capacity in the United Kingdom will grow by 8,000 MW by 2015. Similarly, German offshore development is expected to reach 5,600 MW by 2014 (BSH; BWEA).

In the United States, nine offshore project proposals in state and federal waters are in various stages of development. Proposed projects on the Outer Continental Shelf are under the jurisdiction of the Minerals Management Service (MMS) with their authority established by the Energy Policy Act (EPAct) of 2005 (MMS). Several states are pursuing competitive solicitations for offshore wind projects approval.

2.5.1 Cost of Energy

The current installed capital cost of offshore projects is estimated in the range of $2,400 to $5,000 per kW (Black & Veatch 2007; Pace Global 2007). Because offshore wind energy tends to take advantage of extensive land-based experience and mature offshore oil and gas practices, offshore cost reductions are not expected to be as great as land-based reductions spanning the past two decades. However, offshore wind technology is considerably less mature than land-based wind energy, so it does have significant potential for future cost reduction. These cost reductions are achievable through technology development and innovation, implementation and customization of offshore oil and gas practices, and learning-curve reductions that take advantage of more efficient manufacturing and deployment processes and procedures.

2.5.2 Current Technology

Today's baseline technology for offshore wind turbines is essentially a version of the standard land-based turbine adapted to the marine environment. Although turbines of up to 5 MW have been installed, most recent orders from Vestas (Randers, Denmark) and Siemens (Munich, Germany), the two leading suppliers of offshore wind turbines, range from 2.0 MW to 3.6 MW.

The architecture of the baseline offshore turbine and drivetrain comprises a three-bladed upwind rotor, typically 90 m to 107 m in diameter. Tip speeds of offshore turbines are slightly higher than those of land-based turbines, which have speeds of 80 m/s or more. The drivetrain consists of a gearbox generally run with variable-speed torque control that can achieve generator speeds between 1,000 and 1,800 rpm. The offshore tower height is generally 80 m, which is lower than that of land-based towers, because wind shear profiles are less steep, tempering the advantage of tower height.

The offshore foundation system baseline technology uses monopiles at nominal water depths of 20 m. Monopiles are large steel tubes with a wall thickness of up to 60 mm and diameters of 6 m. The embedment depth varies with soil type, but a typical North Sea installation must be embedded 25 m to 30 m below the mud line. The monopile extends above the surface where a transition piece with a flange to fasten the tower is leveled and grouted. Its foundation requires a specific class of installation equipment for driving the pile into the seabed and lifting the turbine and tower into place. Mobilization of the infrastructure and logistical support for a large offshore wind plant accounts for a significant portion of the system cost.

Turbines in offshore applications are arranged in arrays that take advantage of the prevailing wind conditions measured at the site. Turbines are spaced to minimize aggregate power plant energy losses, interior plant turbulence, and the cost of cabling between turbines.

The power grid connects the output from each turbine, where turbine transformers step up the generator and the power electronics voltage to a distribution voltage of about 34 kilovolts (kV). The distribution system collects the power from each turbine at a central substation where the voltage is stepped up and transmitted to shore through a number of buried, high-voltage subsea cables. A shore-based interconnection point might be used to step up the voltage again before connecting to the power grid.

Shallow water wind turbine projects have been proposed and could be followed by transitional and finally deepwater turbines. These paths should not be considered as mutually exclusive choices. Because there is a high degree of interdependence among them, they should be considered a sequence of development that builds from a shallow water foundation of experience and knowledge to the complexities of deeper water.

2.5.3 Technology Needs and Potential Improvements

Offshore, wind turbine cost represents only one-third of the total installed cost of the wind project, whereas on land, the turbine cost represents more than half of the total installed cost. To lower costs for offshore wind, the focus must be on lowering the balance-of-station costs. These costs, which include those for foundations, electrical grids, O&M, and installation and staging costs, dominate the system COE. Turbine improvements that make turbines more reliable, more maintainable, more rugged, and larger, will still be needed to achieve cost goals. Although none of these improvements are likely to lower turbine costs, the net result will lower overall system costs.

Commercialization of offshore wind energy faces many technical, regulatory, socioeconomic, and political barriers, some of which may be mitigated through targeted short- and long-range RD&D efforts. Short-term research addresses impediments that prevent initial industry projects from proceeding and helps sharpen the focus for long-term research. Long-term research involves a more complex development process resulting in improvements that can help lower offshore life-cycle system costs.

Short-Term RD&D Options

Conducting research that will lead to more rapid deployment of offshore turbines should be an upfront priority for industry. This research should address obstacles to today's projects, and could include the following tasks:

- **Define offshore resource exclusion zones:** A geographically based exclusion study using geographic information system (GIS) land use overlays would more accurately account for all existing and future marine uses and sensitive areas. This type of exclusion study could be part of a regional programmatic environmental impact statement and is necessary for a full assessment of the offshore resource (Dhanju, Whitaker, and Kempton 2006). Currently, developers bear the burden of siting during a pre-permitting phase with very little official guidance. This activity should be a jointly funded industry project conducted on a regional basis.
- **Develop certification methods and standards:** MMS has been authorized to define the structural safety standards for offshore wind turbines on the OCS. Technical research, analysis, and testing are needed to build confidence that safety will be adequate, and to prevent overcautiousness that will increase costs unnecessarily. Developing these standards will require a complete evaluation and harmonization of the existing offshore wind standards and the American Petroleum Institute (API) offshore oil and gas standards. MMS is currently determining the most relevant standards.
- **Develop design codes, tools, and methods:** The design tools that the wind industry uses today have been developed and validated for

land-based utility-scale turbines, and the maturity and reliability of the tools have led to significantly higher confidence in today's wind turbines. By comparison, offshore design tools are relatively immature. The development of accurate offshore computer codes to predict the dynamic forces and motions acting on turbines deployed at sea is essential for moving into deeper water. One major challenge is predicting loads and the resulting dynamic responses of the wind turbine's support structure when it is subjected to combined wave and wind loading. These offshore design tools must be validated to ensure that they can deal with the combined dominance of simultaneous wind and wave load spectra, which is a unique problem for offshore wind installations. Floating system analysis must be able to account for additional turbine motions as well as the dynamic characterization of mooring lines.

- **Site turbines and configure arrays:** The configuration and spacing of wind turbines within an array have a marked effect on power production from the aggregate wind plant, as well as for each individual turbine. Uncertainties in power production represent a large economic risk factor for offshore development. Offshore wind plants can lose more than 10% of their energy to array losses, but improvements in array layout and array optimization models could deliver substantial recovery (SEAWIND 2003). Atmospheric boundary layer interaction with the turbine wakes can affect both energy capture and plant-generated turbulence. Accurate characterization of the atmospheric boundary layer behavior and more accurate wake models will be essential for designing turbines that can withstand offshore wind plant turbulence. Wind plant design tools that are able to characterize turbulence generated by wind plants under a wide range of conditions are likely necessary.
- **Develop hybrid wind-speed databases:** Wind, sea-surface temperatures, and other weather data are housed in numerous satellite databases available from the National Oceanic and Atmospheric Administration (NOAA), NASA, the National Weather Service (NWS), and other government agencies. These data can be combined to supplement the characterization of coastal and offshore wind regimes (Hasager et al. 2005). The limitations and availability of existing offshore data must be understood. Application of these data to improve the accuracy of offshore wind maps will also be important.

Long-Term R&D Options

Long-term research generally requires hardware development and capital investment, and it must take a complex development path that begins early enough for mature technology to be ready when needed. Most long-term research areas relate to lowering offshore life-cycle system costs. These areas are subdivided into infrastructure and turbine-specific needs. Infrastructure to support offshore wind development represents a major cost element. Because this is a relatively new technology path, there are major opportunities for reducing the cost impacts. Although land-based wind turbine designs can generally be used for offshore deployment, the offshore environment will impose special requirements on turbines. These requirements must be taken into account to optimize offshore deployment. Areas where industry should focus efforts include:

2

- **Minimize work at sea:** There are many opportunities to lower project costs by reallocating the balance between work done on land and at sea. The portion of labor devoted to project O&M, land-based installation and assembly, and remote inspections and diagnostics can be rebalanced with upfront capital enhancements, such as higher quality assurance, more qualification testing, and reliable designs. This rebalancing might enable a significant life-cycle cost reduction by shifting the way wind projects are designed, planned, and managed.

- **Enhance manufacturing, installation and deployment strategies:** New manufacturing processes and improvements in existing processes that reduce labor and material usage and improve part quality have high potential for reducing costs in offshore installations. Offshore wind turbines and components could be constructed and assembled in or near seaport facilities that allow easy access from the production area to the installation site, eliminating the necessity of shipping large components over inland roadways. Fabrication facilities must be strategically located for mass-production, land-based assembly, and for rapid deployment with minimal dependence on large vessels. Offshore system designs that can be floated out and installed without large cranes can reduce costs significantly. New strategies should be integrated into the turbine design process at an early stage (Lindvig 2005; Poulsen and Skjærbæk 2005).

- **Incorporate offshore service and accessibility features:** To manage O&M, predict weather windows, minimize downtime, and reduce the equipment needed for up-tower repairs, operators should be equipped with remote, intelligent, turbine condition monitoring and self-diagnostic systems. These systems can alert operators to the need for operational changes, or enable them to schedule maintenance at the most opportune times. A warning about an incipient failure can alert the operators to replace or repair a component before it does significant damage to the system or leaves the machine inoperable for an extended period of time. More accurate weather forecasting will also become a major contributor in optimizing service schedules for lower cost.

- **Develop low-cost foundations, anchors, and moorings:** Current shallow-water foundations have already reached a practical depth limit of 30 m, and anchor systems beyond that are derived from conservative and expensive oil and gas design practices. Cost-saving opportunities arise for wind power plants in deeper water with both fixed-bottom and floating turbine foundations, as well as for existing shallow-water designs in which value-engineering cost reductions can be achieved. Fixed-bottom systems comprising rigid lightweight substructures, automated mass-production fabrication facilities, and integrated mooring and piling deployment systems that minimize dependence on large sea vessels are possible low-cost options. Floating platforms will require a new generation of mooring designs that can be mass produced and easily installed.

- **Use resource modeling and remote profiling systems:** Offshore winds are much more difficult to characterize than winds over land. Analytical models are essential for managing risk during the initial siting of offshore projects, but are not very useful by themselves for micrositing (Jimenez et al. 2005). Alternative methods are needed to measure wind speed and wind shear profiles up to elevations where wind turbines operate. This will require new equipment such as sonic detection and ranging (SODAR), light detection and ranging (LIDAR), and coastal RADAR-based systems that must be adapted to measure offshore wind from more stable buoy systems or from fixed bases. Some systems are currently under development but have not yet been proven (Antoniou et al. 2006). The results of an RD&D measurement program on commercial offshore projects could generate enough confidence in these systems to eliminate the requirement for a meteorological tower.

- **Increase offshore turbine reliability:** The current offshore service record is mixed, and as such, is a large contributor to high risk. A new balance between initial capital investment and long-term operating costs must be established for offshore systems. This new balance will have a significant impact on COE. Offshore turbine designs must place a higher premium on reliability and anticipation of on-site repairs than their land-based counterparts. Emphasis should be placed on avoiding large maintenance events that require expensive and specialized equipment. This can be done by identifying the root causes of component failures, understanding the frequency and cost of each event, and appropriately implementing design improvements (Stiesdal and Madsen 2005). Design tools, quality control, testing, and inspection will need heightened emphasis. Blade designers must consider strategies to offset the impacts of marine moisture, corrosion, and extreme weather. In higher latitudes, designers must also account for ice flows and ice accretion on the blades. Research that improves land-based wind turbine reliability now will have a direct impact on the reliability of future offshore machines.

- **Assess the potential of ultra-large offshore turbines:** Land-based turbines may have reached a size plateau because of transportation and erection limits. Further size growth in wind turbines will largely be pushed by requirements unique to offshore turbine development. According to a report on the EU-funded UpWind project, "Within a few years, wind turbines will have a rotor diameter of more than 150 m and a typical size of 8 MW–10 MW" (Risø National Laboratory 2005). The UpWind project plans to develop design tools to optimize large wind turbine components, including rotor blades, gearboxes, and other systems that must perform in large offshore wind plants. New size-enabling technologies will be required to push wind turbines beyond the scaling limits that constrain the current fleet. These technologies include lightweight composite materials and composite manufacturing, lightweight drivetrains, modular pole direct-drive generators, hybrid space frame towers, and large gearbox and bearing designs that are tolerant of slower speeds and larger scales. All of the weight-reducing features of the taller land-based tower systems will have an even greater value for very large offshore machines (Risø National Laboratory 2005).

2

RD&D Summary

The advancement of offshore technology will require the development of infrastructure and technologies that are substantially different from those employed in land-based installations. In addition, these advances would need to be tailored to U.S. offshore requirements, which differ from those in the European North Sea environment. Government leadership could accelerate baseline research and technology development to demonstrate feasibility, mitigate risk, and reduce regulatory and environmental barriers. Private U.S. energy companies need to take the technical and financial steps to initiate near-term development of offshore wind power technologies and bring them to sufficient maturity for large-scale deployment. Musial and Ram (2007) and Bywaters and colleagues (2005) present more detailed analyses of actions for offshore development.

2.6 Distributed Wind Technology

Distributed wind technology (DWT) applications refer to turbine installations on the customer side of the utility meter. These machines range in size from less than 1 kW to multimegawatt, utility-scale machines, and are used to offset electricity consumption at the retail rate. Because the WinDS deployment analysis does not currently segregate DWT from utility deployment, DWT applications are part of the land-based deployment estimates in the 20% Wind Energy Scenario.

Historically, DWT has been synonymous with small machines. The DWT market in the 1990s focused on battery charging for off-grid homes, remote telecommunications sites, and international village power applications. In 2000, the industry found a growing domestic market for behind-the-meter wind power, including small machines for residential and small farm applications and multimegawatt-scale machines for larger agricultural, commercial, industrial, and public facility applications. Although utility-scale DWT requirements are not distinguishable from those for other large-scale turbines, small machines have unique operating requirements that warrant further discussion.

2.6.1 Small Turbine Technology

Until recently, three-bladed upwind designs using tail vanes for passive yaw control dominated small wind turbine technology (turbines rated at less than 10 kW). Furling, or turning the machine sideways to the wind with a mechanical linkage, was almost universally used for rotor overspeed control. Drivetrains were direct-drive, permanent-magnet alternators with variable-speed operation. Many of these installations were isolated from the grid. Today, there is an emerging technology trend toward grid-connected applications and nonfurling designs. U.S. manufacturers are world leaders in small wind systems rated at 100 kW or less, in terms of both market and technology.

Turbine technology begins the transition from small to large systems between 20 kW and 100 kW. Bergey Windpower (Norman, Oklahoma) offers a 50 kW turbine that uses technology commonly found in smaller machines, including furling, pultruded blades, a direct-drive, permanent-magnet alternator, and a tail vane for yaw control. Distributed Energy Systems offers a 100 kW turbine that uses a direct-drive, variable-speed synchronous generator. Although most wind turbines in the 100 kW range have features common to utility-scale turbines, including gearboxes, mechanical brakes, induction generators, and upwind rotors with active yaw control,

Endurance Windpower (Spanish Fork, Utah) offers a 5 kW turbine with such characteristics.

For small DWT applications, reliability and acoustic emissions are the prominent issues. Installations usually consist of a single turbine. Installations may also be widely scattered. So simplicity in design, ease of repair, and long maintenance and inspection intervals are important. Because DWT applications are usually close to workplaces or residences, limiting sound emissions is critical for market acceptance and zoning approvals. DWT applications are also usually located in areas with low wind speeds that are unsuitable for utility-scale applications, so DWT places a premium on low-wind-speed technologies.

The cost per kW of DWT turbines is inversely proportionate with turbine size. Small-scale DWT installation costs are always higher than those for utility-scale installations because the construction effort cannot be amortized over a large number of turbines. For a 1 kW system, hardware costs alone can be as high as $5,000 to $7,000/kW. Installation costs vary widely because of site-specific factors such as zoning and/or permitting costs, interconnection fees, balance-of-station costs, shipping, and the extent of do-it-yourself participation. Five-year warranties are now the industry standard for small wind turbines, although it is not yet known how this contributes to turbine cost. The higher costs of this technology are partially offset by the ability to compete with retail electricity rates. In addition, small turbines can be connected directly to the electric distribution system, eliminating the need for an expensive interconnection between the substation and the transmission.

Tower and foundation costs make up a larger portion of DWT installed cost, especially for wind turbines of less than 20 kW. Utility-scale turbines commonly use tapered tubular steel towers. However, for small wind turbines, multiple types, sources, and heights of towers are available.

2.6.2 Technology Trends

Recent significant developments in DWT systems less than 20 kW include the following:

- **Alternative power and load control strategies:** Furling inherently increases sound levels because the cross-wind operation creates a helicopter-type chopping noise. Aerodynamic models available today cannot accurately predict the rotor loads in the highly skewed and unsteady flows that occur during the furling process, complicating design and analysis. Alternative development approaches include soft-stall rotor-speed control, constant-speed operation, variable-pitch blades, hinged blades, mechanical brakes, and centrifugally actuated blade tips. These concepts offer safer, quieter turbines that respond more predictably to high winds, gusts, and sudden wind direction changes.
- **Advanced blade manufacturing methods:** Blades for small turbines have been made primarily of fiberglass by hand lay-up manufacturing or pultrusion. The industry is now pursuing alternative manufacturing techniques, including injection, compression, and reaction injection molding. These methods often provide shorter fabrication time, lower parts costs, and increased repeatability and uniformity, although the tooling costs are typically higher.

- **Rare-earth permanent magnets:** Ferrite magnets have long been the staple in permanent-magnet generators for small wind turbines. Rare-earth permanent magnets are now taking over the market with Asian suppliers offering superior magnetic properties and a steady decline in price. This enables more compact and lighter weight generator designs.
- **Reduced generator cogging:** Concepts for generators with reduced cogging torque (the force needed to initiate generator rotation) are showing promise to reduce cut-in wind speeds. This is an important advancement to improve low-wind-speed turbine performance and increase the number of sites where installation is economical.
- **Induction generators:** Small turbine designs that use induction generators are under development. This approach, common in the early 1980s, avoids the use of power electronics that increase cost and complexity, and reduce reliability.
- **Grid-connected inverters:** Inverters used in the photovoltaics market are being adapted for use with wind turbines. Turbine-specific inverters are also appearing in both single- and three-phase configurations. Another new trend is obtaining certification of most inverters by Underwriters Laboratories and others for compliance with national interconnection standards.
- **Reduced rotor speeds:** To reduce sound emissions, turbine designs with lower tip-speed ratios and lower peak-rotor speeds are being pursued.
- **Design standards and certification:** The industry is increasing the use of consensus standards in its turbine design efforts for machines with rotor swept areas under 200 m^2 (about 65 kW rated power). In particular, IEC Standard 61400-2 Wind Turbines – Part 2: Design Requirements of Small Wind Turbines. Currently, however, a limited number of wind turbines have been certified in compliance with this standard because of the high cost of the certification process. To address this barrier, a Small Wind Certification Council has been formed in North America to certify that small wind turbines meet the requirements of the draft AWEA standard that is based on the IEC standard (AWEA 1996–2007).

2.7 Summary of Wind Technology Development Needs

Wind technology must continue to evolve if wind power is to contribute more than a few percentage points of total U.S. electrical demand. Fortunately, no major technology breakthroughs in land-based wind technology are needed to enable a broad geographic penetration of wind power into the electric grid. However, there are other substantial challenges (such as transmission and siting) and significant costs associated with increased penetration, which are discussed in other chapters of this report. No improvement in cost or efficiency for a single component can achieve the cost reductions or improved capacity factor that system-level advances can achieve.

The wind capacity factor can be increased by enlarging rotors and installing them on taller towers. This would require advanced materials, controls, and power systems that can significantly reduce the weight of major components. Capital costs would also be brought down by the manufacturing learning curve that is associated with continued technology advancement and by a nearly fivefold doubling of installed capacity.

The technology development required to make offshore wind a viable option poses a substantial potential risk. Offshore wind deployment represents a significant fraction of the total wind deployment necessary for 20% wind energy by 2030. Today's European shallow-water technology is still too expensive and too difficult to site in U.S. waters. Deepwater deployment would eliminate visual esthetics concerns, but the necessary technologies have yet to be developed, and the potential environmental impacts have yet to be evaluated. To establish the offshore option, work is needed to develop analysis methods, evaluate technology pathways, and field offshore prototypes.

Figure 2-17. Types of repairs on wind turbines from 2.5 kW to 1.5 MW

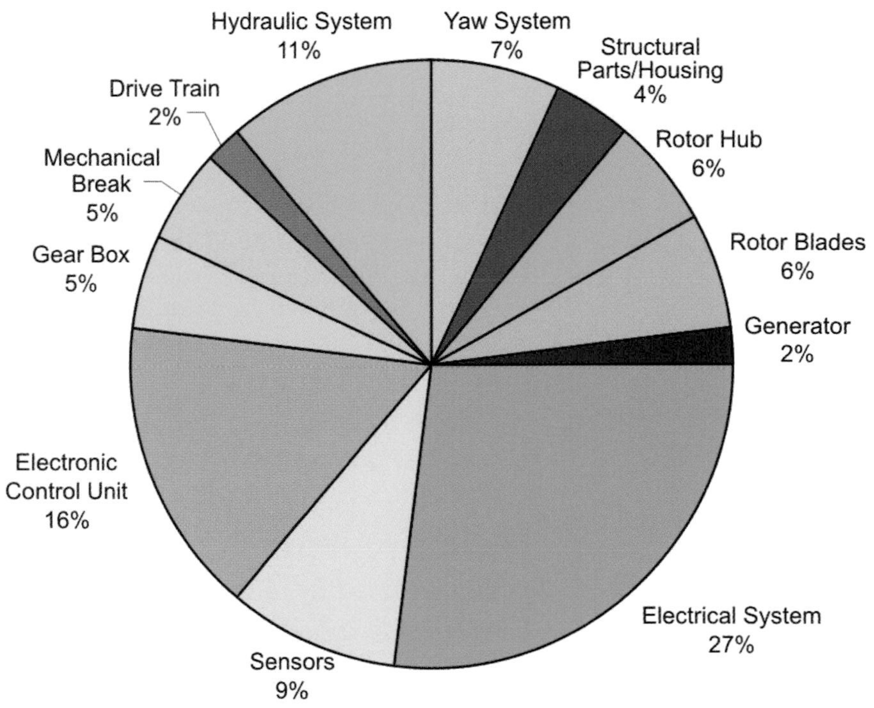

Today's market success is the product of a combination of technology achievement and supportive public policy. A 20% Wind Scenario would require additional land-based technology improvements and a substantial development of offshore technology. The needed cost and performance improvements could be achieved with innovative changes in existing architectures that incorporate novel advances in materials, design approaches, control strategies, and manufacturing processes. Risks are mitigated with standards that produce reliable equipment and full-scale testing that ensures the machinery meets the design requirements.

The 20% Wind Scenario assumes a robust technology that will produce cost-competitive generation with continued R&D investment leading to capital cost reduction and performance improvement. Areas where industry can focus RD&D efforts include those which require the most frequent repairs (see Figure 2-17). Such industry efforts, along with government-supported RD&D efforts, will support progress toward achieving two primary wind technology objectives:

- Increasing capacity factors by placing larger rotors on taller towers (this can be achieved economically only by using lighter components and load-mitigating rotors that reduce the integrated tower-top mass and structural loads; reducing parasitic losses throughout the system can also make gains possible), developing advanced controls, and improving power systems.

- Reducing the capital cost with steady learning-curve improvements driven by innovative manufacturing improvements and a nearly fivefold doubling of installed capacity

2.8 References and Other Suggested Reading

Antoniou, I., H.E. Jørgensen, T. Mikkelsen, S. Frandsen, R. Barthelmie, C. Perstrup, and M. Hurtig. 2006. "Offshore Wind Profile Measurements from Remote Sensing Instruments." Presented at the European Wind Energy Conference, February 27–March 2, Athens, Greece.

Ashwill, T. 2004. *Innovative Design Approaches for Large Wind Turbine Blades: Final Report*. Report No. SAND2004-0074. Albuquerque, NM: Sandia National Laboratories.

AWEA (American Wind Energy Association). 1996–2007. IEC Wind Turbine Standards. http://www.awea.org/standards/iec_stds.html#WG4.

Behnke, Erdman, and Whitaker Engineering (BEW Engineering). 2006. *Low Wind Speed Technology Phase II: Investigation of the Application of Medium-Voltage Variable-Speed Drive Technology to Improve the Cost of Energy from Low Wind Speed Turbines*. Report No. FS-500-37950, DOE/GO-102006-2208. Golden, CO: National Renewable Energy Laboratory (NREL). http://www.nrel.gov/docs/fy06osti/37950.pdf.

Black & Veatch. 2007. *20% Wind Energy Penetration in the United States: A Technical Analysis of the Energy Resource*. Walnut Creek, CA.

Bossanyi, E.A. 2003. "Individual Blade Pitch Control for Load Reduction," *Wind Energy*, 6(2): 119–128.

Brown, A. 2007. "Very Light and Fast." *Mechanical Engineering,* January. http://www.memagazine.org/jan07/features/verylight/verylight.html.

BSH (Bundesamt für Seeschifffahrt und Hydrographie.). *Wind Farms*. http://www.bsh.de/en/Marine%20uses/Industry/Wind%20farms/index.jsp.

BTM Consult. 2005. *World Market Update 2005*. Ringkøbing, Denmark: BTM Consult ApS. http://www.btm.dk/Pages/wmu.htm.

BWEA (British Wind Energy Association). "Offshore Wind." http://www.bwea.com/offshore/info.html.

Bywaters, G., V. John, J. Lynch, P. Mattila, G. Norton, J. Stowell, M. Salata, O. Labath, A. Chertok, and D. Hablanian. *2005. Northern Power Systems WindPACT Drive Train Alternative Design Study Report; Period of Performance: April 12, 2001 to January 31, 2005.* Report No. SR-500-35524. Golden, CO: NREL. http://www.nrel.gov/publications/

Cohen, J., T. Schweizer, A. Laxson, S. Butterfield, S. Schreck, L. Fingersh, P. Veers, and T. Ashwill. 2008. *Technology Improvement Opportunities for Low Wind Speed Turbines and Implications for Cost of Energy Reduction*. Report No. NREL/SR-500-41036. Golden, CO: NREL.

Cotrell, J., W.D. Musial, and S. Hughes. 2006. *The Necessity and Requirements of a Collaborative Effort to Develop a Large Wind Turbine Blade Test Facility in North America.* Report No. TP-500-38044. Golden, CO: NREL

Dhanju A., P. Whitaker, and W. Kempton. 2006. "Assessing Offshore Wind Resources: A Methodology Applied to Delaware." Presented at the AWEA Conference & Exhibition, June 4–7, Pittsburgh, PA.

DOE (U.S. Department of Energy). 2000. *World's Most Advanced Gas Turbine Ready to Cross Commercial Threshold.* Washington, DC: DOE. http://www.fossil.energy.gov/news/techlines/2000/tl_ats_ge1.html.

EIA (Energy Information Administration). 2006. "State Electricity Sales Spreadsheet." http://www.eia.doe.gov/cneaf/electricity/epa/sales_state.xls.

EUI (Energy Unlimited Inc.). 2003. *Variable Length Wind Turbine Blade*. Report No. DE-FG36-03GO13171. Boise, ID: EUI. http://www.osti.gov/bridge/servlets/purl/841190-OF8Frc/

Griffin, D.A. 2001. *WindPACT Turbine Design Scaling Studies Technical Area 1 – Composite Blades for 80- to 120-Meter Rotor.* Report No. SR-500-29492.Golden, CO: NREL.

Hasager, C.B., M.B. Christiansen, M. Nielsen, and R. Barthelmie. 2005. "Using Satellite Data for Mapping Offshore Wind Resources and Wakes." Presented at the Copenhagen Offshore Wind Proceedings, October 26–28, Copenhagen, Denmark.

IEC (International Electrotechnical Commission). 2007. "Technical Committee 88: Wind turbines, Standards 61400-x." http://nettedautomation.com/standardization/IEC_TC88/index.html

ISET (Institut fuer Solare Energieversorgungstechnik). 2003. *Experience Curves: A Tool for Energy Policy Programmes Assessment (EXTOOL).* Lund, Sweden: ISET. http://www.iset.uni-kassel.de/extool/Extoolframe.htm.

Jimenez, B., F. Durante, B. Lange, T. Kreutzer, and L. Claveri. 2005. "Offshore Wind Resource Assessment: Comparative Study between MM5 and WAsP." Presented at the Copenhagen Offshore Wind Proceedings, October 26–28, Copenhagen, Denmark.

Knight, S., and L. Harrison. 2005. "A More Conservative Approach." *Windpower Monthly,* November.

Kühn, P. 2006. "Big Experience with Small Wind Turbines (SWT)." Presented at the 49th IEA Topical Expert Meeting, September, Stockholm, Sweden.

Lindvig, K. 2005. "Future Challenges for a Marine Installation Company." Presented at the Copenhagen Offshore Wind Proceedings, October 26–28, Copenhagen, Denmark.

MMS (Minerals Management Service). Alternative Energy and Alternate Use Program. http://www.mms.gov/offshore/RenewableEnergy/RenewableEnergyMain.htm.

Musial, W. 2007. "Offshore Wind Electricity: A Viable Energy Option for the Coastal United States." *Marine Technology Society Journal*, 42 (3), 32-43.

2

Musial, W. and B. Ram. 2007. *Large Scale Offshore Wind Deployments: Barriers and Opportunities*, NREL Technical Report No. NREL/TP-500-40745,. Golden, CO: Draft.

Northern Power Systems. 2006. *Low Wind Speed Technology Phase I: Advanced Power Electronics for Low Wind Speed Turbine Applications*. Report No. FS-500-37945, DOE/GO-102006-2205. Golden, CO: NREL. http://www.nrel.gov/docs/fy06osti/37945.pdf.

NREL. 2002. Addendum to *WindPACT Turbine Design Scaling Studies Technical Area 3 – Self-Erecting Tower and Nacelle Feasibility*: Report No. SR-500-29493-A. Golden, CO: NREL.

Pace Global Energy Services, Aug. 2007, Assessment of Offshore Wind Power Resources, http://www.lipower.org/newscenter/pr/2007/pace_wind.pdf

Peregrine Power. 2006. *Low Wind Speed Technology Phase II: Breakthrough in Power Electronics from Silicon Carbide*. Report No. FS-500-37943, DOE/GO-102006-2203. Golden, CO: NREL. http://www.nrel.gov/docs/fy06osti/37943.pdf

Poulsen, S.F., and P.S. Skjærbæk. 2005. "Efficient Installation of Offshore Wind Turbines: Lessons Learned from Nysted Offshore Wind Farm." Presented at the Copenhagen Offshore Wind Proceedings, October 26–28, Copenhagen, Denmark.

Risø National Laboratory. 2005. *Association Euratom - Risø National Laboratory Annual Progress Report 2005.* Report No. Risø-R-1579(EN). Roskilde, Denmark: Risø National Laboratory. http://www.risoe.dk/rispubl/ofd/ofdpdf/ris-r-1579.pdf.

SEAWIND, Altener Project, 2003. (Per Nielsen) "Offshore Wind Energy Projects Feasibility Study Guidelines," Denmark.

Stiesdal, H., and P.H. Madsen. 2005. "Design for Reliability." Presented at the Copenhagen Offshore Wind Proceedings, October 26–28, Copenhagen, Denmark.

Walford, C.A. 2006. *Wind Turbine Reliability: Understanding and Minimizing Wind Turbine Operation and Maintenance Costs.* Report No. SAND2006-1100. Albuquerque, NM: Sandia National Laboratories.

Wiser, R., and M. Bolinger. 2007. *Annual Report on U.S. Wind Power Installation, Cost, and Performance Trends: 2006.* DOE/GO–102007-2433. Golden, CO: NREL. http://www.osti.gov/bridge/product.biblio.jsp?query_id=0&page=0&osti_id=908214

Chapter 3. Manufacturing, Materials, and Resources

A 20% Wind Energy Scenario would support expansion of domestic manufacturing and related employment. Production of several key materials for wind turbines would require substantial but achievable growth.

Stakeholders and decision makers need to know whether the effort to achieve a generation mix with 20% wind energy by 2030 might be constrained by raw materials availability, manufacturing capability, or labor availability. This chapter examines the adequacy of these critical resources.

Over the past five years, the wind industry in the United States has grown by an average of 22% annually. In 2006 alone, America's wind power generating capacity increased by 27%.

The U.S. wind energy industry invested approximately $4 billion to build 2,454 MW of new generating capacity in 2006, making wind the second largest source of new power generation in the nation—surpassed only by natural gas—for the second year in a row. Recently installed wind farms increased cumulative installed U.S. wind energy capacity to 13,884 MW—well above the 10,000 MW milestone reached in August 2006 (AWEA 2007). On average, 1 MW of wind power produces enough electricity to power 250 to 300 U.S. homes.

Based on estimates released by the U.S. Department of Energy (DOE) Energy Information Administration (EIA 2006), annual electricity consumption in the United States is expected to grow at a rate of 1.3% annually—from 3.899 billion megawatt-hours (MWh) in 2006 to about 5.368 billion MWh in 2030. Although wind energy supplied approximately 0.8% of the total electricity in 2006, more and larger wind turbines can help to meet a growing demand for electricity. (See the Glossary in Appendix E for explanations of wind energy capacity and measurement units.)

The most common large turbines currently in use have a rated capacity of between 1 MW and 3 MW, with rotor diameters between 60 m and 90 m, tower heights between 60 m and 100 m, and capacity factors between 30% and 40% (capacity factor is an indicator of annual energy production). Although currently installed machines are expected to operate through 2030, larger turbines (with capacity factors that increase over time, as discussed in Chapter 2) are expected to become more common as offshore technology advances are transferred to land-based turbines. These larger turbines could reach rated power between 4 MW and 6 MW with capacity factors between 40% and 50%.

To estimate the raw materials and investments needed to support the 20% Wind Scenario, industry leaders have assumed that most of the wind turbines used in the next two to three decades will be in the 1 MW to 3 MW class, with a modest contribution of the larger-sized machines (see Chapter 2). Today, approximately 2,000 turbines are installed each year, but that figure is expected to rise and to level out at about 7,000 turbines per year by 2017.

3

3.1 RAW MATERIALS REQUIREMENTS

Wind turbines are built in many sizes and configurations, with the larger sizes utilizing a wide range of materials. Reducing the weight and cost of the turbines is key to making wind energy competitive with other power sources. Throughout the next few decades, business opportunities are expected to expand in wind turbine components and materials manufacturing. To reach the high levels of wind energy associated with the 20% Wind Scenario, materials usage will also need to increase considerably, even as new technologies that improve component performance are introduced.

To estimate the raw materials required for the 20% Wind Scenario, this analysis focuses on the most important materials used in building a wind turbine today (such as steel and aluminum) and on main turbine components. Table 3-1 shows the percentage of different materials used in each component and each component's percentage of total turbine weight. The table applies to 1.5 MW turbines MW and larger.

Table 3-2 uses the materials consumption model in Table 3-1 to further describe the raw materials required to reach manufacturing levels of about 7,000 turbines per year. This analysis assumes that turbines will become lighter, annual installation rates will level off to roughly 7,000 turbines per year by 2017, and installation will continue at that rate through 2030. Approximately 100,000 turbines will be required to produce 20% of the nation's electricity in 2030.

No single component dominates a wind turbine's total cost, which is generally split evenly among the rotor, electrical system, drivetrain, and tower. The technological progress described in Chapter 2, however, could significantly reduce costs (e.g., through the use of lighter weight components for blades and towers).

The availability of critical resources is crucial for large-scale manufacturing of wind turbines. The most important resources are steel, fiberglass, resins (for composites and adhesives), blade core materials, permanent magnets, and copper. The production status of these materials is reviewed in the following list:

- **Steel:** The steel needed for additional wind turbines is not expected to have a significant impact on total steel production. (In 2005, the United States produced 93.9 million metric tons of steel, or 8% of the worldwide total.) Although steel will be required for any electricity generation technology installed over the next several decades, it can be recycled. As a result, replacing a turbine after 20+ years of service would not significantly affect the national steel demand because recycled steel can be used in other applications where high-quality steel is not required (Laxson, Hand, and Blair 2006).

Table 3-1. Main components and materials used in a wind turbine (%)

1.5 MW	Weight %	Permanent Magnet	Concrete	Steel	Aluminum	Copper	GRP	CRP	Adhesive	Core	TOTAL
Rotor											
Hub	6.0			100							100.0
Blades	7.2			2			78		15	5	100.0
Nacelle											
Gearbox	10.1			96	2	2					100.0
Generator	3.4			65		35					100.0
Frame	6.6			85	9	3	3				100.0
Tower	66.7		2	98							
	100.0	**0.0**	**1.3**	**89.1**	**0.8**	**1.6**	**5.8**	**0.0**	**1.1**	**0.4**	**100.0**
4 MW		**Permanent Magnet**	**Concrete**	**Steel**	**Aluminum**	**Copper**	**GRP**	**CRP**	**Adhesive**	**Core**	
Rotor											
Hub	6.00			100							100.0
Blades	7.6			2			68	10	15	5	100.0
Nacelle											
Gearbox	10.10			96	2	2					100.0
Generator	2.7	3		93		4					100.0
Frame	6.60			85	9	3	3				100.0
Tower	67.00		2	98							
	100.0	**0.08**	**1.34**	**89.63**	**0.80**	**0.51**	**5.37**	**0.76**	**1.14**	**0.38**	**100.0**

Notes: Tower includes foundation. GRP = glass-fiber-reinforced plastic. CRP = carbon fiber reinforced plastic
Source: Sterzinger and Svrcek (2004)

Table 3-2. Yearly raw materials estimate (thousands of metric tons)

Year	kWh/kg	Permanent Magnet	Concrete	Steel	Aluminum	Copper	GRP	CRP	Adhesive	Core
2006	65	0.03	1,614	110	1.2	1.6	7.1	0.2	1.4	0.4
2010	70	0.07	6,798	464	4.6	7.4	29.8	2.2	5.6	1.8
2015	75	0.96	16,150	1,188	15.4	10.2	73.8	9.0	15.0	5.0
2020	80	2.20	37,468	2,644	29.6	20.2	162.2	20.4	33.6	11.2
2025	85	2.10	35,180	2,544	27.8	19.4	156.2	19.2	31.4	10.4
2030	90	2.00	33,800	2,308	26.4	18.4	152.4	18.4	30.2	9.6

Notes: kg = kilograms; GRP = glass-fiber-reinforced plastic. CRP = carbon fiber reinforced plastic
Source: Sterzinger and Svrcek (2004)

- **Fiberglass:** Additional fiberglass furnaces would be needed to build more wind turbines. Primary raw materials for fiberglass (sand) are in ample supply, but availability and costs are expected to fluctuate for resins, adhesives, and cores made from the petroleum-based chemicals that are used to impregnate the fiberglass (Laxson, Hand, and Blair 2006).
- **Core:** End-grain balsa wood is an alternative core material that can replace the low-density polymer foam used in blade construction. Availability of this wood might be an issue based on the growth rate of balsa trees relative to the projected high demand.
- **Carbon fiber:** Current global production of commercial-grade carbon fiber is approximately 50 million pounds (lb) per year. The use of carbon fiber in turbine blades in 2030 alone would nearly double this demand. To achieve such drastic industry scale-up, changes to carbon fiber production technologies, production facilities, packaging, and emissions-control procedures will be required.

- **Permanent magnets:** By eliminating copper from the generator rotor and using permanent magnets, which are becoming more economically feasible, it is possible to build smaller and lighter generators. World magnet production in 2005 was about 40,000 metric tons, with about 35,000 metric tons produced in China. Although supply is not expected to be restricted, significant additions to the manufacturing capability would be required to meet the demand for wind turbines and other products (Trout 2002; Laxson, Hand, and Blair 2006).
- **Copper:** Although wind turbines use significant amounts of copper, the associated level of demand still equates to less than 4% of the available copper. This demand level, would not have a significant impact on national demand (U.S. refined copper consumption was 2.27 million metric tons in 2005). Although copper ranks third after steel and aluminum in world metals consumption, global copper production is adequate to satisfy growing demands from the wind industry. However, in recent years copper prices have escalated more quickly than inflation, which could affect turbine costs.

Despite the demand and supply status of these materials, new component developments are expected to significantly change material requirements. Generally, trends are toward using lighter-weight materials, as long as the life-cycle costs are low. In addition to the findings of Ancona and McVeigh (2001; described in the Materials Usage Analysis sidebar), other trends in turbine components are outlined in the subsections that follow.

Material Usage Analysis
(Ancona and McVeigh 2001)

- Turbine material usage is, and will continue to be, dominated by steel.
- Opportunities exist for introducing aluminum or other lightweight composites, provided that cost, strength, and fatigue requirements can be met.
- GRP is expected to continue to be used for blades.
- The use of carbon fiber might help reduce weight and cost.
- Low costs and high reliability remain the primary drivers.
- Variable-speed generators will become more common.
- Permanent-magnet generators on larger turbines will increase the need for magnetic materials.
- Simplification of the nacelle machinery might reduce raw material costs and also increase reliability.

Evolution of Rotors

Most rotor blades in use today are built from glass-fiber-reinforced plastic (GRP). Steel and various composites such as carbon filament-reinforced plastic (CFRP) are also used. As the rotor size increases for larger machines, the trend will be toward high-strength, fatigue-resistant materials. Composites involving steel, GRP, CFRP, and possibly other new materials will likely come into use as turbine designs evolve.

Changes to Machine Heads

The machine head contains an array of complex machinery including yaw drives, blade-pitch-change mechanisms, drive brakes, shafts, bearings, oil pumps and coolers, controllers, a bedplate, the drivetrain, the gearbox, and an enclosure. Design simplifications and innovations are anticipated in each element of the machine head.

3.2 Manufacturing Capability

In principle, a sustainable level of annual wind turbine installation would be best supported by a substantial domestic manufacturing base. However, if installation rates fluctuate greatly from one year to the next, manufacturing capability may not be able to grow or shrink as necessary. The National Renewable Energy Laboratory (NREL) created a simple model to explore sustainable installation rates that would maintain wind energy production at specific levels spanning several decades (Laxson, Hand, and Blair 2006).

Figure 3-1. a. Annual installed wind energy capacity to meet 20% of energy demand. b. Cumulative installed wind energy capacity to meet 20% of energy demand.

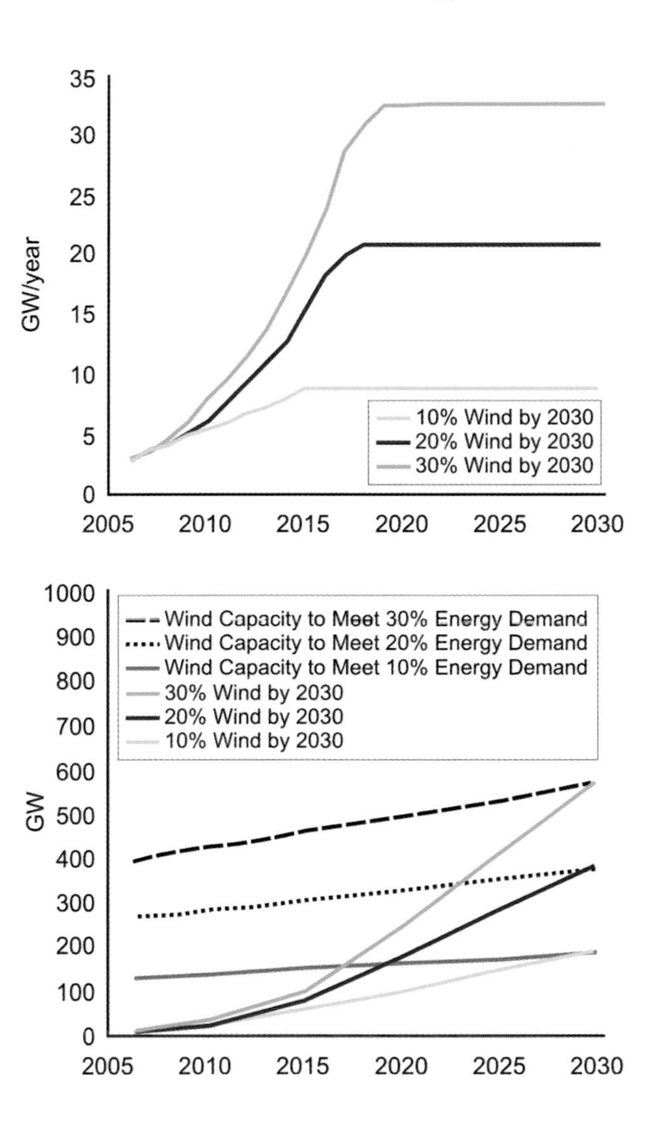

NREL's study explored a number of alternative scenarios for annual wind power capacity expansion to understand their potential impact on wind energy installation and manufacturing rates. The results indicate that achieving the 20% Wind Scenario by 2030 would not overwhelm U.S. industry (Laxson, Hand, and Blair 2006).

NREL's study assessed potential barriers that would prohibit near-term high wind penetration levels, such as manufacturing rates or resource limitations. To reach 20% electric generation from wind by 2030 in the United States, the authors noted, an annual installed capacity increase of about 20% would need to be sustained for a decade (Laxson, Hand, and Blair 2006). Figure 3-1 compares the installation rates required to meet three energy supply goals of 10%, 20%, and 30% of total national electrical energy production from wind by 2030. Figure 3-1(a) shows the annual rates and Figure 3-1(b) shows the cumulative capacity attained in each case. A manufacturing production level of 20 gigawatts (GW) per year by 2017—and maintained at this value thereafter—would reach levels close to 400 GW of wind energy capacity by 2030.

NREL's study assumed that the wind plant capacity factor would not change from year to year or from location to location. This assumption provided an upper bound on the annual installation rate and cumulative capacity required to produce 20% of electricity demand. Alternatively, the 20% Wind Scenario evaluation assumes that plant capacity factors will increase modestly with experience and technology improvements (see Chapter 2). The 20% Scenario also accounts for regional variations in wind resources, as explained in Appendix A's detailed description of the analytic modeling approach employed. Note that when these refinements are included, the 20% curve in Figure 3-1(a) shifts downward, somewhat similar to that shown in Figure 3-2 on the next page.

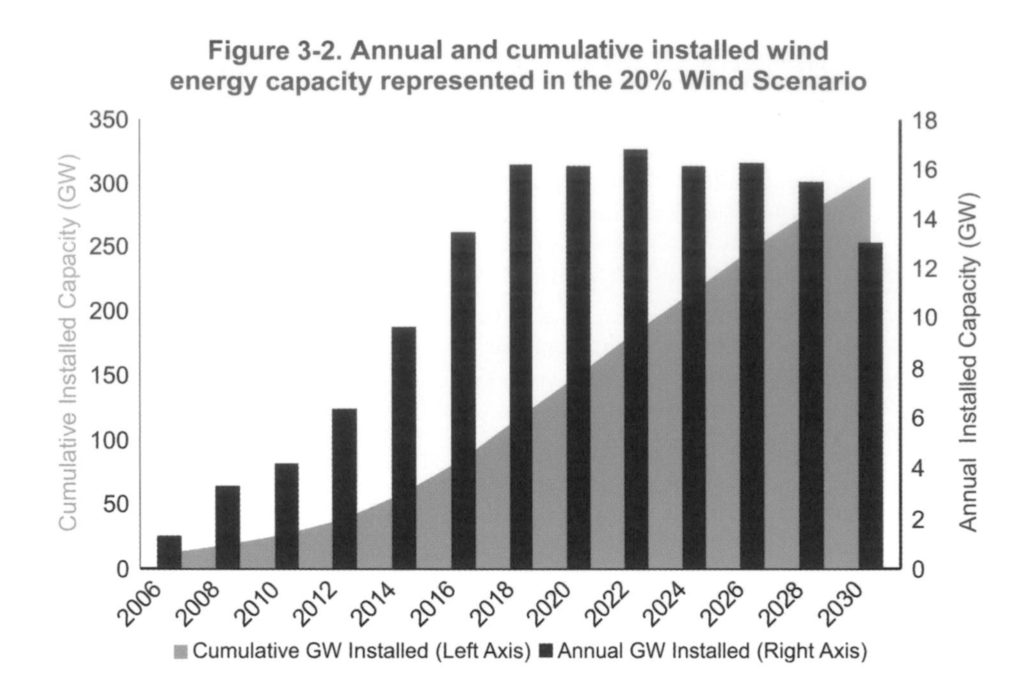

Figure 3-2. Annual and cumulative installed wind energy capacity represented in the 20% Wind Scenario

This chapter discusses the materials and manufacturing needed to pursue the 20% Wind Scenario from 2007 through 2030 to meet the annual and cumulative installed capacity shown in Figure 3-2. This figure shows the forecasts for annual and cumulative installed wind energy capacity, which also forms the basis for estimates of new wind turbines and the raw materials required to produce them. In this scenario, annual installations climb more than 16 GW per year, and the total installed wind capacity increases to 305 GW by 2030. Between 2007 and 2030, 293 GW are installed. (For more details on the modeling approach used, see Appendix A.)

3.2.1 Current Manufacturing Facilities

A growing number of states and companies in the United States are ramping up capacity to manufacture wind turbines, or have the ability to do so. Jobs are expected to remain in the United States, but only if investments are made in certain components and in advanced manufacturing technologies. Appendix C describes the jobs and economic impacts associated with wind energy, including manufacturing, construction, and operational sectors of the wind industry.

A useful perspective on growing manufacturing requirements is provided by a non-government organization study released in 2004 called *Wind Turbine Development: Location of Manufacturing Activity* (Sterzinger and Svrcek 2004). This study investigated the current and future U.S. wind manufacturing industry, both to determine the location of companies involved in wind turbine production and to examine limitations to a rapidly expanding wind business. The report covered four census regions (the Midwest, Northeast, South, and West) and divided turbine manufacturing into 20 separate components. These components were grouped into five categories, as shown in Table 3-3. The table also shows the locations of U.S. wind turbine component manufacturers in 2004, broken down by region. Among the 106 companies surveyed, about 90 companies directly manufacture components for utility-scale wind turbines, with utility scale being roughly defined as 1 MW or greater.

Table 3-3. Locations of U.S. wind turbine component manufacturers

Region	Division	Rotor	Nacelle and Controls	Gearbox & Drivetrain	Generator & Power Electronics	Tower	Division Total
Midwest	East North Central	6	5	8	1	2	22
	West North Central	1	0	1	1	8	11
Northeast	Middle Atlantic	3	4	4	5	1	17
	New England	0	6	0	2	0	8
South	East South Central	0	0	0	0	2	2
	South Atlantic	3	2	1	1	2	9
	West South Central	4	5	0	1	6	16
West	Mountain	1	0	0	1	0	2
	Pacific	5	4	2	4	4	19
	Component Total:	**23**	**26**	**16**	**16**	**25**	**106**

(Sterzinger and Svrcek 2004)

Figure 3-3 on the next page shows the locations of a number of the current manufacturers of wind turbines and components. These firms are widely distributed around the country and some are located in regions with, as yet, little wind power development.

A large national investment in wind would likely spread beyond these active companies. To identify this potential, the North American Industrial Classification System (NAICS; http://www.census.gov/epcd/www/naics.html) was searched to identify companies operating under relevant industry codes. The manufacturing activity related to wind power development is substantial and widely dispersed (Sterzinger and Svrcek 2004). As Table 3-4 shows, more than 16,000 firms are currently producing products under one or more of the NAICS codes that include

Table 3-4. U.S. Manufacturing firms with technical potential to enter wind turbine component market

NAICS Code	Code Description	Total Employees	Annual Payroll ($1000s)	Number of Companies
326199	All Other Plastics Products	501,009	15,219,355	8,174
331511	Iron Foundries	75,053	3,099,509	747
332312	Fabricated Structural Metal	106,161	3,975,751	3,033
332991	Ball and Roller Bearings	33,416	1,353,832	198
333412	Industrial and Commercial Fans and Blowers	11,854	411,979	177
333611	Turbines, and Turbine Generators, and Turbine Generator Sets	17,721	1,080,891	110
333612	Speed Changer, Industrial	13,991	539,514	248
333613	Power Transmission Equip.	21,103	779,730	292
334418	Printed Circuits and Electronics Assemblies	105,810	4,005,786	716
334519	Measuring and Controlling Devices	34,499	1,638,072	830
335312	Motors and Generators	62,164	2,005,414	659
335999	Electronic Equipment and Components, NEC	42,546	1,780,246	979
Total		**1,025,327**	**35,890,079**	**16,163**

Figure 3-3. Examples of manufacturers supplying wind equipment across the United States

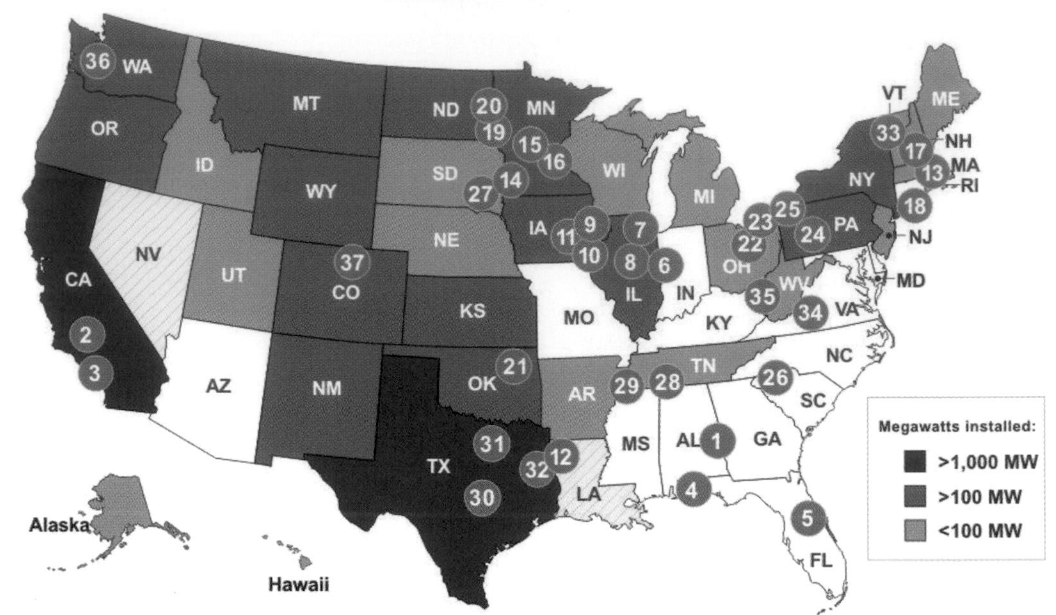

Wind power creates manufacturing jobs even in regions like the Southeast that do not have a large wind resource.

1 Vectorply. Phenix City, AL (composites for blades)
2 GE Energy, Tehachapi, CA (wind turbine manufacturing facility)
3 Bragg Crane & Rigging Service, Long Beach, CA (cranes, rigging, transportation)
4 GE Energy, Pensacola, FL (blade technology development)
5 Mitsubishi Power Systems, Lake Mary, FL (gear boxes)
6 White Construction Inc., Clinton, IN (construction services)
7 Winergy Drive Systems Corporation, Elgin, IL (gear units, generators, power converters)
8 Trinity Industries, Clinton, IL (towers)
9 Clipper Windpower, Cedar Rapids, IA (turbine manufacturing, assembly)
10 Siemens, Fort Madison, IA (blades)
11 Acciona Energia, West Branch, IA (planned) (turbine manufacturing)
12 Beaird Industries, Shreveport, LA (towers, tower flanges and bolts)
13 Second Wind Inc., Somerville, MA (anemometers, electronic controllers, sensors/data loggers)
14 Suzlon Wind Energy, Pipestone, MN (blade manufacture, turbine assembly)
15 D.H. Blattner & Sons, Avon, MN (construction)
16 M.A. Mortenson Co., Minneapolis, MN (construction)
17 Hendrix Wire & Cable Inc., Milford, NH (cables to substations)
18 Hailo LLC, Holbrook, NY (ladder and lift systems)
19 DMI Industries, West Fargo, ND (towers)
20 LM Glasfiber, Grand Forks, ND (blades)
21 Trinity Structural Towers, Tulsa, OK (towers)
22 Owens Corning Composites, Granville, OH (composites for blades)
23 Hamby Young, Aurora, OH (substations and high voltage applications)
24 Gamesa, Ebensburg, PA (blade, nacelle, tower manufacturing)
25 GE Energy, Erie, PA (wind turbine components)
26 GE Energy, Greenville, SC (turbine assembly plant)
27 Knight & Carver, Howard, SD (blade manufacturing)
28 Aerisyn Inc, Chattanooga, TN (towers)
29 Thomas & Betts Corp., Memphis, TN (towers, tower flange and bolts)
30 DeWind, Inc./TECO Westinghouse, Round Rock, TX (wind turbine manufacturing)
31 Trinity Structural Towers, Fort Worth, TX (towers)
32 CAB Incorporated, Nacogdoches, TX (blade extender, hub, nacelle frame, tower flange and bolts)
33 NRG Systems, Hinesburg, VT (anemometers, sensors/data loggers)
34 GE Energy, Salem, VA (wind turbine components)
35 Tower Logistics, Huntington, WV (lifts for turbines)
36 PowerClimber, Seattle, WA (traction hoists, rigging equipment)
37 Vestas, Windsor, CO (planned) (blade and turbine manufacturing)

manufacture of wind components. These firms are spread across all 50 states. They are concentrated, however, in the most populous states and the states that have suffered the most from loss of manufacturing jobs. The 20 states that would likely receive the most investment and the most new manufacturing jobs from wind power expansion account for 75% of the total U.S. population, and 76% of the manufacturing jobs lost in the last 3.5 years.

A 2006 NGO report entitled "*Renewable Energy Potential: A Case Study of Pennsylvania* (Sterzinger and Stevens 2006) identified the bottlenecks in the component supply chain. Bottlenecks were identified for various components, but obtaining gearbox components was particularly problematic. Currently, only a few manufacturers in the world deliver gearboxes for large wind turbines. Additional investments will be required to support the development of a gearbox industry specifically for large wind applications. Investments will also be needed to expand the manufacture of large bearings and large castings.

The wind equipment manufacturing sector also faces trade-offs between using domestic or foreign manufacturing facilities. An advantage to domestic operations is a reduction reducing the significant transportation costs of moving large components such as blades and towers. Manufacturing many significant wind turbine components is also a labor-intensive process. With U.S. labor wage rates at higher levels than those paid in many other countries, manufacturers have naturally been drawn to setting up their factories outside the United States (e.g., in Mexico and China). One wind blade manufacturer with significant international manufacturing experience estimates that, to make a U.S. factory competitive, the labor hours per blade would need to be reduced by a factor of 30%–35%. To ensure that the bulk of these manufacturing jobs stay in the United States, automation and productivity gains through the development of advanced manufacturing technology are needed. These gains will allow the higher U.S. wage rates to be competitive.

To attract these jobs, a number of U.S. states have set aside funds for RD&D, with plans to collaborate with industry and the federal government on a cost-shared basis. Collaboration among state, industry, and federal programs on advanced manufacturing technology can create competitive U.S. factories and provide better job security for U.S. employees.

3.2.2 Ramping Up Energy Industries

In the United States, several industries have experienced large rates of growth over a short period of time. The power plants most commonly used to produce electricity around the world—such as thermal power stations fired with coal, gas or oil, or nuclear reactors—are large in scale. Nuclear power stations, developed mainly since the middle of the twentieth century, have now reached a penetration of 17.1% in the world's power supply. Worldwide, nuclear power plant installations saw a 17% annual growth rate between 1960 and 1997 (BTM 1999). Despite a halt in new nuclear plant licensing in the early 1980s, U.S. nuclear plants generate about 20% of the nation's electrical energy, and have done so for the last decade or more. The history of nuclear power shows that it is possible to achieve substantial levels of penetration over two to three decades with a new technology.

Even though the time horizon of the 20% Wind Scenario is consistent with the historical development of nuclear power, it is nonetheless difficult to directly compare penetration patterns for nuclear power that is typically about 1,000 MW and wind power technology. A wind turbine is a smaller-scale technology that has a current typical commercial unit size of 2 MW–3 MW. Despite the smaller scales of wind power, its modularity makes it ideal for all sizes of installations—from a single unit (2 MW–3 MW) to a large utility-scale wind farm (1,000 MW). On the supply side, serial production of large numbers of similar units can reduce manufacturing costs. These factors suggest that manufacturing ramp-up for wind turbines should be less daunting than ramp-up for nuclear power plant equipment.

Experiences with natural-gas-fired power plants over the past decade also provide important perspectives on the ability to rapidly expand manufacturing capability for wind power. From the early 1990s through the first half of the current decade, the U.S. electric sector experienced a rush toward new gas combined-cycle and combustion-turbine generation. This growth was driven by the expectation—now discounted—of continuing low natural gas prices. From 1999 through 2005, tens of gigawatts of natural gas power plants were manufactured and installed in the United States each year, with installations peaking in 2002 at more than 60 GW (Black & Veatch 2007). The experience with natural gas demonstrates that huge amounts of power generation equipment can be manufactured in the United States if sufficient market demand exists.

3

As Table 3-5 shows, Toyota North America exemplifies the manufacturing scale-up of a modular technology and capability that is possible in the United States. Toyota has continued to establish U.S. manufacturing capability since the mid-1980s, and automobiles, like wind turbines, require large quantities of steel, plastics, and electronic components. There is no indication that Toyota's domestic expansion caused any strain on the nation's manufacturing or materials-supply sectors. Today, the majority of vehicles Toyota sells in the U.S. are produced in this country.

Table 3-5. Toyota North America vehicle production and sales

Direct U.S. Employment (2005)	32,003 employees
2005 Payroll	$2,244,946,444
Cumulative U.S. Production	12,374,062 vehicles
Cumulative Sales	$272,390,226,806
U.S. Vehicle Sales (2005)	2,269,296 vehicles
U.S. Vehicle Production (2005)	1,393,100 vehicles
Average Engine Power 2004-2005	227 horsepower or 0.17 MW
2005 U.S. Production in Power Output Terms	275 million horsepower 236 million kW or 236 GW
2005 U.S. Sales in Power Output Terms	448 million horsepower 384 million kW or 384 GW

Source: Adapted from Toyota website data
http://www.toyota.com/about/operations/manufacturing/

Table 3-5 shows that Toyota's annual U.S. production, when expressed in terms of engine power output, increased to 236 GW by 2005. This annual production begins to approach in power capability the total amount of wind generation installed between 2007 and 2030 through realization of the 20% Wind Scenario.

3.3 Labor Requirements

Beyond the raw material and manufacturing facilities required to create wind turbines and components, a skilled labor force would be required. This staff would need a range of skills and experience to fill many new employment opportunities. The likely outcome from developing new capabilities and capacity would be expansion of manufacturing in areas currently capable of competing or development in locations where logistic advantages exist.

3.3.1 Maintaining and Expanding Relevant Technical Strength

Major expansion of wind power in the United States would require substantial numbers of skilled personnel available to design, build, operate, maintain, and advance wind power equipment and technology. Toward this end, a number of educational programs are already offered around the nation, including those shown in Table 3-6.

Table 3-6. Wind technology-related educational programs around the United States today

School	Location	Degree or Program
Wind Energy Applications Training Symposium	Boulder, Colorado	Workshops for industry
Colorado State University	Fort Collins, Colorado	65 MW turbine on campus for research (engineering, environmental, etc.)
Advanced Technology Environmental Education Center: Sustainable Energy Education and Training	Bettencourt, Iowa	Workshops for upper level high school and community college technology instructors
Iowa Lakes Community College	Estherville, Iowa	One-year diploma for wind technician; two-year associate in applied science degree for wind technician
University of Massachusetts at Amherst: College of Engineering, and Renewable Energy Research Laboratory (becoming University of Massachusetts Wind Energy Center in late 2008)	Amherst, Massachusetts	MS and Ph.D. level engineering programs specializing in wind energy
Minnesota West Community and Technical College	Canby, Maine	Associate of applied science degree program in wind energy technology; diploma for wind energy mechanic; online certificate program for "windsmith"
Southwestern Indian Polytechnic Institute	Albuquerque, New Mexico	Under development: Integration of renewable energy technology experiential learning into the electronics technology, environmental science, agricultural science, and natural resources certificate and degree programs
Mesalands Community College: North American Wind Research and Training Center	Tucumcari, New Mexico	Under development: Curriculum for operations and maintenance technician; two-year associate degree in wind farm management
Wayne Technical and Career Center	Williamson, New York	New Vision Renewable Energy Program for high school seniors
Columbia Gorge Community College	Hood River, Oregon	One-year certificate and two-year degree for renewable energy technician
Lane Community College	Eugene, Oregon	Two-year associate of applied science degree for energy management technician; two-year associate of applied science option for renewable energy technician
Texas Tech and other American universities: Wind Science & Engineering Research Center	Lubbock, Texas	Integrative graduate education and research traineeship
Lakeshore Technical College	Cleveland, Wisconsin	Associate degree in applied science; electromechanical technology with a wind system Technician track
Fond du Lac Tribal and Community College	Fond du Lac, Wisconsin	Clean Energy Technician Certificate Program

Although this is an excellent beginning, many more programs of a similar nature will be needed nationwide to satisfy the needs stemming from the 20% Wind Scenario. One concern is that the number of students in power engineering programs has been dropping in recent years. Currently, U.S. graduate power engineering programs produce about 500 engineers per year; in the 1980s, this number approached 2,000. In addition, the number of wind engineering programs in U.S. graduate schools is significantly lower than in Europe. This concern is echoed in Figure 3-4 below, which shows that the number of college graduates receiving

Figure 3-4. Projected percentage of 22-year-olds with a bachelor's degree in science and engineering through 2050

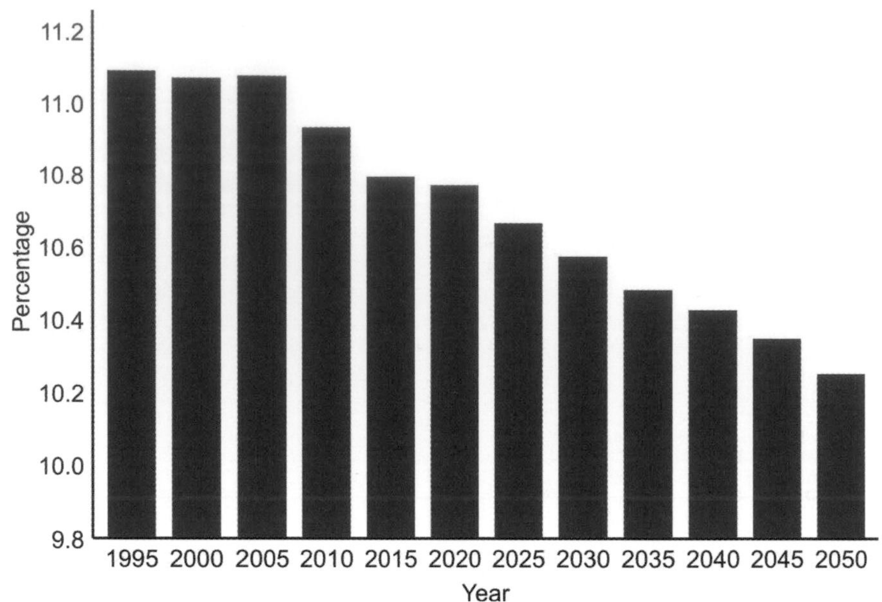

degrees in science and engineering has been declining, and that this trend is projected to continue for the foreseeable future (NSTC 2000).

Even the level of U.S. graduate programs is well below similar graduate programs in Europe (Denmark, Germany, etc). At this rate, the United States will be unable to provide the necessary trained talent and manufacturing expertise. Unless this trend is reversed, even with major new wind installations in the United States, most of the technology will be imported, and a significant portion of the economic gains will be foreign rather than domestic.

3.4 Challenges to 20% Wind Energy by 2030

3.4.1 Challenges

Materials

Several key materials are crucial to the production of a wind turbine. The availability of some key raw materials—including fiberglass (about 9 metric tons required per megawatt of wind turbine capacity), resins, and permanent magnets—might potentially constrain the ability to develop an infrastructure producing high levels of wind power. To give perspective, the glass fiber requirements would be about half the level used domestically for roofing shingles (which is currently the largest consumer of fiberglass) and about double the amount now used in boat building.

Manufacturing

The 20% Wind Scenario would demand installations at a sustained growth rate of 20% annually for nearly a decade and then require maintaining that level of annual installations through 2030. For turbine companies, it is no longer simply a matter of where to establish new manufacturing capacity. Investment decisions must now address strategies for building out and securing supply lines on a global basis; a proactive stance is essential to operate successfully in an environment of rapidly growing and shifting demand for wind turbines (Hays, Robledo, and Ambrose 2006). Fortunately, the 20% Wind Scenario could be feasible even with the potential challenges related to the availability of raw material or increased manufacturing demands. For rapid growth of manufacturing capacity to be achieved, stable and consistent policies that encourage investment in these new sectors of activity are needed.

3

Labor

One potential gap in achicving high rates of wind energy development is the availability of a qualified work force. In a report published by the National Science and Technology Council (NSTC), as noted above, the percentage of 22-year-olds earning degrees in science and engineering will continue to drop in the next 40 years (NSTC 2000). More support from industry, trade organizations, and various levels of government could foster university programs in wind and renewable energy technology, preparing the work force to support the industry's efforts.

3.5 References and Other Suggested Reading

Ancona and McVeigh. 2001. Princeton Energy Resources International, LLC. Rockville, MD http://www.generalplastics.com/uploads/technology/WindTurbine-MaterialsandManufacturing_FactSheet.pdf

AWEA (American Wind Energy Association). 2007. *Wind Power Capacity In U.S. Increased 27% in 2006 and Is Expected To Grow an Additional 26% in 2007.* Washington, DC: AWEA. http://www.awea.org/newsroom/releases/Wind_Power_Capacity_012307.html

BTM Consult. 1999. *Wind Force 10: A Blueprint to Achieve 10% of the World's Electricity from Wind Power by 2010.* Ringkøbing, Denmark: BTM Consult ApS. http://www.inforse.dk/doc/Windforce10.pdf

EIA (Energy Information Administration). February 2006. *Annual Energy Outlook 2006*. Report No. DOE/EIA-0383.Washington, DC: EIA.

Hays, K., C. Robledo, and W. Ambrose. 2006. *Wind Power at a Crossroads, Supply Shortages Spark Industry Restructuring*, Strategy White Paper. Cambridge, MA: Emerging Energy Research.

Laxson, A., M.M. Hand, and N. Blair. 2006. *High Wind Penetration Impact on W.S. Wind Manufacturing Capacity and Critical Resources*. Report No. NREL/TP-500-40482. Golden, CO: National Renewable Energy Laboratory (NREL).

NSTC (National Science and Technology Council). 2000. *Ensuring a Strong U.S. Scientific, Technical and Engineering Workforce in the 21st Century*. Washington, DC:NSTC.

Black & Veatch. 2007 20 *% Wind Energy Penetration in the United States: A Technical Analysis of the Energy Resource*. Walnut Creek, CA

Sterzinger, G., and M. Svrcek. September 2004. *Wind Turbine Development: Location of Manufacturing Activity*. Washington, DC: Renewable Energy Policy Project (REPP).

Sterzinger, G., and M. Svrcek. 2005. *Component Manufacturing: Ohio's Future in the Renewable Energy Industry*. Washington, DC: REPP.

Sterzinger, G., and J. Stevens. October 2006. *Renewable Energy Potential: A Case Study of Pennsylvania*. Washington, DC: REPP.

Trout, S.R. 2002. "Rare Earth Magnet Industry in the USA: Current Status and Future Trends." Presented at the XVII Rare Earth Magnet Workshop, August 18–22, Newark, New Jersey.

Chapter 4. Transmission and Integration into the U.S. Electric System

The ever-increasing sophistication of the operation of the U.S. electric power system—if it continues on its current path—would allow the 20% Wind Scenario to be realized by 2030. The 20% Wind Scenario would require the continuing evolution of transmission planning and system operations, in addition to expanded electricity markets.

There are two separate and distinct power system challenges to obtaining 20% of U.S. electric energy from wind. One challenge lies in the need to reliably balance electrical generation and load over time when a large portion of energy is coming from a variable power source such as wind, which, unlike many traditional power sources, cannot be accessed on demand or is "nondispatchable." The other challenge is to plan, build, and pay for the new transmission facilities that will be required to access remote wind resources. Substantial work already done in this field has outlined scenarios in which barriers to achieving the 20% Wind Scenario could be removed while maintaining reliable service and reasonable electricity rates.

This chapter begins with an examination of several detailed studies that have looked at the technical and economic impacts of integrating high levels of wind energy into electric systems. Next, this chapter examines how wind can be reliably accommodated into power system operations and planning. Transmission system operators must ensure that enough generation capacity is operating on the grid at all times, and that supply meets demand, even through the daily and seasonal load cycles within the system. To accommodate a nondispatchable variable source such as wind, operators must ensure that sufficient reserves from other power sources are available to keep the system in balance. However, overall it is the net system load that must be balanced, not an individual load or generation source in isolation. When seen in this more systemic way, wind energy can play a vital role in diversifying the power system's energy portfolio.

As the research discussed in this chapter demonstrates, wind's variability need not be a technical barrier to incorporating it into the broader portfolio of available options. Although some market structures, generation portfolios, and transmission rules accommodate much more wind energy than others, reforms already under consideration in this sector can better accommodate wind energy. Experience and studies suggest that with these reforms, wind generation could reliably supply 20% of U.S. electricity demand.

Finally, this chapter assesses the feasibility and cost of building new transmission lines and facilities to tap the remote wind resources that would be needed for the 20% Wind Scenario. Many challenges are inherent in building transmission systems to accommodate wind energy. If electric loads keep growing as expected, however, extensive new transmission will be required to connect new generation to loads. Over the coming decades, this will be true regardless of the power sources that dominate, whether they are fossil fuels, wind, hydropower, or others. The U.S. power industry has renewed its commitment to a robust transmission system, and support continues to grow for cleaner generation options. In this environment, designers and engineers must find ways to build transmission at a reasonable cost and take a closer look at the alternatives to conventional power generation in a carbon-constrained future.

Wind Penetration Levels

At least three different measures are used to describe wind penetration levels: energy penetration, capacity penetration, and instantaneous penetration. They are defined and related as follows:

Energy penetration is the ratio of the amount of energy delivered from the wind generation to the total energy delivered. For example, if 200 megawatt-hours (MWh) of wind energy are supplied and 1,000 MWh are consumed during the same period, wind's energy penetration is 20%.

Capacity penetration is the ratio of the nameplate rating of the wind plant capacity to the peak load. For example, if a 300 MW wind plant is operating in a zone with a 1,000 MW peak load, the capacity penetration is 30%. The capacity penetration is related to the energy penetration by the ratio of the system load factor to the wind plant capacity factor. Say that the system load factor is 60% and the wind plant capacity factor is 40%. In this case, and with an energy penetration of 20%, the capacity penetration would be 20% × 0.6/0.4, or 30%.

Instantaneous penetration is the ratio of the wind plant output to load at a specific point in time, or over a short period of time.

4.1 LESSONS LEARNED

4.1.1 WIND PENETRATION EXPERIENCES AND STUDIES

The needs of system operators—reflected in grid codes—ensure that wind power will continue to be integrated in ways that guarantee the continued reliable operation of the power system. Grid codes are regulations that govern the performance characteristics of different aspects of the power system, including the behavior of wind plants during steady-state and dynamic conditions. Grid codes around the world are also changing to incorporate wind plants; the Federal Energy Regulatory Commission (FERC) Order 661-A in the United States is an example.

Several U.S. utilities are approaching 10% wind capacity as a percentage of their peak load, including the Public Service Company of New Mexico (PNM) and Xcel

Energy (which serves parts of Colorado, Michigan, Minnesota, New Mexico, North Dakota, South Dakota, Texas, and Wisconsin). Xcel Energy could actually exceed 13% by the end of 2007. MidAmerican Energy in Iowa has already exceeded 10%, and Puget Sound Energy (PSE) in Washington expects to reach 10% capacity penetration shortly after 2010.

4.1.2 Power System Studies Conclude that 20% Wind Energy Penetration Can Be Reliably Accommodated

Rapid growth in wind power has led a number of utilities in the United States to undertake studies of the technical and economic impacts of incorporating wind plants, or high levels of wind energy, into their electric systems. These studies are yielding a wealth of information on the expected impacts of wind plants on power system operations.

General Electric International (GE), for example, has conducted a comprehensive study for New York state that examines the impact of 10% capacity penetration of wind by 2008 (Piwko et al. 2005). The state of California has set the ambitious goal of achieving 20% of its electrical energy from renewable sources by 2010 and 30% by 2020 (CEC 2007). The state of Minnesota has studied wind energy penetration of up to 25%, to be implemented statewide by 2020 (EnerNex Corporation 2006). The Midwest ISO (independent system operator) has examined the impact of achieving a wind energy penetration of 10% in the region by 2020, with 20% in Minnesota (Midwest ISO 2006).

U.S. experience with studies on wind were reviewed in a special issue of the Institute of Electrical and Electronics Engineers (IEEE) *Power & Energy Magazine* (IEEE 2005). The Utility Wind Integration Group (UWIG) also summarized these studies in cooperation with the three large utility trade associations—the Edison Electric Institute (EEI), the American Public Power Association (APPA), and the National Rural Electric Cooperative Association (NRECA). The UWIG (2006) summary came to the following conclusions:

- "Wind resources have impacts that can be managed through proper plant interconnection, integration, transmission planning, and system and market operations."
- "On the cost side, at wind penetrations of up to 20% of system peak demand, system operating cost increases arising from wind variability and uncertainty amounted to about 10% or less of the wholesale value of the wind energy. These conclusions will need to be reexamined as results of higher-wind-penetration studies—in the range of 25%–30% of peak balancing-area load—become available. However, achieving such penetrations is likely to require one or two decades."
- "During that time, other significant changes are likely to occur in both the makeup and the operating strategies of the nation's power system. Depending on the evolution of public policies, technological capabilities, and utility strategic plans, these changes can be either more or less accommodating to the natural characteristics of wind power plants."

4

4

- "A variety of means—such as commercially available wind forecasting and others discussed below—can be employed to reduce these costs."
- "There is evidence that with new equipment designs and proper plant engineering, system stability in response to a major plant or line outage can actually be improved by the addition of wind generation."
- "Since wind is primarily an energy—not a capacity—source, no additional generation needs to be added to provide back-up capability provided that wind capacity is properly discounted in the determination of generation capacity adequacy. However, wind generation penetration may affect the mix and dispatch of other generation on the system over time, since non-wind generation is needed to maintain system reliability when winds are low."
- "Wind generation will also provide some additional load carrying capability to meet forecasted increases in system demand. This contribution is likely to be up to 40% of a typical project's nameplate rating, depending on local wind characteristics and coincidence with the system load profile. Wind generation may require system operators to carry additional operating reserves. Given the existing uncertainties in load forecasts, the studies indicate that the requirement for additional reserves will likely be modest for broadly distributed wind plants. The actual impact of adding wind generation in different balancing areas can vary depending on local factors. For instance, dealing with large wind output variations and steep ramps over a short period of time could be challenging for smaller balancing areas, depending on the specific situation."

Load, Wind Generation, and Reserves

The first phase in determining how to integrate wind energy into the power grid is to conduct a wind integration study, which begins with an analysis of the impact of the wind plant profiles relative to the utility load curve. By way of illustration, Figure 4-1 shows a two-week period of system loads in the spring of 2010 for the Xcel system in Minnesota. This system has 1,500 MW of wind capacity on a 10,000 MW peak-load system (Zavadil et. al. 2004). Because both load and wind generation vary, it is the resulting variability—load net of wind generation—that system operators must manage, and to which the non-wind generation must respond.

Although wind plants exhibit significant variability and uncertainty in their output, electric system operators already deal with these factors on similar time scales with current power system loads. It is critical to understand that output variability and uncertainty are not dealt with in isolation, but rather as one component of a large, complex system. The system must be operated with balance and reliability, taking into account the aggregate behavior of all of its loads and generation operating together.

To maintain system balance and security, the electric system operator analyzes the regulation and load-following requirements of wind relative to other resources. Wind energy contributes some net increase in variability above that already imposed by cumulative customer loads. This increase, however, is less than the isolated variability of the wind alone on all time scales of interest. Although specific details

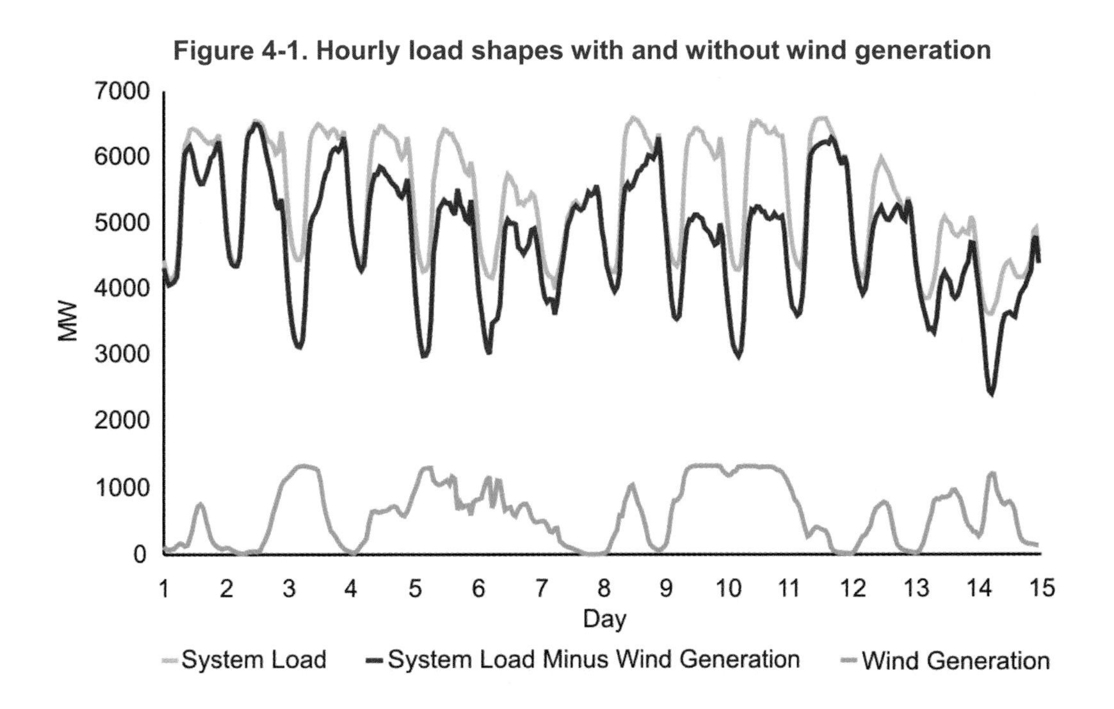

Figure 4-1. Hourly load shapes with and without wind generation

vary, distribution of changes in the load net flattens and broadens when large-scale wind is added to the system. The resulting reserve requirements can be predicted with statistical analysis. It is not necessary, or economically feasible, to counter each movement of wind with a corresponding movement in a traditional energy source. As a result, the load net of wind requires fewer reserves than would be required to balance the output of individual wind plants, or all the wind plants aggregated together, in isolation from the load. In the very short time frame, the additional regulation burden has been found to be quite small, typically adding less than $0.50/MWh to the cost of the wind energy (Zavadil, et. al. 2004).

Operational impacts of nondispatchable variable resources can occur in each of the time scales managed by power system operators. Figure 4-2 below illustrates these time scales, which range from seconds to days. "Regulation" is a service that rapid-response maneuverable generators deliver on short time scales, allowing operators to maintain system balance. This typically occurs over a few minutes, and is provided by generators using automatic generation control (AGC). "Load following" includes both capacity and energy services, and generally varies from 10 minutes up to several hours. This time scale incorporates the morning load pick-up and evening load drop-off. The "scheduling" and "unit-commitment" processes ensure that sufficient generation will be available when needed over several hours or days ahead of the real time schedule.

A statistical analysis of the load net of wind indicates the amount of reserves needed to cope with the combination of wind and load variability. The reserve determination starts with the assumption that wind generation and load levels are independent variables. The resultant variability is the square root of the sum of the squares of the individual variables (rather than the arithmetic sum). This means that the system operator, who must balance the total system, needs a much smaller amount of reserves to balance the load net of wind. Higher reserves would be needed if that operator were to try to balance the output of individual wind plants, or all the wind plants aggregated together in isolation from the load.

Figure 4-2. Time scales for grid operations

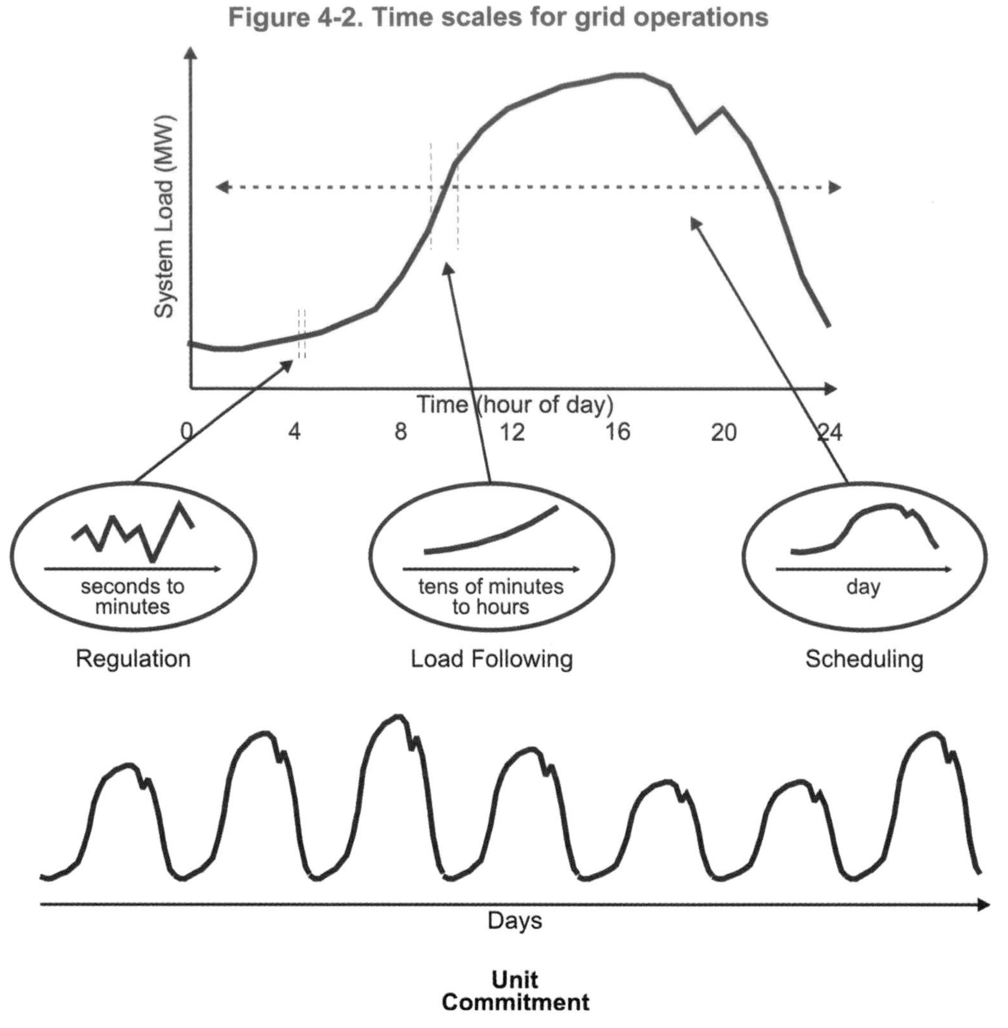

Source: Milligan et al. (2006)

Some suggest that hydropower capacity, or energy storage in the form of pumped hydro or compressed air, should be dedicated to supply backup or firming and shaping services to wind plants. Given an ideally integrated grid, this capacity would not be necessary because the pooling of resources across an electric system eliminates the need to provide costly backup capacity for individual resources. Again, it is the net system load that needs to be balanced, not an individual load or generation source in isolation. Attempting to balance an individual load or generation source is a suboptimal solution to the power system operations problem

Reserve Requirements Calculation

A hypothetical example is offered to calculate reserve requirements. Say that system peak load for tomorrow is projected at 1,000 MW with a 2% forecast error, which makes the forecast error (i.e., expected variability of peak load) equal to 20 MW. Wind generation for a 200 MW wind plant in that balancing area is predicted at a peak hour output of 100 MW with an error band of 20%. The expected variability of peak wind generation, then, is 20 MW. Assuming that these are independent variables, the total error is calculated as the square root of the sum of the squares of the individual variables (which is the square root of (2 × 20) squared, or 1.41 × 20, which equals 28 MW). Adding the two variables to estimate reserve requirements would result in an incorrect value of 40 MW.

because it introduces unnecessary extra capacity and an associated increase in cost. Hydro capacity and energy storage are valuable resources that should be used to balance the system, not just the wind capacity.

Figure 4-3 illustrates the incremental load-following impact of wind on an electrical system, as determined in the work of Zavadil and colleagues (2004). The histograms show more high-ramp requirements with wind than without wind, and a general reduction in small-ramp requirements compared to the no wind case. For these illustrative summer and winter hours, following load alone entails relatively fewer large-megawatt changes in generation (ramps). Following load net of wind generation, however, creates a wider variability in the magnitude of load change between two adjacent hours. A system with wind generation needs more active load-following generation capability than one without wind, or more load-management capability to offset the combined variability of load net of wind.

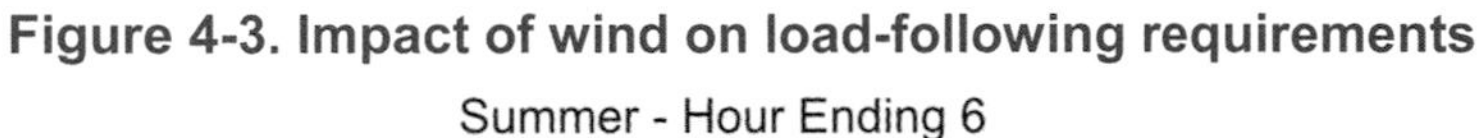

Figure 4-3. Impact of wind on load-following requirements

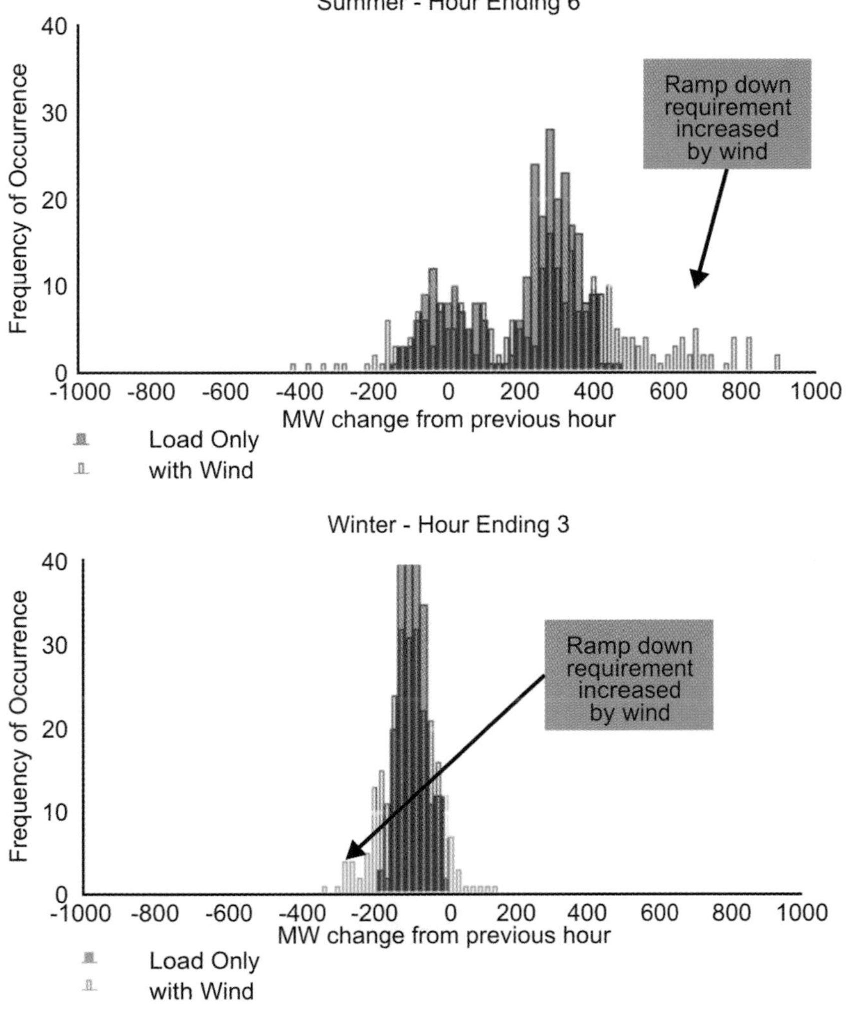

Wind Integration Cost

One impact of the variability that wind imposes on the system is an increase in the uncertainty introduced into the day-ahead unit-commitment process. Specifically, despite improvements in wind generation forecasting, greater uncertainty remains about what the next day's load net of wind and resulting generation requirements will be. The impact of these effects has been shown to increase system operating cost by up to $5.00/MWh of wind generation at wind capacity penetrations up to 20%. These figures are shown in the Unit-Commitment Cost column of Table 4-1. These day-ahead cost impacts are significantly higher than the others, reflecting the high cost of starting up generating units on a daily basis—even when they might not be needed.

The impact of wind's variability depends on the nature of the dispatchable generation sources, their fuel cost, the market and regulatory environment, and the characteristics of the wind generation resources. The most recent study conducted for Minnesota, for example, examined up to 25% energy penetration in the Midwest ISO market context (EnerNex 2006). The study found that the cost of wind integration is similar to that found in a study done two years earlier for a 15% wind capacity penetration in a vertically integrated market (Zavadil et al. 2004). A comparison of these results illustrates the beneficial effect of regional energy markets, namely that large operational structures reduce variability, contain more load-following resources, and offer more useful financial mechanisms for managing the costs of wind integration. Handling large output variations and steep ramps over short time periods (e.g., within the hour), though, can be challenging for smaller balancing areas.

Table 4-1 shows the integration cost results from recent U.S. studies. The wind integration issue is primarily a matter of cost, but the costs in the 20% Wind Scenario are expected to be less than 10% of the wholesale cost of energy (COE).

Table 4-1. Wind integration costs in the U.S.

Date	Study	Wind Capacity Penetration (%)	Regulation Cost ($/MWh)	Load Following Cost ($/MWh)	Unit Commit-ment Cost ($/MWh)	Gas Supply Cost ($/MWh)	Total Operating Cost Impact ($/MWh)
May 03	Xcel-UWIG	3.5	0	0.41	1.44	na	1.85
Sep 04	Xcel-MNDOC	15	0.23	na	4.37	na	4.60
Nov 06	MN/MISO	35 (25% energy)	0.15	na	4.26	na	4.41
July 04	CA RPS Multi-year Analysis	4	0.45	na	na	na	na
June 03	We Energies	4	1.12	0.09	0.69	na	1.90
June 03	We Energies	29	1.02	0.15	1.75	na	2.92
2005	PacifiCorp	20	0	1.6	3.0	na	4.6
April 06	Xcel-PSCo	10	0.20	na	2.26	1.26	3.72
April 06	Xcel-PSCo	15	0.20	na	3.32	1.45	4.97

Source: Adapted from IEEE (2005)

Wind Penetration Impacts

U.S. studies for capacity penetrations in the range between 20% and 35% have found that the additional reserves required to meet the intrahour variability are within the capabilities of the existing stack of units expected to be committed. In the high-penetration Minnesota study (EnerNex 2006), changes in total reserve requirements amounted to 7% of the wind generation needed to reach 25% wind energy penetration (5,700 MW). These reserves included 20 MW of additional regulating reserve, 24 MW of additional load-following reserve, and 386 MW maximum of additional operating reserve to cover next-hour errors in the wind forecast. Existing capacity is expected to cover these reserve needs, although over time, load growth could reduce this spare capacity if new dispatchable power plants are not constructed. Because wind and load are generally uncorrelated over short time scales, the regulation impact of wind is modest. The system operator will schedule sufficient spinning and nonspinning reserves so that unforeseen events do not endanger system balance, and so that control performance standards prescribed by the North American Electric Reliability Corporation (NERC) are met.

4.1.3 Wind Turbine Technology Advancements Improve System Integration

As described in more detail in the Wind Turbine Technology chapter, wind turbine technology has advanced dramatically in the last 20 years. From a performance point of view, modern wind power plants have much in common with conventional utility power plants, with the exception of variability in plant output. In the early days of wind power applications, wind plants were often thought of as a curiosity or a nuisance. Operators were often asked to disconnect from the system during a disturbance and reconnect once the system was restored to stable operation. With the increasing penetration of wind power, most system operators recognize that wind plants can and should contribute to stable system operation during a disturbance, as do conventional power plants.

As grid codes are increasingly incorporating wind energy, new plants are now capable of riding through a serious fault at the point of interconnection and are able to contribute to the supply of reactive power and voltage control, just like a conventional power plant. The supply of reactive power is a critical aspect of the design and operation of an interconnected power system. Modern wind plants can perform this function and supply voltage support for secure grid operations.

In addition, modern wind plants can be integrated into a utility's supervisory control and data acquisition (SCADA) system. They can provide frequency response similar to that of other conventional machines and participate in plant output control functions and ancillary service markets. Figure 4-4 illustrates the ability of a wind power plant to increase its output (grey line) in response to a drop in system frequency (red line). Figure 4-5 illustrates various control modes possible via

Figure 4-4. GE turbine frequency response

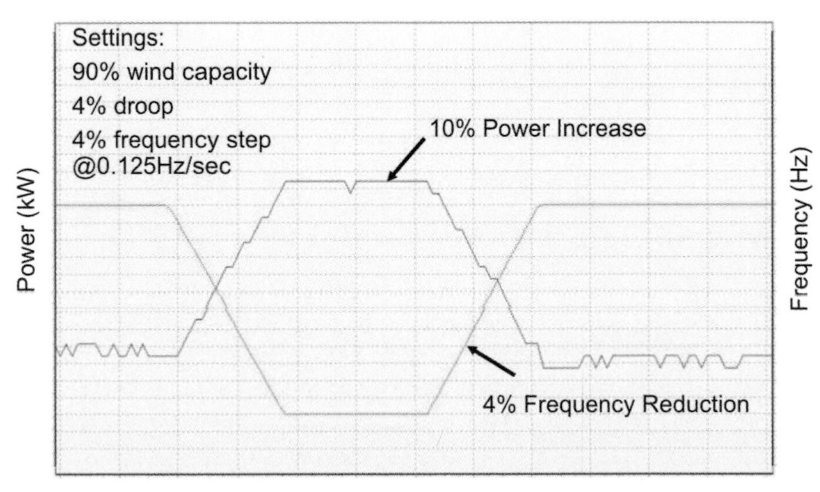

10% increase in plant watts with 4% under-frequency

Figure 4-5. Vestas wind turbine control capability

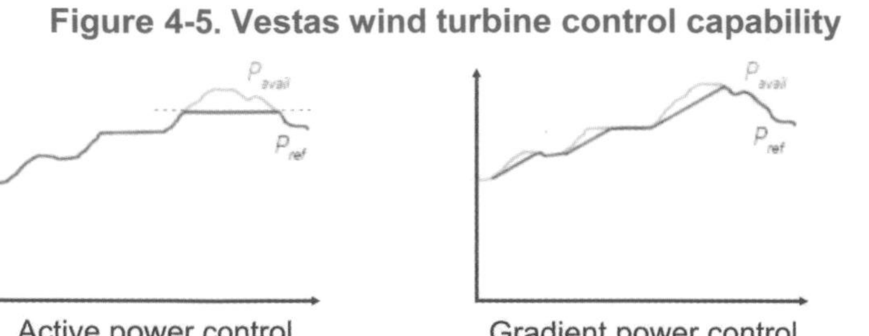

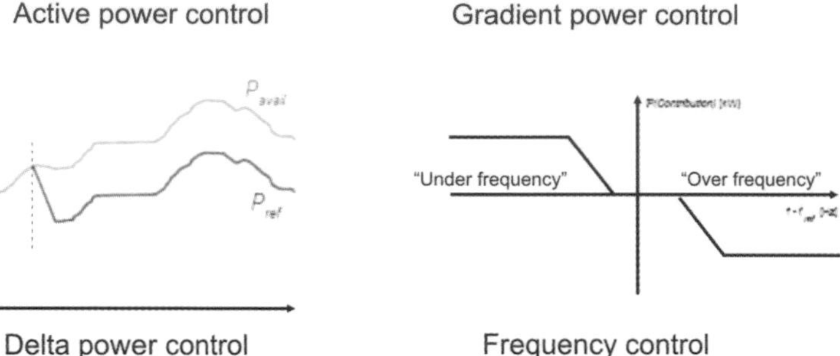

SCADA participation, including the ability to limit plant output power at any given time, control ramp rate in moving up or down, and carry spinning reserves as ordered (Saylors 2006). These plants also have the ability to tap frequency-responsive reserves. These control features come at a cost, however, which is that of "spilling" wind, a free energy resource. In any given geographic area, the cost of operating wind units in this manner so as to provide ancillary services would have to be compared with the cost of furnishing such services by other means.

Wind plant control systems offer another mechanism for dealing with the variability of the wind resource. Controllers can hold system voltage constant at a remote bus, even under widely varying wind speed conditions. Figure 4-6 shows an example of

Figure 4-6. GE wind plant controls

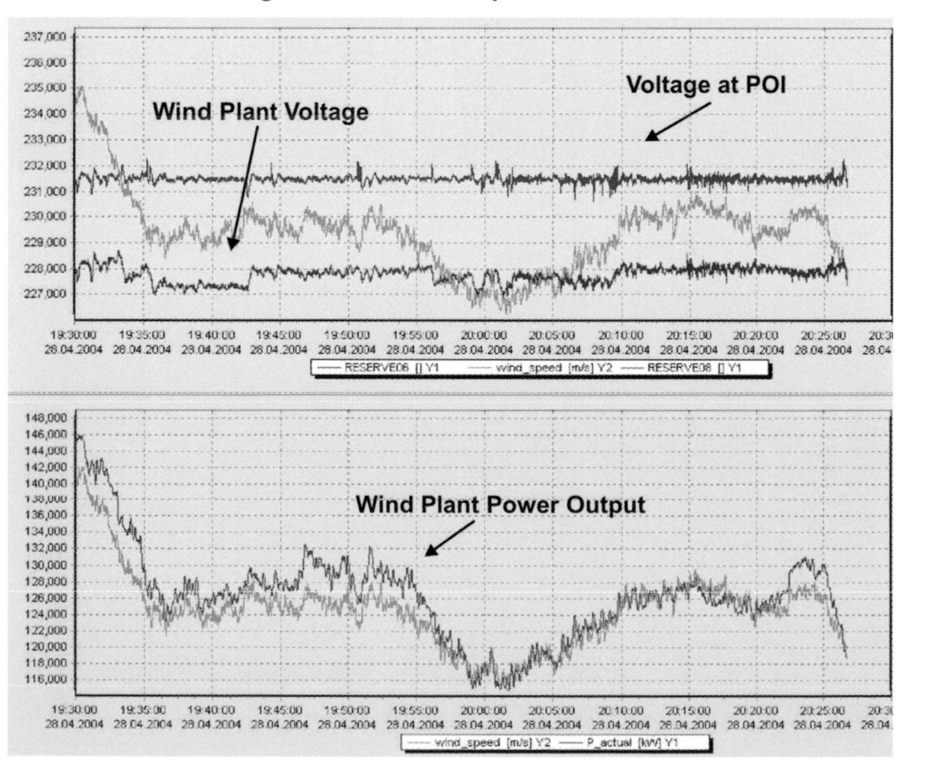

the voltage control features on a GE wind plant built recently in Colorado. In this system, voltage can be controlled across a broad range of wind conditions and power plant output. Voltage disturbances at the point of interconnection (POI) on the remote bus trigger offsetting changes in the wind plant voltage, controlling variations in the bus voltage.

Modern wind plants can be added to a power grid without degrading system performance. In fact, they can contribute to improvements in system performance. A severe test of the reliability of a system is its ability to recover from a three-phase fault at a critical point in the system. (For definitions of faults, see the Glossary in Appendix E.) System stability studies have shown that modern wind plants—equipped with power electronic controls and dynamic voltage support capabilities—can improve system performance by supporting postfault voltage recovery and damping power swings.

This performance is illustrated in Figure 4-7, which simulates a normally cleared three-phase fault on a critical 345 kV bus in the Marcy substation in central New York state (Piwko et al. 2005). The simulation assumed a 10% wind penetration (3,300 MW on a 33,000 MW system) of wind turbines with doubly fed induction

Figure 4-7. Impact of wind generation on system dynamic performance

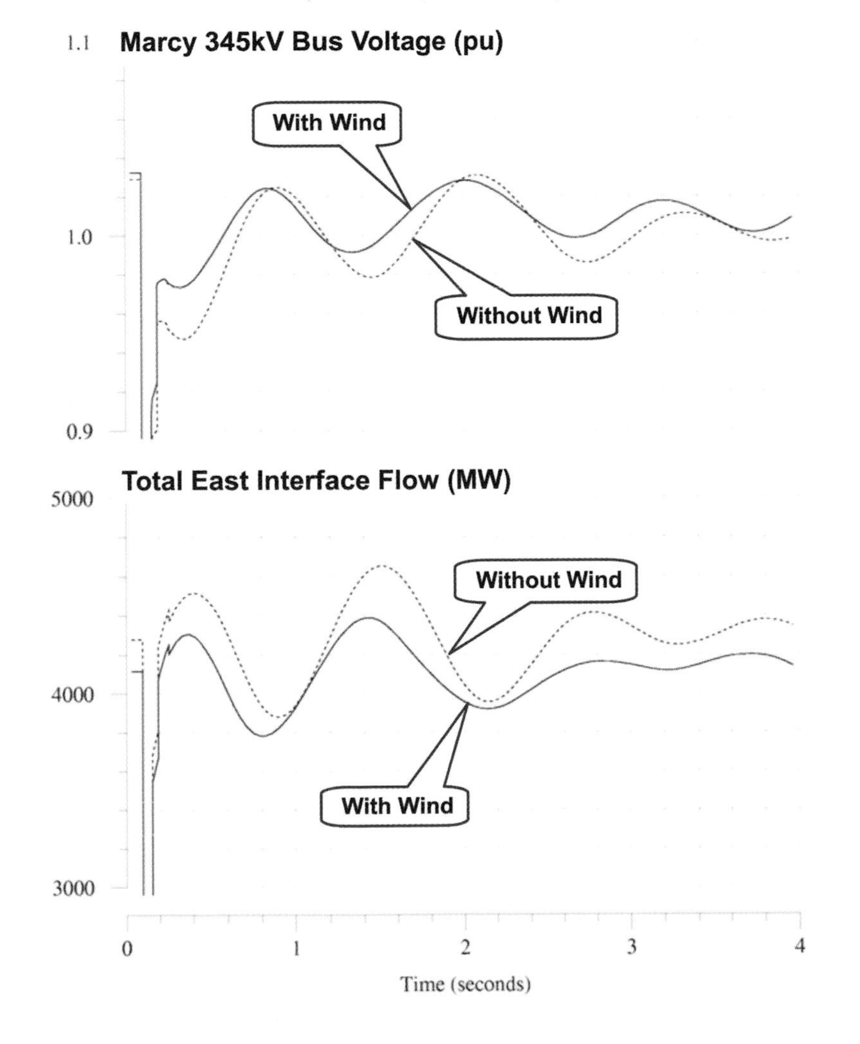

generators. It incorporated power electronics that allowed for independent control of real and reactive power. The top half of the figure shows the quicker recovery and increased damping in the system voltage transient at the Marcy 345 kV bus. The bottom half of the figure similarly shows that the flow on the east interface has less overshoot and is more highly damped with wind. And because the power electronics capabilities of these wind turbines remain connected to the grid and respond to grid conditions with or without real power generation, they manage voltage on the grid even when the turbine is not generating power.

Utility planners use models to understand and represent the capabilities and performance of generators and transmission system assets. Detailed wind plant models that incorporate today's sophisticated wind turbine and plant control features are being used to study future system configurations, as well as to improve the power system performance of conventional technology. Wind turbine manufacturers and developers are giving a high priority to the development of improved models in response to the leadership of utility organizations such as the Western Electricity Coordinating Council (WECC). The models are critical tools that enable planners to understand wind plant capabilities and accurately determine the impact of wind plants on power system behavior.

Improved performance features are likely to be incorporated into wind models as the utility interface and control characteristics of wind turbines and wind plants continue to evolve. Variable-speed designs with power electronic controls are improving real and reactive power control within wind turbines under both transient and steady-state conditions.

4.1.4 Wind Forecasting Enhances System Operation

System operators can significantly reduce the uncertainty of wind output by using wind forecasts that incorporate meteorological data to predict wind production. Such systems yield both hour-ahead and day-ahead forecasts to support real-time operations. They also inform the scheduling and market decisions necessary for day-ahead planning.

Forecasting allows operators to anticipate wind generation levels and adjust the remainder of generation units accordingly. Piwko and colleagues (2005) found that a perfect wind forecast reduced annual variable production costs by $125 million. And a state-of-the-art forecast delivered 80% of the benefit of a perfect forecast. Improved short-term wind production forecasts let operators make better day-ahead market operation and unit-commitment decisions, help real-time operations in the hour ahead, and warn operators about severe weather events. Advanced forecasting systems can also help warn the system operator if extreme wind events are likely so that the operator can implement a defensive system posture if needed. The operating impact with the largest cost is found in the unit-commitment time frame. The seamless integration of wind plant output forecasting—into both power market operations and utility control room operations—is a critical next step in accommodating large penetrations of wind energy in power systems.

4.1.5 Flexible, Dispatchable Generators Facilitate Wind Integration

Studies and actual operating experience indicate that it is easier to integrate wind energy into a power system where other generators are available to provide balancing energy and precise load-following capabilities. In 2005, Energinet.dk published the preliminary results of a study of the impact of meeting 100% of western Denmark's annual electrical energy requirement from wind energy (Pedersen 2005). The study showed that the system could absorb about 30% energy from wind without any excess (wasted) wind production, assuming no transmission ties to outside power systems. Surplus wind energy starts to grow substantially after the wind share reaches 50%. And if wind generates 100% of the total energy demand of 26 terawatt-hours (TWh), 8 TWh of the wind generation would be surplus because it would be produced during times that do not match customer energy-use patterns. Other energy sources, such as thermal plants, would supply the deficit, including the balancing energy. In the Pedersen study, the cost of electricity doubled when wind production reached 100% of the load. The study made very conservative assumptions, however, of no external ties or market opportunities for the excess wind energy.

4.1.6 Integrating an Energy Resource in a Capacity World

Wind energy has characteristics that differ from those of conventional energy sources. Wind is an *energy* resource, not a *capacity* resource. Capacity resources are those that can be available on demand, particularly to meet system peak loads. Because only a fraction of total wind capacity has a high probability of running consistently, wind generators have limited capacity value. Traditional planning methods, however, focus on reliability and capacity planning. Incorporating wind energy into power system planning and operation, then, will require new ways of thinking about energy resources.

Traditional system planning techniques use tools that are oriented toward ensuring adequate capacity. Most transmission systems, however, can make room for additional energy resources if they allow some flexibility for interconnection and operation. This flexibility includes choice of interconnection voltage, operation as a price-taker in a spot market, and limited curtailment. Economic planning tools and probabilistic analytical methods must also be used to ensure that a bulk power system has adequate generation and transmission capacity while optimizing its use of energy resources such as wind and hydropower.

Many hydropower generators produce low-cost variable energy. Unlike wind energy, most hydropower energy can be scheduled and delivered at peak times, so it contributes greater capacity value to the system. But because the reality of droughts

Effective Load Carrying Capability (ELCC)

The ELCC is the amount of additional load that can be served at the target reliability level with the addition of a given amount of generation (wind in this case). For example, if the addition of 100 MW of wind could meet an increase of 20 MW of system load at the target reliability level, it would have an ELCC of 20 MW, or a capacity value of 20% of its nameplate value.

Consider the following example: There are 1,000 MW of wind capacity in a concentrated geographic area, with an ELCC of 200 MW or a capacity value of 20%. The peak load of the system is 5,000 MW. On the peak-load day of the year, there is a dead calm over the area, and the output of the wind plant is 0. The lost capacity is 200 MW (20% of 1,000 MW). If this system were planned with a nominal 15% reserve margin, it would have a planning reserve of 750 MW that would well exceed the reserves needed to replace the loss of the wind capacity at system peak load.

causes hydropower capacity to vary from year to year, the capacity value of this energy resource (effective load-carrying capacity [ELCC]) must be calculated using industry-standard reliability models. The capacity value is used for system planning purposes on an annual basis, not on a daily operating basis. Some combination of existing market mechanisms and utility unit-commitment processes must be used to plan capacity for day-to-day reliability.

Planning techniques for a conventional power system focus on the reliable capacity offered by the units that make up the generation system. This is essential for meeting the system planning reliability criterion, such as the loss of load probability (LOLP) of 1 day in 10 years. The ELCC of a generation unit is the metric used to determine its contribution to system reliability. It is important to recognize that wind does offer some additional planning reserves to the system, which can be calculated with a standard reliability model. The ELCC of wind generation, which can vary significantly, depends primarily on the timing of the wind energy delivery relative to times of high system risk. The capacity value of wind has been shown to range from approximately 5% to 40% of the wind plant rated capacity, as shown in Table 4-2. In some cases, simplified methods are used to approximate the rigorous reliability analysis.

Table 4-2. Methods to estimate wind capacity value in the United States

Region/Utility	Method	Note
CA/CEC	ELCC	Rank bid evaluations for RPS (20%-25%)
PJM	Peak Period	Jun-Aug HE 3 -7 p.m., capacity factor using 3-year rolling average (20%, fold in actual data when available)
ERCOT	10%	May change to capacity factor for the hours between 4 -6 p.m. in July (2.8%)
MN/DOC/Xcel	ELCC	Sequential Monte Carlo (26%-34%)
GE/NYSERDA	ELCC	Offshore/land-based (40%/10%)
CO PUC/Xcel	ELCC	PUC decision (10%), Full ELCC study using 10-year data gave average value of 12.5%
RMATS	Rule of thumb	20% for all sites in RMATS
PacifiCorp	ELCC	Sequential Monte Carlo (20%). New Z-method 2006
MAPP	Peak Period	Monthly 4-hour window, median
PGE		33% (method not stated)
Idaho Power	Peak Period	4 p.m. -8 p.m. capacity factor during July (5%)
PSE and Avista	Peak Period	The lesser of 20% or 2/3 of January Capacity Factor
SPP	Peak Period	Top 10% loads/month; 85th percentile

Reliability planning entails determining how much generation capacity of what type is needed to meet specified goals. Because wind is not a capacity resource, it does not require 100% backup to ensure replacement capacity when the wind is not blowing. Although 12,000 MW of wind capacity have been installed in the United States, little or no backup capacity for wind energy has been added to date. Capacity in the form of combustion turbines or combined cycle units has been added to meet system reliability requirements for serving load. It is not appropriate to think in terms of "backing up" the wind because the wind capacity was installed to generate, low-emissions energy, but not to meet load growth requirements. Wind power cannot replace the need for many "capacity resources," which are generators and dispatchable load that are available to be used when needed to meet peak load. If wind has some capacity value for reliability planning purposes, that should be viewed as a bonus, but not a necessity. Wind is used when it is available, and system reliability planning is then conducted with knowledge of the ELCC of the wind plant. Nevertheless, in some areas of the nation where access to generation and markets that span wide regions has not developed, the wind integration process could be more challenging. (For more information on capacity terminology, see the Glossary in Appendix E.)

Plant capacity factors illustrate the roles that different power technologies play in a bulk power system. The capacity factor (CF) of a unit measures its actual energy production relative to its potential production at full utilization over a given time period. Table 4-3 shows the capacity factors of different power plant types within the Midwest ISO for a year. The units with the highest capacity factors—nuclear (75% CF) and coal (62% and 71% CF)—are the workhorses of the system because they produce relatively low-cost baseload energy and are fully dispatchable. Wind (30% CF) and hydro (27% CF) generate essentially free energy, so the wind is taken whenever it is available (subject to transmission availability) and the hydro is scheduled to deliver maximum value to the system (to the extent possible). The plants with the lowest capacity factors (combined cycle, combustion turbines, and oil- and gas-fired steam boilers) are operated as peaking and load-following plants and essential capacity resources. As illustrated in Table 4-3, many resources in the system operate at far less than their rated capacity for much of the year, but all are necessary components of an economic and reliable system.

Table 4-3. Midwest ISO plant capacity factor by fuel type (June 2005–May 2006)

Fuel Type	Number of Units	Max Capacity (MW)	Possible Energy (MWh)	Actual Energy (MWh)	Capacity Factor (%)
Combined Cycle	50	12,130	106,257,048	11,436,775	11
Gas Combustion Turbine (CT)	275	21,224	185,924,868	14,749,450	8
Oil CT	187	7,488	65,595,756	2,292,288	3
Hydro	113	2,412	21,129,120	5,696,734	27
Nuclear	17	11,895	104,200,200	77,764,757	75
Coal Steam Turbine (ST; <300 MW)	230	25,432	222,786,948	137,771,172	62
Coal ST Coal (≥300 MW)	113	51,155	448,116,048	320,014,108	71
Gas ST	20	1,673	14,651,976	1,256,756	9
Oil ST	12	1,790	15,676,896	560,910	4
Other ST	10	345	3,021,324	1,722,434	57
Wind	28	1,103	9,658,776	2,882,459	30
Total	**1055**	**136,646**	**1,197,018,960**	**576,147,844**	

4.1.7 Aggregation Reduces Variability

The greater the number of wind turbines operating in a given area, the less their aggregate production variability. This is shown in Table 4-4, which gives an analysis of wind production variability as a function of an increasing number of aggregated wind turbines in a large wind plant in the Midwest (Wan 2005). Table 4-4 shows the average and standard deviation of step changes in wind plant output for different numbers of turbines over different time periods. These results indicate that wind production changes very little over short time periods. As the time period increases from seconds to minutes to hours, the output variability increases because it is driven by changes in weather patterns. In addition, as a general trend, the more wind

Table 4-4. Wind generation variability as a function of the number of generators and time interval

		14 Turbines (%)	61 Turbines (%)	138 Turbines (%)	250+Turbines (%)
1-Second Interval					
	Average	0.4	0.2	0.1	0.1
	Std. Dev.	0.5	0.3	0.2	0.1
1-Minute Interval					
	Average	1.2	0.8	0.5	0.3
	Std. Dev.	2.1	1.3	0.8	0.6
10-Minute Interval					
	Average	3.1	2.1	2.2	1.5
	Std. Dev.	5.2	3.5	3.7	2.7
1-Hour Interval					
	Average	7.0	4.7	6.4	5.3
	Std. Dev.	10.7	7.5	9.7	7.9

Note: This table compares output at the start and end of the indicated time period in terms of the percentage of total generation from each turbine group. Std. Dev. is the abbreviation for standard deviation.

turbines that are operating in a given period, the lower the production variability during that period. Simply put, system operators in the United States have found that as more wind generating capacity is installed, the combined output becomes less variable.

A careful evaluation of integrating wind into current operations should include a determination of the magnitude and frequency of occurrence of changes in the net load on the system during the time frames of interest (seconds, minutes, and hours). This analysis, which should be conducted both before and after the wind generation is added, will help determine the additional requirements on the balance of the generation mix.

Similarly, as more wind turbines are installed across larger geographic areas, the aggregated wind generation becomes more predictable and less variable. The benefits of geographical diversity can be seen in Figure 4-8, which shows the change in wind plant hourly capacity factor over one year for four different levels of wind plant aggregation. This figure shows the operational capacity factor of wind turbines aggregated over successively larger areas—first over southwest Minnesota, then across southwest and southeast Minnesota, then across the entire state, and finally across both Minnesota and central North Dakota. There is a decrease in the number of occurrences of very high and very low hourly capacity factors in the tails of the distribution as the degree of aggregation increases. A considerable benefit is also realized across a broad mid-range of capacity factors from 20% to 80% (EnerNex 2006).

4.1.8 Geographic Dispersion Reduces Operational Impacts

Actual wind production data and sophisticated mesoscale weather modeling techniques have shown that a sudden and simultaneous loss of all wind power on a system is not a credible event. This scenario would be prevented by spatial variations of wind from turbine to turbine in a wind plant, and to a greater degree, from plant to plant. Because of the higher capacities of existing thermal plants and

Figure 4-8. Annual hourly capacity factor

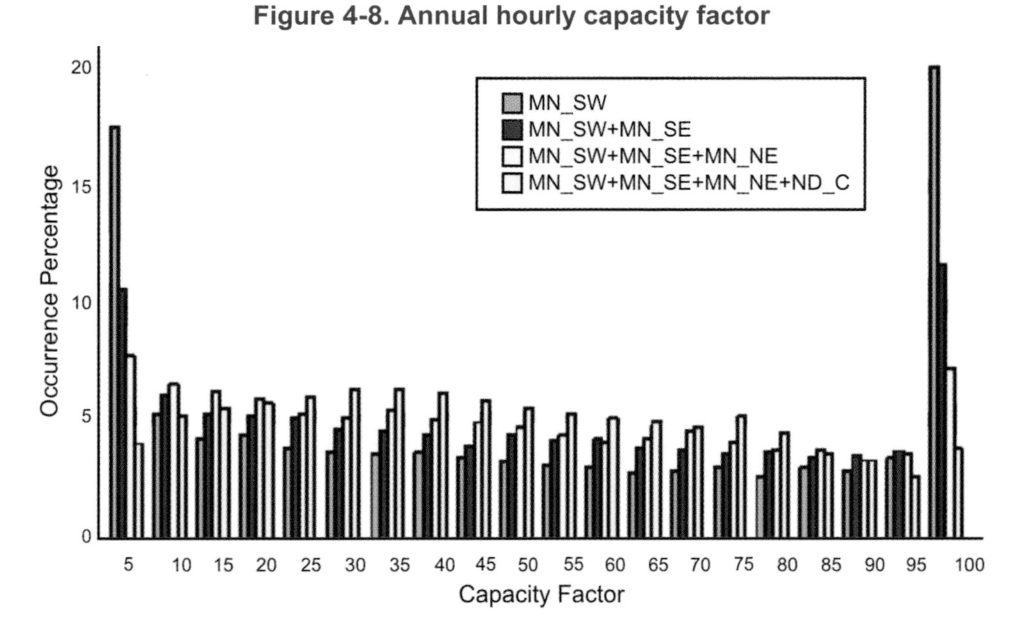

transmission lines, the loss of a wind plant will seldom be the single largest first contingency event for planning purposes. Severe weather events can lead to the loss of wind plant output as individual turbines trip off-line and/or restart as a storm front passes through. This kind of event happens on the time scale of tens of minutes to hours, however, rather than seconds.

4.1.9 Large Balancing Areas Reduce Impacts

To maintain the stable operation of the electric system, the system must instantaneously balance the amount of generation supplied and the load. If the generation and load are not in balance, the system could potentially suffer a loss of either, or lose stability and collapse. The system-balancing function is performed by authorities who operate a portion of the system called a "balancing area." (For more information on balancing areas, see the Glossary in Appendix E.) Today there are about 130 balancing areas in the U.S. grid. The largest balancing area is the PJM grid, which is part of the Eastern Interconnection, with a peak load of 145,000 MW. A small balancing area, in contrast, might be a small utility with a peak load of a few hundred MW. Balancing areas are an outgrowth of the evolution of power systems. In some areas, the current patchwork nature of the grid resulted when a number of small, isolated systems were combined into a single balancing area such as PJM.

Systems became interconnected for a number of reasons, mostly having to do with reliability and economics. Consider this example: If three adjacent systems, each with a peak load of 3,000 MW, had a single largest contingency (loss of a line or generator) of 300 MW, each would carry 300 MW of reserves. If the three systems were interconnected, and the single largest contingency was still 300 MW, each system would need only 100 MW of reserves to cover contingency reserve requirements. In this example, and as another advantage, the peak load of the combined system would be less than 9,000 MW because of diversity in the load of the three systems. Finally, operators can call on the most efficient and lowest-cost producers available across the combined system and shift production away from more-expensive units. This approach ensures that the generation mix used to meet the aggregated system's changing load is always relatively more efficient. Overall, the three interconnected systems are able to operate more efficiently at a reduced operating cost.

Wind units operate in a parallel situation across multiple balancing areas. As indicated previously, geographically dispersed wind units produce electricity more consistently and predictably. Similarly, when a system is operating across a larger area, more wind generators are available to offset customer demands, making the resulting load net of wind less variable and more predictable.

The Energy Policy Act (EPAct) of 2005 created an Electricity Reliability Organization (ERO), overseen by FERC, to enforce mandatory reliability standards, with fines for rule violations. The resulting reliability standards, implemented by NERC, include the following:

- Operator training
- Balancing authority performance criteria
- Control room situational awareness capability
- Control center hardware and software capability

These reliability requirements are likely to increase pressure on small balancing areas to consolidate. In addition to providing reliability benefits, consolidation of balancing areas would offer economic advantages because it would reduce operating costs and lower the cost of increased penetration of wind power. Virtual balancing-area consolidation can deliver the benefits of large-area aggregation without physically merging balancing areas under a single operator. Virtual consolidation can be accomplished through reserve sharing or pooling across a group of utilities, sharing of area control error (ACE) data among several balancing areas, and dynamic scheduling of wind plants from a smaller to a larger balancing area. All of these methods can help deal with the challenges of high penetrations of wind power. (For further explanation of ACE, see Appendix E.)

4.1.10 Balancing Markets Ease Wind Integration

Experience has shown that the use of well-functioning hour-ahead and day-ahead markets and the expansion of access to those markets are effective tools for dealing with wind's variability. A deep, liquid real-time market is the most economical approach to providing the balancing energy required by wind plants with variable outputs (IEA 2005). The absence of a wind production forecast introduces significant costs into the day-ahead market. As a result, wind plant participation in day-ahead markets is important for minimizing total system cost. Price-responsive load markets and associated technologies are helpful components of a well-functioning electricity market, which allows the power system to better deal with increased variability. In some regions of the United States that lack centralized markets, access to balancing and related services is being pursued through instruments such as bilateral contracts and reserve-sharing agreements.

The electricity market allows energy from all generators across the area to be dispatched based on real-time prices. When wind blows strongly, the real-time price falls, signaling more controllable generators to reduce their output and save costly fuel. Conversely, when wind drops off, real-time prices rise and dispatchable generators increase their output. As an example, the Midwest ISO covers a footprint of 15 states, so there is a deep pool of generators that can ramp up and down in response to wind output. The EnerNex (2006) study in Minnesota examined up to 25% energy penetration in the Midwest ISO market context (33% capacity penetration). The integration costs were similar to the results of a study done two years earlier (Zavadil et al. 2004) for a 15% wind capacity penetration in a structure without the regional Midwest ISO balancing market.

4.1.11 Changing Load Patterns Can Complement Wind Generation

To date, the electric system has been planned and operated under the fundamental assumption that the supply system must perfectly meet every customer's energy use, and that demand is relatively uncontrolled. But this assumption is starting to change as policy makers work to create opportunities for customers to manage their energy use in response to price signals. Wider use of price-responsive demand is expected to boost the competitiveness of wholesale electricity markets, enhance grid reliability, and improve the efficiency of resource use. Technology and regulatory options that enable customer energy management are gaining momentum because of increasing support from electricity regulators, regional transmission organizations (RTOs), and retail electricity providers.

Several customer-driven energy trends could have a significant impact on wind development. Much wind generation occurs in hours when energy use is low. Two proposed off-peak electricity uses—the deployment of plug-in hybrid vehicles with off-peak charging and the production of hydrogen to power vehicles—could absorb much of this off-peak, low-cost wind generation. In addition, as more customers gain the ability to practice automated price-responsive demand or to automatically receive and respond to directions to increase or decrease their electricity use, system loads will be able to respond to, or manage, variability from wind and other energy sources.

4.2 Feasibility and Cost of the New Transmission Infrastructure Required for the 20% Wind Scenario

If the considerable wind resources of the United States are to be utilized, a significant amount of new transmission will be required. Transmission must be recognized as a critical infrastructure element needed to enable regional delivery and trade of energy resources, much like the interstate highway system supports the nation's transportation needs. Every era of new generation construction in the United States has been accompanied by new transmission construction. Federal hydropower developments of the 1930s, 1940s, and 1950s, for example, included the installation of integral long-distance transmission owned by the federal government. Construction and grid integration of large-scale nuclear and coal plants in the 1960s and 1970s entailed installing companion high-voltage interstate transmission lines, which were needed to deliver the new generation to loads. Even the natural gas plants of the 1990s, although requiring less new electric transmission, relied on expansion of the interstate gas transportation network. Significant expansion of the transmission grid will be required under any future electric industry scenario. Expanded transmission will increase reliability, reduce costly congestion and line losses, and supply access to low-cost remote resources, including renewables.

Much of the current electric grid was built to deliver power from remote areas to load centers. During the past two decades, however, investment in gas-fired generation units located closer to load centers allowed the power system to grow without investment in major new transmission (Hirst and Kirby 2001). Transmission investment lagged substantially behind that of previous decades because of uncertainty about the outcome of electricity restructuring. The average level of investment for the last half of the 1990s was under $3 billion per year, as illustrated in Figure 4-9. This amount was down from investments of approximately $5.5 billion per year in the mid 1970s (adjusted for inflation). Although transmission investment declined for two decades, it has been steadily climbing since the late 1990s .

Figure 4-9. Annual transmission investments from 1975 through 1999 and projections through 2005

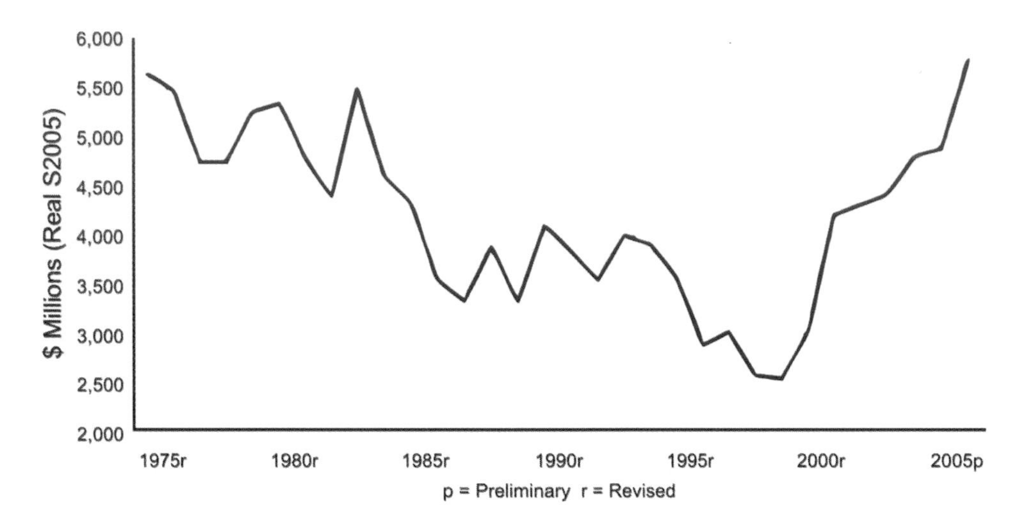

Transmission investment from investor-owned utilities and independent transmission companies climbed from $3.0 billion per year in 2000 to $6.9 billion in 2006 (Eisenbrey 2007). Nearly $8 billion of investment is expected in 2007, with the figure growing to $8.4 billion in 2009. The steady increase in new transmission investment reflects not only a catch-up in local transmission, but new commitments to backbone transmission systems for major new generation, intra- and inter-regional trade, and increased reliability.

The 20% Wind Scenario would require continued transmission investment. Many new transmission infrastructure studies, plans, and projects are already under way. Current or recent activities include the following:

- Planning by the Western Governors' Association's (WGA) Clean and Diversified Energy Advisory Committee (CDEAC 2006)
- The collaboration of Minnesota utilities in the Capital Expansion Plan for 2020 (CapX 2020)
- The creation of Competitive Renewable Energy Zones (CREZ) by the state legislature in Texas (ERCOT 2006)
- The creation of state transmission or infrastructure authorities in Wyoming, Kansas, South Dakota, New Mexico, and Colorado
- The proliferation of large interstate transmission projects in the West (WIEB 2007)
- The SPP "X Plan" and Extra High Voltage analysis (SPP)

- The Midwest ISO *Transmission Expansion Plan 2006* (Midwest ISO 2006).

4.2.1 A New Transmission Superhighway System Would Be Required

Wind energy development requires two types of transmission. Trunk-line transmission runs from areas with high-quality wind resources and often carries a high proportion of energy from wind and other renewable sources. Backbone high-voltage transmission runs across long distances to deliver energy from production areas to load centers. These superhighways mix power from many generating areas, sources, and shippers—just as a highway carries all types of vehicles traveling a range of distances.

To determine how much transmission would be needed for the 20% Wind Scenario, the National Renewable Energy Laboratory's (NREL) Wind Deployment System (WinDS) model was used (see Appendices A and B). The approach, described in Appendices A and B, used the WinDS model to determine distances from the point of production to the point of consumption, as well as the cost-effectiveness of building wind plants close to load or in remote locations and paying the transmission cost. To account for the cost of transmission that would be required by coal and other resources, the analysis added the typical cost of transmission needed to interconnect those resources to the capital cost. This method, although providing balance in the overall cost assessment, is only a first step. More work must be done in regional transmission planning processes to evaluate the transmission required for the desired portfolio of resources.

When determining whether it is more efficient to site wind projects close to load or in higher quality wind resource areas that are remote from load and require transmission, the WinDS optimization model finds that it is often more efficient to site wind projects remotely. In fact, the model finds that it would be cost-effective to build more than 12,000 miles of additional transmission, at a cost of approximately $20 billion in net present value terms. Much of that transmission would be required in later years after an initial period in which generation is able to use the limited remaining capacity available on the existing transmission grid. The transmission required for the 20% Wind Scenario can be seen in the red lines on the map in Figure 4-10. The red lines represent general areas where new transmission capacity would be needed. The existing transmission grid illustrated by green lines. As a point of comparison, more than 200,000 miles of transmission lines are currently operating at 230 kV and above.

This analytical approach is consistent with other recent or current studies and plans, such as the following:

- The CDEAC evaluated a "high renewables" case and found that it would require an additional 3,578 line miles of transmission at a total cost of $15.2 billion (CDEAC 2006). This transmission investment would access 68.4 GW of renewable generation (predominantly wind) and 84.6 GW of new fossil fuel generation. Under the CDEAC analysis, if half of the transmission cost is assigned to wind, the resulting cost would be approximately $120 per new kilowatt of wind developed. This represents about a 7% increase in the capital cost of wind development (based on capital costs for a wind energy facility of about $1,800/kW).

Figure 4-10. Conceptual new transmission line scenario by WinDS region

2030 - New Transmission Lines - WinDS Region Level - Simplified Corridors >= 100 MW

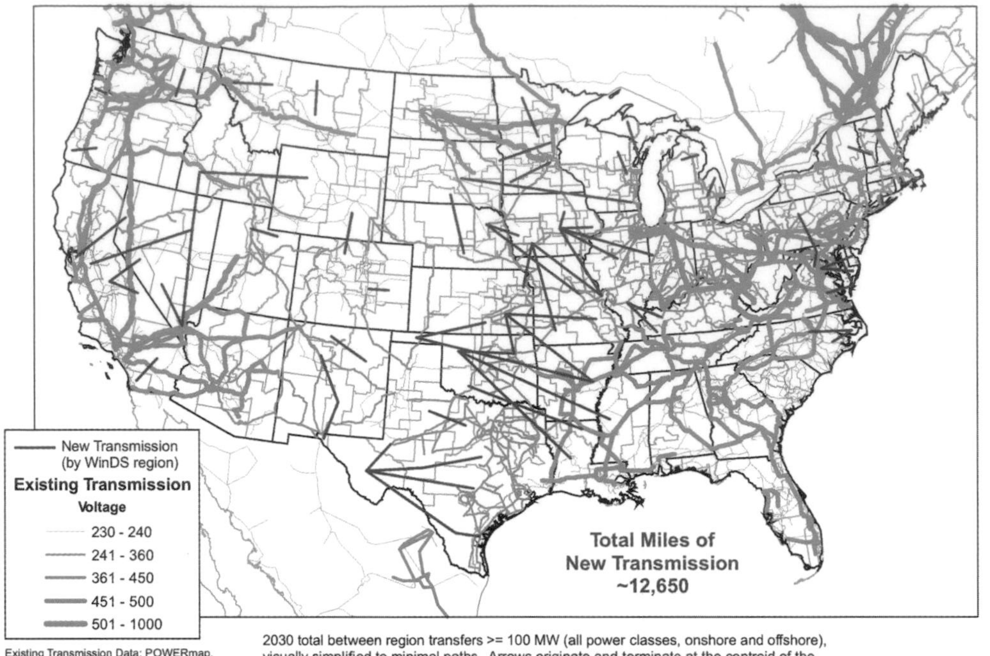

Existing Transmission Data: POWERmap, powermap.platts.com ©2007 Platts, A Division of The McGraw-Hill Companies

2030 total between region transfers >= 100 MW (all power classes, onshore and offshore), visually simplified to minimal paths. Arrows originate and terminate at the centroid of the region for visualization purposes; they do not represent physical locations of transmission lines.

20% Wind 06-19-2007

- The Midwest ISO compared the benefits and costs of bringing 8,640 MW of new wind energy online. Using a natural gas price of $5 per million British thermal units (MMBtu; well below 2007 prices), the annual benefits of reduced natural gas costs from new transmission and development of wind generation were between $444 and $478 million (Midwest ISO 2003). The Midwest ISO recently studied the costs of developing 16,000 MW of wind within its system, along with 5,000 miles of new 765 kV transmission lines to deliver the wind from the Dakotas to the New York City area. Although the overall generation and transmission costs reached an estimated investment of $13 billion, the project produced annual savings of $600 million over its costs. These savings are in the form of lower wholesale power costs and prices in the eastern part of the Midwest ISO footprint—such as Ohio and Indiana—resulting from greater access to lower-cost generation in western states such as Iowa and the Dakotas.
- AEP, a large utility and transmission owner/operator, produced a conceptual transmission plan to integrate 20% electricity from wind. The conceptual plan provides for 19,000 miles of new 765 kV transmission line at a discounted or net present value cost of $26 billion. This estimate is close to the WinDS model estimate (AEP 2007).
- ERCOT, the independent transmission operator for most of Texas, evaluated 12 options to build transmission for additions of 1,000 MW to 4,600 MW of wind energy. ERCOT found that the transmission addition would cost between $15 million and $1.5 billion, depending on the distance required. The transmission cost averages $180/kW of wind energy, or about 10% of the $1,800/kW

capital cost (ERCOT 2006). The benefits available from such transmission are often reported in terms of annual savings to consumers and the reduced cost of energy production. The graph in Figure 4-11 illustrates the cumulative benefits in the Texas study, for the weakest investment of the 12 analyzed by ERCOT. It should be noted that wind transmission cost estimates remain highly uncertain. For example, ERCOT recently updated their earlier study and found that for additions of 5,150 MW to 18,000 MW of wind energy, the transmission addition would cost between $2.95 billion and $6.38 billion, or in the range of $350/kW to $570/kW (ERCOT 2008).

Figure 4-11. Cumulative savings versus total transmission cost for renewable energy zone (worst case)

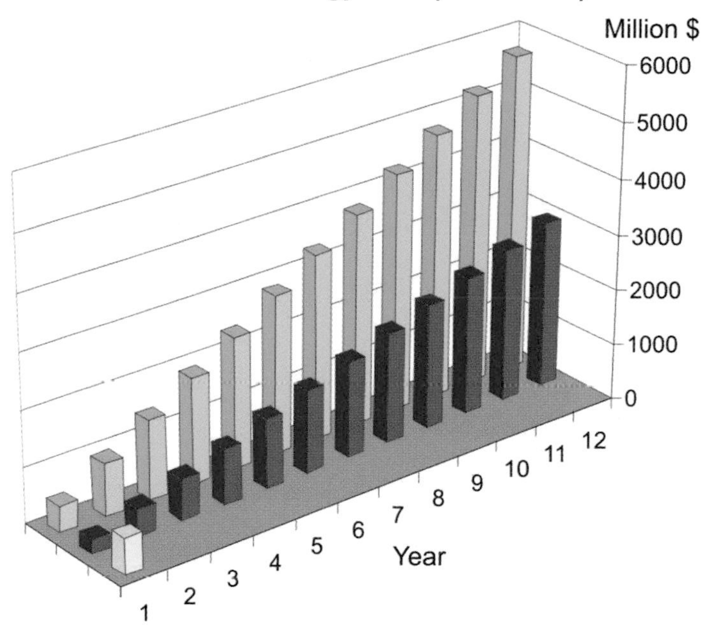

Source: ERCOT (2006)

- In another study analyzing transmission costs, the CDEAC Wind Task Force used NREL's WinDS geographic information system (GIS) database to create wind energy supply curves for many states in the western United States. This analysis showed that the western states can build 30 GW of wind capacity that can be delivered at a price of $50/MWh (counting both generation and transmission costs). Building additional transmission to reach more wind resources and more loads would raise the marginal cost by 20% to $60/MWh. More than 100 GW of new wind capacity could be developed at that price, using 2005 equipment costs (CDEAC 2006).

Clearly, significant additional transmission capacity would be required to integrate high levels of wind across the country. As the studies described here demonstrate, however, meeting this challenge could be economically and technically feasible. In addition, sizable net reductions in the cost of delivering bulk electricity to load centers could be achievable.

Developing any major new generation sources in remote or semiremote locations will require new transmission to deliver the energy to loads. As long as load continues to grow, investment in transmission will be needed as well. Most high-voltage transmission additions serve multiple generation resources, not just wind. Once the marginal transmission cost for wind is balanced against its low energy cost and environmental impacts, the net costs might turn out to be not much greater in the portfolio context than the transmission costs of traditional fossil fuel resources.

An investment of approximately $60 billion (in undiscounted terms) in transmission between now and 2030, as suggested by the NREL analysis, amounts to an expenditure of approximately $3 billion per year over the next 22 years. Current transmission investment level is nearly $8 billion per year and growing. Regardless of wind's role, most analysts believe that this figure will continue to increase as utilities make up for decades of underinvestment in the grid. As long as electricity demands grow, new transmission will be required to serve any new generation developed, and incremental transmission costs will be unavoidable.

4.2.2 Overcoming Barriers to Transmission Investment

Barriers to transmission investment include:

- Transmission planning
- Allocation of the costs of new transmission investments
- Assurance of cost recovery
- Siting of new transmission facilities

More details on each area are given in the following subsections.

Transmission Planning

Generation companies are currently reluctant to commit to a new generation project unless it is clear that transmission will be available, but transmission developers are equally reluctant to step forward until generator interconnection requests have been filed (hence, transmission planning has its own "chicken or the egg" conundrum). Most electric utilities planned generation and transmission in an integrated process until the 1990s, when federal open access rules required the separation of transmission and generation businesses. The effects of this separation on planning can be reduced through open, transparent transmission planning processes, which are now required by FERC's recently enacted ruling, Order No. 890 (FERC 2007).

The 20% Wind Scenario would require a generic change in the way transmission planning is done in many areas of the country. Numerous parties across a wide geographic area would need to collaborate on developing a common plan, instead of individual entities planning in isolation. This approach yields major economies of scale in that all users would benefit by pooling solutions to their needs into a single plan that would be more productive (in regional terms) than simply summing the needs of individual organizations. FERC's Order No. 890 is a large step toward this regional joint planning approach, but success will depend on collaborative follow-through at the regional level.

Cost Allocation

Transmission is often a "public good"— meaning that its benefits are widely dispersed and that some parties can enjoy these benefits without incurring direct costs. In such situations, parties might have incentives to avoid paying their fair share of the costs. Accordingly, public good status cannot be achieved unless some government agency determines how the costs are to be allocated and is able to enforce that allocation.

Under the Federal Power Act, FERC is responsible for determining how transmission costs are to be allocated. For regions with RTOs or ISOs, FERC has typically reviewed generic cost-allocation plans proposed by these organizations and approved the plans with modifications. In areas without RTOs or ISOs, prospective transmission developers propose cost-allocation arrangements to FERC on a project-by-project basis. FERC reviews the proposals; calls for additional information if needed; and either approves them, rejects them, or approves them with certain conditions attached.

Cost Recovery

A new transmission facility, regardless of need or merit, will not be built until the participating utilities (and the financial community) have a very high degree of certainty that the cost of the facility will be recoverable in a predictable manner. FERC and state regulatory approval of a cost-allocation plan and a rate of return on the investment are essential.

Creating Renewable Energy Models for New Transmission

A few states that have good wind resources and RPS laws have decided to expand their states' transmission in advance of generation to enable the modular development of location-constrained, clean, and diversified resource areas to meet state goals. Texas, Minnesota, Colorado, and California, for example, are leaders in renewable energy development, and have created renewable energy models for new transmission. North Dakota, South Dakota, Wyoming, Kansas, and New Mexico have also established new authorities to spur investment in additional transmission infrastructure.

Transmission Siting

Local opposition to proposed transmission lines is often a major challenge to transmission expansion. An AC transmission line typically benefits all users along its path by increasing reliability, allowing for new generation and associated economic development, and providing access to lower-cost resources. Local owners, however, do not always value such benefits and frequently have other concerns that must be addressed. Some transmission companies have been more effective than others at obtaining local input, identifying and dealing with landowners' concerns, and selecting routes. Best practices in this area need to be identified and broadly applied.

State agencies sometimes reject interstate transmission proposals if it appears that they would not result in significant benefits for intrastate residents. This concern led the U.S. Congress to include a provision in the 2005 EPAct that establishes a federal "backstop" transmission siting authority, which can be invoked if the U.S. Department of Energy (DOE) has designated the relevant geographic area as a "national interest electric transmission corridor" (i.e., a "national corridor"), and an affected state has withheld approval of a proposed transmission facility in the national corridor for more than one year.

4.2.3 Making a National Investment in Transmission

The 20% Wind Scenario would require widespread recognition that there is national interest in ensuring adequate transmission. Expanding the country's transmission infrastructure would support the reliability of the power system; enable open, fair, and competitive wholesale power markets; and grant owners and operators access to low-cost resources. Although built to enable access to wind energy, the new transmission infrastructure would also increase energy security, reduce GHG emissions, and enhance price stability through fuel diversity.

4.3 U.S. Power System Operations and Market Structure Evolution

The lessons summarized from research done to date illustrate a number of changes that would facilitate reaching 20% wind energy penetration. Expanding from approximately 12 GW at the time of this writing to over 300 GW will require most or all of these changes. This section summarizes the operational and market features that would support the 20% Wind Scenario. These features are also important to the long-term sustainability of the electric industry.

4.3.1 Expanding Market Flexibility

The 20% Wind Scenario would be aided by the development of or access to energy spot markets where participants who have an excess or shortfall of power could trade at competitive prices that reflect the marginal cost of balancing load. Such markets were recently implemented in the 15-state Midwest ISO region, the mid-Atlantic PJM region, New York, New England, and the Southwest Power Pool (SPP), showing the feasibility of such reforms. It is certainly possible that other regions could pursue such reforms by 2030.

Broad geographical markets and inter-area trading would allow the benefits of geographic dispersion and aggregation of wind plant output to be realized. These benefits have been shown to reduce the variability of wind plant output on a large scale, which makes a market-based approach and trading system all the more worthwhile. The challenge is that energy spot markets have been subject to opposition as market prices have risen because of higher fuel costs.

4.3.2 Enhancing Wind Forecasting and System Flexibility

The 20% Wind Scenario would require highly trained power system operators, equipped with state-of–the-art wind resource forecasting tools that would be fully integrated with power system operations. Forecasting is spreading rapidly and improving significantly, particularly in terms of its adoption and integration within power system operations. Some power system dispatchers, however, still need to be trained to operate systems with high wind penetration and to use forecasting and operations tools that predict and respond to wind plant output fluctuations.

To achieve balance in a power system using wind energy, the 20% Wind Scenario would require the use of the existing fleet of flexible, dispatchable, mainly gas-fired generators designed for frequent and rapid ramping. There would need to be enough dispatchable units to balance the system as fluctuations occur in wind plant output and load.

Transmission services vary across regions in the United States. Regions with RTOs have "financial transmission rights" that are more flexible than capacity reservations and allow for payment based on usage. In addition, under FERC Order 890 (FERC 2007), all regions are now required to develop "conditional firm" services, which would allow for resources such as wind to be better integrated into the grid.

The 20% Wind Scenario would require end users to be able (via price signals and technology) to respond to system needs by shifting or curtailing consumption. Time-shifting of demand would help reduce today's large difference between peak and off-peak loads and encourage more flexible loads (such as plug-in hybrid cars, hydrogen production, and smart appliances) that take energy from the grid during low-load periods. These practices would smooth electricity demand and open a larger market for off-peak wind energy.

The 20% Wind Scenario would require a smarter, more flexible, and more robust high-voltage transmission grid than the one in place today. Greater reliance on flexible AC transmission system (FACTS) devices and wide-area monitoring and control systems would be necessary. Increased flexibility would accommodate variations in technology choices, resource mixes, market rules, and regional characteristics. Greater robustness would help ensure future reliability. Information technologies for distributed intelligence, sensors, smart systems, controls, and distributed energy resources would need to be standardized and integrated with market and customer operations.

4.4 References and Other Suggested Reading

Ackermann, T. 2005. *Wind Power in Power Systems*. Chichester, UK: Wiley Europe Ltd.

AEP (American Electric Power). 2007. Interstate Transmission Vision for Wind Integration. http://www.aep.com/about/i765project/docs/windtransmissionvisionwhitepaper.pdf

Bouillon, H., P. Fösel, J. Neubarth, and W. Winter. 2004. *Wind Report 2004*. Bayreuth, Germany: E.ON Netz. http://www.wind-watch.org/documents/wp-content/uploads/EonWindReport2004.pdf

CapX 2020. CapX 2020 Web site. http://www.capx2020.com/.

Cardinal, M., and N. Miller. 2006. "Grid Friendly Wind Plant Controls: WindCONTROL – Field Test Results." Presented at WindPower 2006, June 4–7, Pittsburgh, PA.

CDEAC (Clean and Diversified Energy Advisory Committee). 2006. *Combined Heat and Power White Paper*. Denver, CO: CDEAC. http://www.westgov.org/wga/initiatives/cdeac/CHP-full.pdf

CEC (California Energy Commission). 2007. *Intermittency Analysis Project: Final Report*. California Energy Commission, July 2007. http://www.energy.ca.gov/pier/final_project_reports/CEC-500-2007-081.html

Deutsche Energie-Agentur GmbH (dena). 2005. *Energy Management Planning for the Integration of Wind Energy into the Grid in Germany, Onshore and Offshore by 2020.* Cologne, Germany: Deutsche Energie-Agentur GmbH. http://www.uwig.org/Dena-2005_English.pdf

Eisenbrey, C. 2007. *EEI Energy Data Alert, Strong Upward Trend in T&D Investment Continued in 2006.* Washington, DC: Edison Electric Institute (EEI).

EnerNex Corporation. 2006. Final Report: *2006 Minnesota Wind Integration Study.* Volumes I and II. Knoxville, TN: EnerNex. http://www.puc.state.mn.us/docs/#electric

ERCOT (Energy Reliability Council of Texas). 2006. *Analysis of Transmission Alternatives for Competitive Renewable Energy Zones in Texas*. Austin: ERCOT. http://www.ercot.com/news/presentations/2006/ATTCH_A_CREZ_Analysis_Report.pdf

ERCOT (Energy Reliability Council of Texas). 2008. ERCOT's Competitive Renewable Energy Zone Transmission Optimization Study. Austin: ERCOT. http://www.ercot.com/news/press_releases/2008/nr04-02-08.html

Eriksen, P., A. Orths, and V. Akhmatov. 2006. "Integrating Dispersed Generation into the Danish Power System – Present Situation and Future Prospects." Presented at the IEEE Power Engineering Society (PES) Meeting, June 20, Montreal.

Erlich, I., W. Winter, and A. Dittrich. 2006. *Advanced Grid Requirements for the Integration of Wind Turbines into the German Transmission System*. Invited panel paper No. 060428. Montreal: IEEE. http://www.uni-duisburg.de/FB9/EAUN/downloads/papers/paper_030706-02.pdf

EWEA (European Wind Energy Association). 2007. *Wind Power Installed in Europe by End of 2006 (Cumulative).* http://www.ewea.org/fileadmin/ewea_documents/documents/publications/statistics/070129_Wind_map_2006.pdf

FERC (Federal Energy Regulatory Commission). 2007. *Order No. 890 Final Rule: Preventing Undue Discrimination and Preference in Transmission Service.* Washington, DC: FERC. http://www.ferc.gov/industries/electric/indus-act/oatt-reform/order-890/fact-sheet.pdf

Gross, R., P. Heptonstall, D. Anderson, T. Green, M. Leach, and J. Skea. 2006. *The Costs and Impacts of Intermittency: An Assessment of the Evidence on the Costs and Impacts of Intermittent Generation on the British Electricity Network.* ISBN 1 90314 404 3. London, UK: Energy Research Centre. http://www.ukerc.ac.uk/ImportedResources/PDF/06/0604_Intermittency_report_final.pdf

Hirst, E., and B. Kirby. 2001. Transmission Planning for a Restructuring U.S. Electric Industry. Washington, DC: EEI. http://www.eei.org/industry_issues/energy_infrastructure/transmission/transmission_hirst.pdf

Holttinen, H. 2004. *The Impact of Large Scale Wind Power Production on the Nordic Electricity System*. ISBN 951-38-6426-X. Espoo, Finland: VTT Technical Research Centre of Finland. http://www.vtt.fi/inf/pdf/publications/2004/P554.pdf

IEA (International Energy Agency). 2005. *Variability of Wind Power and Other Renewables: Management Options and Strategies*. Paris: IEA. http://www.iea.org/Textbase/Papers/2005/variability.pdf

IEEE. 2005. *IEEE Power and Energy Magazine Special Issue: Working with Wind; Integrating Wind into the Power System*, November–December, (3:6).

Milligan, M. and K. Porter. 2006. "Capacity Value of Wind in the United States: Methods and Implementation," *Electricity Journal*, 19(2): 91–99.

Milligan, M., B. Parsons, J.C. Smith, E. DeMeo, B. Oakleaf, K. Wolf, M. Schuerger, R. Zavadil, M. Ahlstrom, and D. Nakafuji. 2006. *Grid Impacts of Wind Variability: Recent Assessments from a Variety of Utilities in the United States*. Report No. NREL/CP-500-39955. Golden, CO: National Renewable Energy Laboratory (NREL). http://www.uwig.org/Ewec06gridpaper.pdf.

Midwest ISO (Independent Transmission System Operator). 2006. *Midwest ISO Transmission Expansion Plan 2006*. http://www.midwestmarket.org

Midwest ISO. 2003. *Midwest ISO Transmission Expansion Plan 2003*. http://www.midwestmarket.org

National Grid. 2006. *Transmission and Wind Energy: Capturing the Prevailing Winds for the Benefit of Customers*. Westborough, MA: National Grid. *http://www.nationalgridus.com/transmission/c3-3_documents.asp*.

Northwest Power and Conservation Council. 2007. *Northwest Wind Integration Action Plan.* Portland, OR: Northwest Power and Conservation Council, Northwest Wind Integration Action Plan Steering Committee. http://www.nwcouncil.org/energy/Wind/library/2007-1.htm

Pedersen, J. 2005. *System and Market Changes in a Scenario of Increased Wind Power Production*. Report No. 238389. Fredericia, Denmark: Energinet.dk.

Pfirrmann, K. 2005. *PJM Testimony Promoting Regional Transmission Planning and Expansion to Facilitate Fuel Diversity Including Expanded Use of Coal-Fired Resources.* FERC Docket No. AD05-3-000. Washington, DC: PJM Interconnection, L.L.C. http://conserveland.org/pp/Transmission/testimony_mtneer.pdf

Piwko, R., B. Xinggang, K. Clark, G. Jordan, N. Miller, and J. Zimberlin. 2005. *The Effects of Integrating Wind Power on Transmission System Planning, Reliability, and Operations*. Schenectady, NY: Prepared for The New York State Energy Research and Development Authority (NYSERDA) by Power Systems Energy Consulting, General Electric International, Inc. http://www.nyserda.com/publications/wind_integration_report.pdf

Saylors, S. 2006. "Wind Park Solutions to Meet Expectations of Grid System Planners and Operators." Presented at WindPower 2006, June 4–7, Pittsburgh, PA.

SPP (Southwest Power Pool). Southwest Power Pool Web site. http://www.spp.org/.

4

The Blue Ribbon Panel on Cost Allocation. 2007. *A National Perspective on Allocating the Costs of New Transmission Investment: Practice and Principles.* A white paper prepared for WIRES (Working Group for Investment in Reliable and Economic Electric Systems). http://www.wiresgroup.com/resources/industry_reports/1Blue%20Ribbon%20Panel%20-%20Final%20Report.pdf

UWIG (Utility Wind Integration Group). 2006. *Utility Wind Integration State of the Art*. Reston, VA: UWIG. http://www.uwig.org/UWIGWindIntegration052006.pdf.

Van Hulle, F. 2005. *Large Scale Integration of Wind Energy in the European Power Supply: Analysis, Issues, and Recommendations*. Brussels, Belgium: EWEA. http://www.ewea.org/fileadmin/ewea_documents/documents/publications/grid/051215_Grid_report.pdf.

Wan, Y. 2005. *Primer on Wind Power for Utility Applications*. Report No. TP-500-36230.Golden, CO: NREL. http://nrelpubs.nrel.gov/Webtop/ws/nich/www/public/SearchForm

WIEB (Western Interstate Energy Board). 2007. *Western Transmission Expansion.* Denver, CO: WIEB. http://www.westgov.org/wieb/electric/Transmission%20Expansion/index.htm

Zavadil, R., J. King., L. Xiadon, M. Ahlstrom, B. Lee, D. Moon, C. Finley, et al. 2004. *Wind Integration Study – Final Report*. Prepared for Xcel Energy and the Minnesota Department of Commerce. Knoxville, TN: EnerNex and WindLogics.

Chapter 5. Wind Power Siting and Environmental Effects

The 20% Wind Scenario offers substantial positive environmental impacts in today's carbon-constrained world. Wind plant siting and approval processes can accommodate increased rates of installation while addressing environmental risks and concerns of local stakeholders.

5

5.1 WIND ENERGY TODAY

Wind energy is one of the cleanest and most environmentally neutral energy sources in the world today. Compared to conventional fossil fuel energy sources, wind energy generation does not degrade the quality of our air and water and can make important contributions to reducing climate-change effects and meeting national energy security goals. In addition, it avoids environmental effects from the mining, drilling, and hazardous waste storage associated with using fossil fuels. Wind energy offers many ecosystem benefits, especially as compared to other forms of electricity production. Wind energy production can also, however, negatively affect wildlife habitat and individual species, and measures to mitigate prospective impacts may be required. As with all responsible industrial development, wind power facilities need to adhere to high standards for environmental protection.

Wind energy generally enjoys broad public support, but siting wind plants can raise concerns in local communities. Successful project developers typically work closely with communities to address these concerns and avoid or reduce risks to the extent possible. Not all issues can be fully resolved, and not every prospective site is appropriate for development, but engaging with local leaders and the public is imperative. Various agencies and stakeholders must also be involved in reviewing and approving projects. If demand increases and annual installations of wind energy approach 10 gigawatts (GW) and more, the wind energy industry and various government agencies would need to scale up their permitting and review capabilities.

To date, hundreds of wind projects have been successfully permitted and sited. Although the wind energy industry must continue to address significant environmental and siting challenges, there is growing market acceptance of wind energy. If challenges are resolved and institutions are adaptive, a 20% Wind Scenario in the United States could be feasible by 2030. As noted by the Intergovernmental Panel on Climate Change (IPCC), under certain conditions, renewable energy could contribute 30% to 35% of the world's electricity supply by 2030 (IPCC 2007).

This chapter reviews environmental concerns associated with siting wind power facilities, public perceptions about the industry, regulatory frameworks, and potential approaches to addressing remaining challenges.

5.1.1 Site-Specific and Cumulative Concerns

About 10% to 25% of proposed wind energy projects are not built—or are significantly delayed—because of environmental concerns. Although public support for wind energy is generally strong, this attitude does not always translate into early support for local projects. Site-specific concerns often create tension surrounding new energy facilities of any kind. Although most wind energy installations around the United States pose only minor risks to the local ecology or communities, some uncertainties remain. Further research and knowledge development will enable some of these uncertainties to be mitigated and make risks more manageable.

Local stakeholders generally want to know how wind turbines might affect their view of their surroundings and their property values. In addition, they might be concerned about the impact on birds and other wildlife. Weighing these risks and benefits raises questions about the best management approaches and strategies.

Wind energy developments usually require permits or approvals from various authorities, such as a county board of supervisors, a public service commission, or another political body (described in more detail in Section 5.5). These entities request information from a project developer—usually in the form of environmental impact studies before construction—to understand potential costs and benefits. The results of these studies guide jurisdictional decisions. A single lead agency might consider the entire life-cycle effects of a wind energy project. This is in contrast to fossil fuel and nuclear projects, in which the life-cycle impacts (e.g., acid rain and nuclear wastes) would be widely dispersed geographically. No single agency considers all impacts.

For many government agency officials, the central issue is whether wind energy projects pose risks to the resources or environments they are required to protect. Officials want to know the net cumulative environmental impact (i.e., emissions reductions versus wildlife impacts) of using 20% wind power in the United States, whether positive or negative. Uncertainty can arise from inadequate data, modeling limitations, incomplete scientific understanding of basic processes, and changing societal or management contexts. Complex societal decisions about risk typically involve some level of uncertainty, however, and very few developers make decisions with complete information (Stern and Fineberg 1996). Because a great deal of experience exists to inform decision making in such circumstances, residual uncertainties about environmental risks need not unduly hinder wind energy project development.

The wind industry may encounter difficulties entering a competitive energy marketplace if it is subject to requirements that competing energy technologies do not face. Risks associated with wind power facilities are relatively low because few of the significant upstream and downstream life-cycle effects that typically characterize other energy generation technologies are realized. Moreover, the potential risks are not commensurate when comparing wind energy and other sources (such as nuclear and fossil fuels), and comparative impact analyses are not readily available. These analyses would need to examine the broader context of the potential adverse effects of wind power on human health and safety (minimal), ecology, visibility, and aesthetics in relation to the alternatives.

The acceptability of risks will vary among communities and sites, so it is important to understand these differences and build broad public engagement. Developing effective approaches to gaining the public's acceptance of risks is a necessary first step toward siting wind energy facilities.

5.2 Environmental Benefits

5.2.1 Global Climate Change and Carbon Reductions

Publicity related to wind power developments often focuses on wind power's impact on birds, especially their collisions with turbines. Although this is a valid environmental concern that needs to be addressed, the larger effects of global climate change also pose significant and growing threats to birds and other wildlife species. The IPCC recently concluded that global climate change caused by human activity is likely to seriously affect terrestrial biological systems, as well as many other natural systems (IPCC 2007). A 2004 study in *Nature* forecast that a mid-range estimate of climate warming could cause 19% to 45% of global species to become extinct. Even with minimal temperature increases and climate changes, the study forecast that extinction of species would be in the 11% to 34% range (Thomas et al. 2004). The future for birds in a world of global climate change is particularly bleak. A recent article found that 950 to 1,800 terrestrial bird species are imperiled by climate changes and habitat loss. According to the study, species in higher latitudes will experience more effects of climate change, while birds in the tropics will decline from continued deforestation, which exacerbates global climate change and land conversion (Jetz, Wilcove, and Dobson 2007). Wind energy, which holds significant promise for reducing these impacts, can be widely deployed across the United States and around the world to begin reducing greenhouse gas emissions (GHGs) now. Although the effects of wind energy development on wildlife should not be minimized, they must be viewed in the larger context of the broader threats posed by climate change.

Compared with the current U.S. average utility fuel mix, a single 1.5 MW wind turbine displaces 2,700 tons of CO_2 per year, or the equivalent of planting 4 square kilometers of forest every year (AWEA 2007).

A primary benefit of using wind-generated electricity is that it can play an important role in reducing the levels of carbon dioxide (CO_2) emitted into the atmosphere. Wind-generated electricity is produced without emitting CO_2, the GHG that is the major cause of global climate change.

Today, CO_2 emissions in the United States approach 6 billion metric tons annually, 39% of which are produced when electricity is generated from fossil fuels (see Figure 5-1; EIA 2006). If the United States obtained 20% of its electricity from wind energy, the country could avoid putting 825 million metric tons of CO_2 annually into the atmosphere by 2030, or a cumulative total of 7,600 million metric tons by 2030 (see assumptions outlined in Appendices A and B).

A relatively straightforward metric used to understand the carbon benefits of wind energy is that a single 1.5 MW wind turbine displaces 2,700 metric tons of CO_2 per year compared with the current U.S. average utility fuel mix, or the equivalent of planting 4 square kilometers of forest every year (AWEA 2007).

5

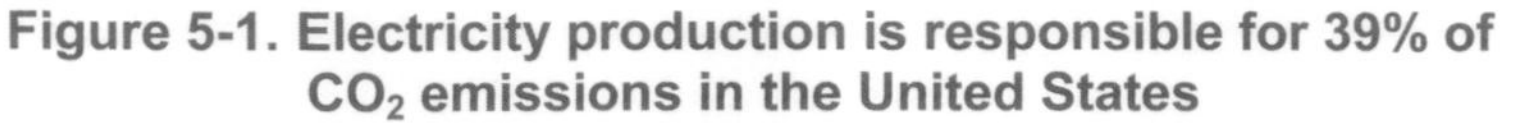

Figure 5-1. Electricity production is responsible for 39% of CO_2 emissions in the United States

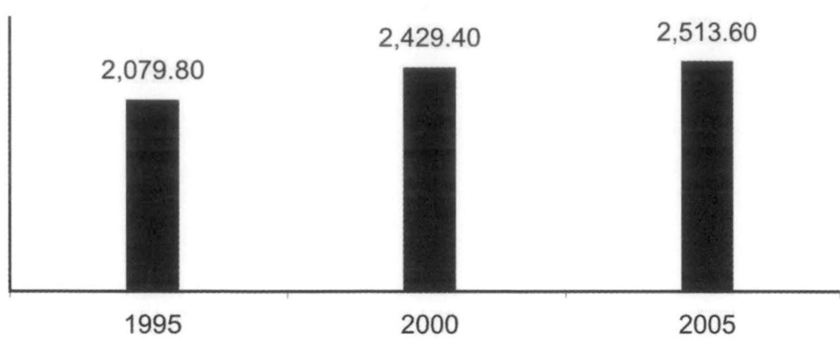

Source: EIA (2006)

The fuel displaced by wind-generated electricity depends on the local grid and the type of generation supply. In most places, natural gas is the primary fuel displaced. Wind energy can displace coal on electric grids with large amounts of coal-fired generation. In the future, wind energy is likely to offset more coal by reducing the need to build new coal plants. Regardless of the actual fuel supplanted, more electricity generated from wind turbines means that other nonrenewable, fossil-based fuels are not being consumed. In New York, for example, a study prepared for the independent system operator (ISO) found that if wind energy provided 10% of the state's peak electricity demand, 65% of the energy displaced would be from natural gas, followed by coal at 15%, oil at 10%, and electricity imported from out of state at 10% (Piwko et al. 2005).

In addition, manufacturing wind turbines and building wind plants together generate only minimal amounts of CO_2 emissions. One university study that examined the issue (White and Kulsinski 1998) found that when these emissions are analyzed on a life-cycle basis, wind energy's CO_2 emissions are extremely low—about 1% of those from coal, or 2% of those from natural gas, per unit of electricity generated. In other words, using wind instead of coal reduces CO_2 emissions by 99%; using wind instead of gas reduces CO_2 emissions by 98%.

5.2.2 Improving Human Health through Reduced Air Emissions

Switching to a zero-emissions energy-generation technology like wind power contributes to cleaner and healthier air. Moreover, wind power generation is not a direct source of regulated pollutants such as nitrogen oxides, sulfur dioxide, and mercury.

Coal-fired power plants are the largest industrial source of mercury emissions in the United States (NESCAUM 2003). The U.S. Environmental Protection Agency (EPA) (EPA 2007) and the American Medical Association (AMA) note that fetal exposure to methylmercury has been linked to problems with neurological development in children (AMA Council on Scientific Affairs 2004).

Furthermore, according the American Lung Association (ALA), almost half of all Americans live in counties where unhealthy levels of smog place them at risk for decreased lung function, respiratory infection, lung inflammation, and aggravation of respiratory illness. And more than 76.5 million Americans are exposed to unhealthful short-term levels of particle pollution, which has been shown to increase heart attacks, strokes, emergency room visits for asthma and cardiovascular disease, and the risk of death. Some 58.3 million Americans suffer from chronic exposure to particle pollution. Even when levels are low, exposure to these particles can also increase the risk of hospitalization for asthma, damage to the lungs, and the risk of premature death (ALA 2005).

5.2.3 Saving Water

The nation's growing communities place greater demands on water supplies and wastewater services, and more electricity is needed to power the expanding water services infrastructure. Future population growth in the United States will heighten competition for water resources. Especially in arid regions, communities are increasingly facing challenges with shortages of water and electric power, resources that are interlinked.

Water is a critical resource for thermoelectric power plants, which use vast quantities. These plants were responsible for 48% of all total water withdrawals in 2000, or about 738 billion liters per day (Hutson et al. 2005). Much of the water withdrawn from streams, lakes, or other sources is returned, but about 9%—totaling about 68 billion liters per day—is consumed in the process. Although regulation will require the majority of new generation plants to use recirculating, closed-loop cooling technologies, which will lessen water withdrawals, this evolution will actually lead to an overall increase in water consumption (DOE 2006).

Even some renewable technologies place a demand on water resources. For example, most ethanol plants have demonstrated a reduction in water use over the past years, but are still in the range of 13.25 to 22.7 liters of water consumed per 3.79 liters of ethanol produced (IATP 2006).

In contrast, wind energy does not require the level of water resources consumed by many other kinds of power generation. As a result, it may offer communities in water-stressed areas the option of economically meeting growing energy needs without increasing demands on valuable water resources. Wind energy can also provide targeted energy production to serve critical local water system needs such as irrigation and municipal systems.

Wind energy has the potential to conserve billions of liters of water in the interior West, which faces declining water reservoirs.

In a nongovernmental organization report entitled *The Last Straw: Water Use by Power Plants in the Arid West,* Baum and colleagues (2003) called attention to water quality and supply issues associated with fossil-fuel power plants in the interior West. Faced with water shortages, the eight states in this region are seeing water for power production compete with other uses, such as irrigation, hydropower, and municipal water supplies. Based on this analysis, the authors estimate that significant savings from wind energy are possible, as illustrated in Table 5-1.

As the United States seeks to lessen the use of foreign oil for fuel, water use and consumption is high among other energy production methods. Most ethanol plants have demonstrated a reduction in water use in recent years, but are still in the range of 13.25 to 22.7 liters of water consumed per 3.79 liters of ethanol produced (IATP 2006). An issue brief, prepared by the World Resources Institute, stated that coal-to-

Table 5-1. Estimated water savings from wind energy in the interior West (Baum et al. 2003)

Wind Energy (MW)	**Water Savings** (billion gallons withdrawn)	**Water Savings** (billion gallons consumed)
1,200	3.15	1.89
3,000	7.88	4.73
4,000	10.51	6.31

Adapted from *The Wind/Water Nexus: Wind Powering America* (DOE 2006)

liquid fuel production is a water-intensive process, requiring about 10 gallons of water use for every gallon of coal-to-liquid product (Logan and Venezia 2007).

5

Global climate change is also expected to impact water supplies. Mountains in the western United States will have less snowpack, more winter flooding, and reduced flows in the summer, all of which worsen the already fierce competition for diminished water resources (IPCC 2007). Because of increasing demand for water and decreasing supplies, some tough decisions will be needed about how this valuable resource should be allocated—especially for the West and Great Plains. Although wind energy cannot solve this dilemma, an increased reliance on wind energy would alleviate some of the increased demand in the electricity sector, thereby reducing water withdrawals for the other energy sources.

5.3 POTENTIAL ENVIRONMENTAL IMPACTS

5.3.1 HABITAT DISTURBANCE AND LAND USE

Fuel extraction and energy generation affect habitat and land use, regardless of the type of fuel. Traditional electricity generation requires mining for coal or uranium and drilling for natural gas, all of which can destroy habitat for many species and cause irreversible ecological damage. With the global and national infrastructure required to move fuel to generating stations—and the sites needed to store and treat the resulting waste—processing fossil fuel and nuclear energy is also a highly land-intensive endeavor.

Coal mining is estimated to disturb more than 400,000 hectares[11] of land every year for electricity generation in the United States, and it destroys rapidly disappearing wildlife habitat. In the next 10 years, more than 153,000 hectares of high-quality mature deciduous forest are projected to be lost to coal mining in West Virginia, Tennessee, Kentucky, and Virginia, according to the National Wildlife Federation (Price and Glick 2002).

Wind development also requires large areas of land, but the land is used very differently. The 20% Wind Scenario (305 GW) estimates that in the United States, about 50,000 square kilometers (km^2) would be required for land-based projects and more than 11,000 km^2 would be needed for offshore projects. However, the footprint of land that will actually be disturbed for wind development projects under the 20% Wind Scenario ranges from 2% to 5% of the total amount (representing land needed

[11] One hectare = 2.47 acres

for the turbines and related infrastructure). Thus the amount of land to be disturbed by wind development under the 20% Wind Scenario is only 1,000 to 2,500 km^2 (100,000 to 250,000 hectares)—an amount of dedicated land that is slightly smaller than Rhode Island. For scale comparisons, available data for existing coal mining activities indicate that about 1,700,000 hectares of land is permitted or covered and about 425,000 hectares of land are disturbed (DOI 2004). An important factor to note is that wind energy projects use the same land area each year; coal and uranium must be mined from successive areas, with the total disturbed area increasing each year. In agricultural areas, land used for wind generation projects has the potential to be compatible with some land uses because only a few hectares are taken out of production, and no mining or drilling is needed to extract the fuel.

Although wind energy may be able to coexist with land uses such as farming, ranching, and forestry, wind energy development might not be compatible with land uses such as housing developments, airport approaches, some radar installations, and low-level military flight training routes. Wind turbines are tall structures that require an otherwise undisturbed airspace around them. The need for relatively large areas of undisturbed airspace can also directly or indirectly affect wildlife habitat.

In a presentation to the National Wind Collaborative Committee, wildlife biologists describe direct construction impacts that include building wind turbines, service roads, and other infrastructure (such as substations). Estimates of temporary construction impacts range from 0.2 to 1.0 hectare per turbine; estimates of permanent habitat spatial displacement range from 0.3 to 0.4 hectare per turbine (Strickland and Johnson 2006). Indirect impacts can include trees being removed around turbines, edges in a forest being detrimental to some species, and the presence of turbines causing some species or individuals to avoid previously viable habitats. For example, a grassland songbird study on Buffalo Ridge in Minnesota found species displacement of 180 meters (m) to 250 m from the wind turbines (Strickland and Johnson 2006).

Indirect habitat impacts on grassland species are a particular concern, especially because extensive wind energy development could take place in grassy regions of the country. Peer-reviewed research has concluded, however, that one species, the Lesser Prairie Chicken, actively avoids electricity infrastructure such as transmission lines and frequent vehicle activity by as much as 0.4 km, and fossil fuel power plants by more than 1 km (Robel et al. 2004). Displacements of already declining local populations are likely, but the magnitude of these effects is uncertain because data specific to wind energy are not yet available. The extent of unknowns surrounding this issue led the National Wind Collaborative Committee (NWCC) Wildlife Workgroup to form the Grassland/Shrub-Steppe Species Collaborative (GS3C), a four-year research program to study the effects of wind turbines on grassland birds (NWCC 2006). Like the Bat and Wind Energy Cooperative (BWEC) discussed later, the GS3C provides a vehicle for public and private funding and for third-party peer-reviewed studies. Issues regarding the conservation of sensitive habitats will need to be addressed over time. Strategic planning and siting to conserve and improve potentially high-value habitat can be constructive and beneficial for both wind energy and wildlife.

5.3.2 WILDLIFE RISKS

Wildlife—and birds in particular–are threatened by numerous human activities, including effects from climate change. Relative to other human causes of avian mortality, wind energy's impacts are quite small. Figure 5-2 puts the wind industry's

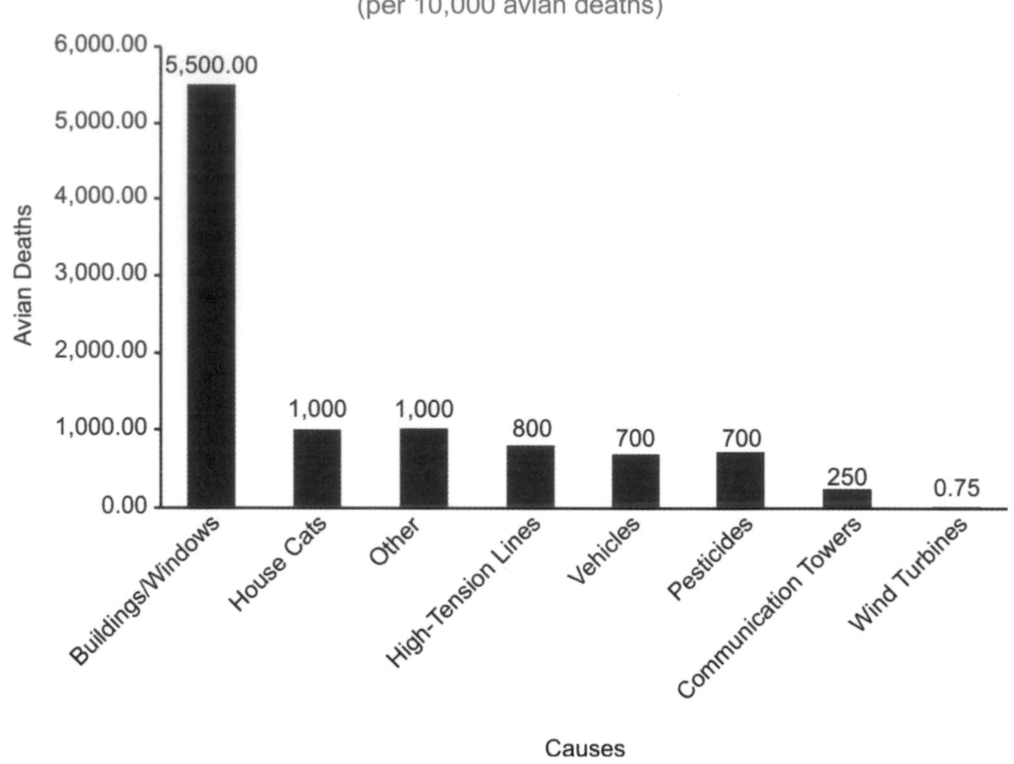

Figure 5-2. Anthropogenic causes of bird mortality
(per 10,000 avian deaths)

Source: Erickson et al. (2002)

impacts into context and illustrates that many human (and some feline) activities pose risks to birds.

As Figure 5-2 shows, anthropogenic causes of bird fatalities range from 100 million to 1 billion annually. Currently, it is estimated that for every 10,000 birds killed by all human activity, less than one death is caused by wind turbines. In fact, a recent National Research Council (NRC 2007) study concluded that current wind energy generation is responsible for 0.003% of human-caused avian mortality. Even with 20% wind energy, turbines are not expected to be responsible for a significant percentage of avian mortality as long as proper precautions are taken in siting and design.

Further comparative analyses are needed to better understand the trade-offs with other energy sources. Avian mortality is also caused, for example, by oil spills, oil platforms built on bird migration routes along the Gulf Coast, acid rain, and mountaintop mining. Wind energy will likely continue to be responsible for a comparatively small fraction of total avian mortality risks, although individual sites can present more-localized risks. Some data relative to specific sites are offered in the list that follows:

- The first large-scale commercial wind resource area developed in the world was Altamont Pass in California's Bay Area in the 1980s. The Altamont Pass development has seen high levels of bird kills, specifically raptors. Although this facility has been problematic, it remains an anomaly relative to other wind energy projects. In January 2007, a number of the parties involved agreed to take steps to reduce raptor fatalities and upgrade the project area with newer technology.

- An NWCC fact sheet (2004) reviewed the mortality figures from 12 comparable postconstruction monitoring studies and found that the fatality rate averaged 2.3 bird deaths per turbine per year and 3.1 birds per megawatt per year of capacity in the United States (outside California). Fatality rates have ranged from a low of 0.63 per turbine and 1 per megawatt at an agricultural site in Oregon to 10 per turbine and 15 per megawatt at a fragmented mountain forest site in Tennessee (NWCC 2004). This information, which is shown in Table 5-2, will be updated in 2008 to incorporate newly available data.

Table 5-2. Estimated avian fatalities per megawatt per year

Wind Project and Location	Total Fatalities
Stateline, OR/WA	2.92
Vansycle, OR	0.95
Combine Hills, OR	2.56
Klondike, OR	0.95
Nine Canyon, WA	2.76
Foote Creek Rim, WY (Phase 1)	2.50
Foote Creek Rim, WY (Phase 2)	1.99
Wisconsin	1.97
Buffalo Ridge, MN (Phase 1)	3.27
Buffalo Ridge, MN (Phase 2)	3.03
Buffalo Ridge, MN (Phase 3)	5.93
Top of Iowa	1.44
Buffalo Mountain, TN	11.67
Mountaineer, WV	2.69

Source: Data adapted from Strickland and Johnson (2006)

- Before 2003, bat kills at wind farms studied were also generally low. The frequency of bat deaths in 2003 at a newly constructed wind farm in West Virginia, though, led researchers to estimate that 1,700 to 2,900 bats had been killed, and that additional bats had probably died a few weeks before and after the six-week research period (Arnett et al. 2005). According to a USGS biologist, bat mortality has also been higher than expected at a number of sites in the United States and Canada (Cryan 2006).

Wildlife collisions with wind turbines are a significant concern, particularly if they affect species populations. To date, no site or cumulative impacts on bird or bat populations have been documented in the United States or Europe. But that does not mean that impacts are nonexistent. This is a particular worry with bats because they are relatively long-lived mammals with low reproduction rates, according to a peer-reviewed study (Arnett et al. 2005). BWEC is currently conducting the necessary research to understand the risks to bats.

Concerns about uncertain risks to birds and bats can lead permitting agencies and developers to conduct lengthy and costly studies that may or may not answer the wildlife impact questions raised. More research is necessary to more clearly understand the link between preconstruction surveys and postconstruction monitoring results. Well-designed research programs can, however, be costly for many projects and require care in assessing the appropriate levels of analysis.

Addressing these uncertainties through additional, focused research would be necessary if the United States is to increase wind development. Although many factors influence decisions to build wind projects, wildlife and environmental concerns can cause site exclusion because of the following:

- Concerns about potential wildlife impacts
- Costly study requirements
- Future risk mitigation requirements
- Conflicts with other resources

The long term viability of the wind industry will be helped by acknowledging and addressing the challenges raised by these uncertain risks. Collaborative efforts such as BWEC and GS3C offer constructive models for this undertaking.

5.3.3 Managing Environmental Risk and Adaptive Management Principles

Dealing with uncertainties associated with siting wind power facilities is a challenge for some institutions because it requires a management structure with high levels of social trust and credibility. As a result, various stakeholders are investigating how adaptive management principles might be applied to assess and manage wildlife and habitat risks at wind power sites. Under these models, developers and operators of the wind site, along with permitting agencies, could adjust the management of the site and the level of required monitoring studies to the potential challenges that arise over the life of the project.

Although the term is used often, "adaptive management" is not always well defined. Here adaptive management refers to an evolutionary management approach that purposely seeks to adapt management and decision-making processes to evolving knowledge of the technology or environmental risks in question (Holling 1978; Walters 1986; Lee 1993). "Social learning" is a centerpiece in this approach, with management seeking to enhance its capability to learn from experience and from an expanding body of knowledge. Management solutions are regarded more as experiments than as definitive solutions to the challenges involved. Valuable experience with this approach exists in such areas as watershed planning (Lee 1993; NRC 2004), fisheries (Walters 1986), and forestry (Holling 1978).

An adaptive management approach contrasts with the more typical regulatory approaches, which assume that sufficient knowledge exists at the outset to define environmental risks and effects. The basic differences between two decision-making approaches—a linear approach commonly called "command and control," and the adaptive management approach—are quite apparent in Figure 5-3. In the figure,

Figure 5-3. Linear decision strategy (command and control) and interactive model with adaptive management principles

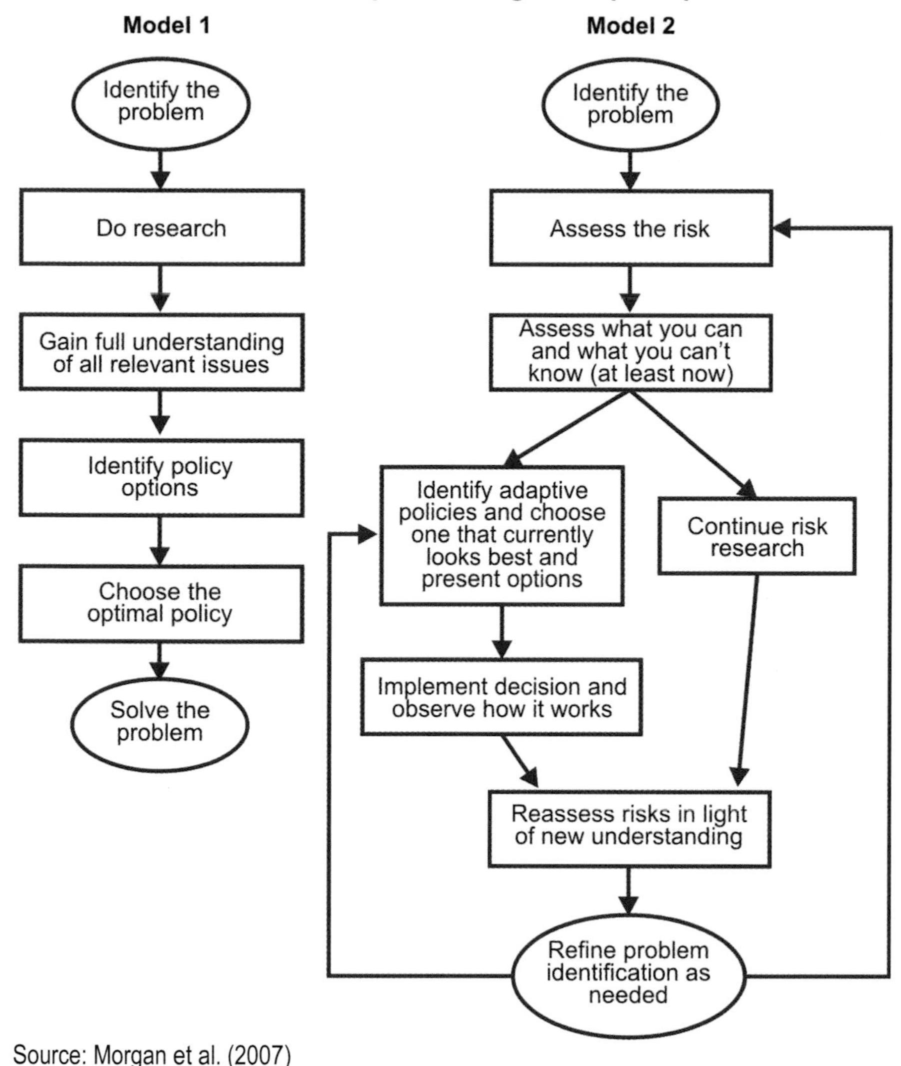

Source: Morgan et al. (2007)

Model 1 assumes that sufficient research and assessment can be done before the technology or management system is deployed, allowing an appropriate management system and the needed regulatory requirements to be put in place at the outset. Risk analysis plays a critical role in this process, with the assumption that major risks can be identified and assessed and appropriate mitigation systems instituted. In Model 2, the assumptions address different types of situations—the risks are uncertain and unlikely to be resolved in the near future; the risks can only be partially assessed at the outset; and surprises are likely as experience unfolds. This model emphasizes the importance of flexible, rapid response to new knowledge or events.

Accordingly, this risk management approach might well be suited for a technology such as wind energy, where experience and knowledge are still growing and where documented effects are strongly site-specific. Guiding principles and applications for this approach are still evolving, but adaptive management seems particularly well suited for situations of high uncertainties or conflict in the political process.

5.4 PUBLIC PERCEPTION AND ENGAGEMENT

Because the environmental benefits of wind energy are significant, public support for expanding wind energy development is widespread. The impacts of wind projects, however, are predominately local and can concern some individuals in the affected communities and landscapes. A primary challenge in achieving 20% of U.S. electricity from wind is to maximize the overall benefits of this form of energy without disrupting or alienating specific communities, especially prospective communities that do not have experience with wind turbines.

5.4.1 PUBLIC ATTITUDES

Wind energy development receives considerable general support among the U.S. population. Of those polled in a study conducted by Yale University in 2005, more than 87% want expanded wind energy development (Global Strategy Group 2005). Only a minority of the U.S. population appears to oppose wind energy, but that opposition can strengthen when particular sites are proposed. Some evidence indicates that, over time, opposition might decrease and support might grow. Surveys commissioned in the United Kingdom and Spain have found, for example, that local support for a wind project increased once it was installed and operating.

Selected Public Opinion Surveys

- In April 2002, RBA Research conducted a study for the British Wind Energy Association of people living near a small project in the United Kingdom. It found that 74% of participants supported the wind farm—37% strongly—and only 8% were opposed. Of those opposed to the project, about 25% remained opposed after the project was constructed. Sixty percent later supported the wind farm (RBA Research 2002).
- Although polls show broad statewide support for the Cape Wind offshore project in Massachusetts, some opponents have been very vocal. When asked, however, some opponents say they might support the project if it were part of a broader strategy to combat global climate change. More information on this topic can be found at www.mms.gov/offshore/alternativeenergy.

Communities must be consulted about the global impacts of wind, and this must include addressing their concerns early on. Involving affected communities early is critical to identifying concerns and addressing them proactively. Stakeholder concerns must be taken seriously, and a long-term commitment to understanding stakeholder interactions must be made.

5.4.2 VISUAL IMPACTS

Wind turbines can be highly visible because of their height and locations (e.g., ridgelines and open plains). Reactions to wind turbines are subjective and varied. The best areas for siting wind turbines tend to be those with lower population densities. Although this can minimize the number of people affected, less populated areas may also be prized for tranquility, open space, and expansive vistas. Some people feel that turbines are intrusive; others see them as elegant and interesting. In either case, the visual impacts of wind energy projects may well be a factor in gauging site acceptability.

Discourse with communities about the expected impacts is important. Wind project developers can conduct visual simulations from specific vantage points and produce maps of theoretical visibility across an affected community (Pasqualetti 2005). With this information, a developer can make technical adjustments to the project layout to accommodate specific concerns, relocate wind turbines, reduce the tower height, or even propose screening devices (such as trees) to minimize visual impact. All of these steps can, of course, affect the economic feasibility of a proposed project, so they should be weighed carefully in siting and development decisions.

Because almost all commercial-scale wind turbines rise more than 60 m above the ground, proposed wind projects must be reviewed by the Federal Aviation Administration (FAA). In February 2007, the FAA updated an advisory circular (FAA 2007) dealing with obstruction lighting and marking, including new uniform recommendations for lighting wind energy projects. The new FAA suggestions are designed to allow pilots flying too low to be warned of obstructions and minimize intrusion to neighbors. The guidance recommends that wind energy projects should be lit at night, but now the lights can be up to 0.8 km apart and be placed only around the project perimeter, reducing the number of lights needed overall. The guidelines recommend red lights, which are less annoying than white lights to people nearby. No daytime lighting is necessary if the turbines and blades are painted white or off-white.

5.4.3 Sound

All machinery with moving parts make some sound, and wind turbines are no exception, though advances in engineering and insulation ensure that modern turbines are relatively quiet; concerns about sound are primarily associated with older technology, such as the turbines of the 1980s, which were considerably louder. The primary sound is aerodynamic noise from the blades moving through the air—the "whoosh-whoosh" sound heard as the blades pass the tower. Less commonly heard in modern turbines are the mechanical sounds from the generator, yaw drive, and gearbox. When the wind picks up and the wind turbines begin to operate, the sound from a turbine (when standing at or closer than 350 m) is 35 to 45 decibels (dB; see Figure 5-4). This sound is equivalent to a running kitchen refrigerator.

Figure 5-4. Decibel levels of various situations

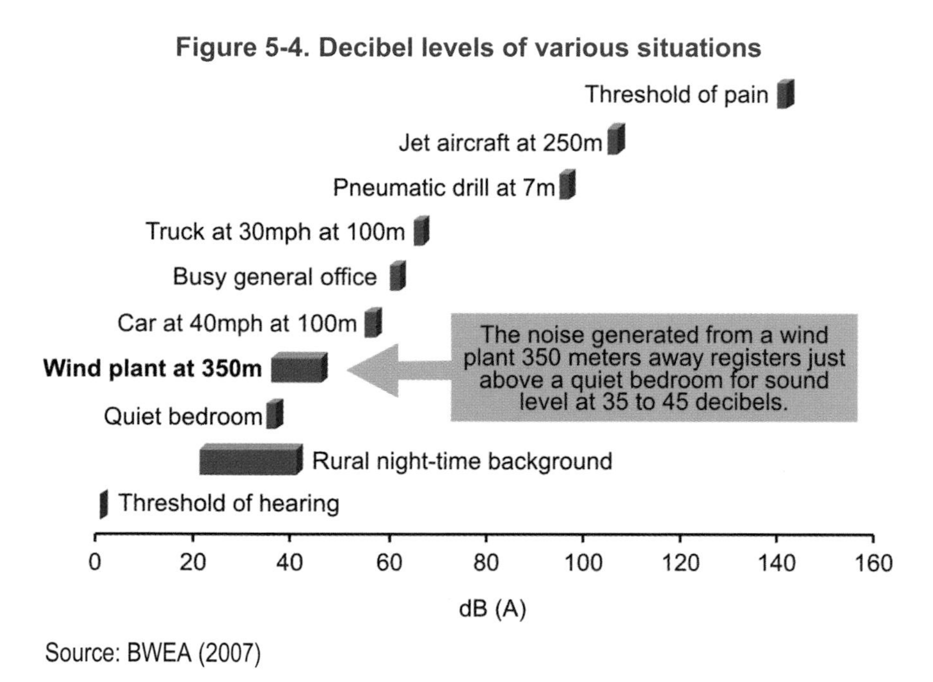

Source: BWEA (2007)

When proposing a wind energy project, wind developers can conduct studies to predict sound levels in various places, including in nearby buildings or homes. Turbines noise might be more obtrusive if, for example, they are located on a windy ridge or if houses are located downwind in a sheltered valley. Changes can usually be made to a project if the sound levels at a particular location are deemed too high. In general, standard setbacks from residences and other buildings appear to reasonably ensure that sound levels from a wind project will be low and nonintrusive.

5.4.4 LAND VALUE

The primary asset for many families is their home, so property values are a serious concern. Residents can become particularly concerned about possible declines in local property values when wind energy projects are proposed in their community. To ascertain what effects they are likely to experience, they may look to other communities with existing wind facilities.

Studies of the effects of wind projects on local property values should be done with great care, even though extensive studies have already been conducted on other energy facilities, such as nuclear plants. Because home values are a composite of many factors, isolating the effects of proximity to a wind project is important (though only a part of the full picture). Wind projects also tend to be located in areas of low residential density, which further compounds the difficulties of controlling the impact on property value. To date, two studies (see "Wind Energy and Home Values" sidebar) have examined these issues in the United States. Though neither is definitive and additional work in this area is needed, both studies found little evidence to support the claim that home values are negatively affected by the presence of wind power generation facilities.

Individuals with turbines on their properties might actually see an increase in their property values because of the lease payments paid by the wind project owner. Lease payments tend to be $2,000 to $5,000 (US$2006) per turbine per year, either through fixed payments or as a small share of the revenue.

Wind Energy and Home Values

In 2003, the Renewable Energy Policy Project (REPP) conducted a study of 24,000 home sales surrounding 11 wind projects in the United States. It compared the average selling price over time of homes near the wind project with a nearby control area that was at least 8 km from the project. No clear evidence of adverse effects on property values was found. In some communities, home values near the facilities rose faster than properties in the control group (Sterzinger, Fredric, and Kostiuk 2003).

In April 2006 a Bard College study focused on a 20-turbine wind project in Madison County, New York. Researchers visited each home, measured the distance to the nearest turbine, and ascertained to what degree the home could see the wind facility. This study also concluded that there was no evidence that the facility affected home values in a measurable way, even when concentrating on homes that sold near to the facility or those with a prominent view of the turbines (Hoen 2006).

5.5 SITING/REGULATORY FRAMEWORK

Currently, wind energy projects are governed by a complex set of laws. Projects are subject to the input of a diverse set of decision makers and different permitting regulations apply in different parts of the country. Authorities at local, state, and federal levels make siting decisions. These authorities have different responsibilities relative to a project, and there can be inconsistencies among them and even within the same agency. In some places, primary decisions rest with the local jurisdiction, although federal and state requirements may still apply. A wide diversity of requirements means that projects across the country must adhere to different standards, and different information is often required before permits are issued. Differing levels of public involvement also occur in these processes. A dramatic increase in development is likely to make this situation even more complex for developers and decision makers alike. Increased uniformity of regulatory requirements across regions would greatly facilitate the increased deployment of wind projects necessary to reach the 20% Wind Scenario.

5.5.1 LOCAL

Locally-elected officials make siting decisions at the county level. This allows the community to maintain control of local land use decisions, which is especially attractive in states where local authority is highly prized. Responsibilities differ among local bodies, but local commissions are often responsible for property assessments, rural road maintenance, economic development, zoning, and water quality (NACO 2003). Local commissions typically are concerned with protecting the environment, enhancing tax revenue, and preserving the local quality of life.

In some cases, local authorities may feel ill-equipped to weigh the highly technical information presented by a wind project developer. They can be easily influenced by proponents or opponents armed with incomplete or inaccurate information. In communities where wind development has a history, decision makers are more comfortable rendering considered permit decisions.

Most wind energy projects go through the local conditional use permit process and must spell out the conditions under which a project will operate. For example, a project permit might limit the sound level or require a setback distance from roads, houses, or property lines. Counties can also create ordinances to permit wind energy facilities: In Pike County, Illinois, the County Board created a permitted use ordinance that lays out standard conditions for wind projects; decision makers in Klickitat County, Washington, designated specific areas to encourage and guide wind energy development; and the local authorities in Kern County, California, conducted a county-wide environmental impact review to enable development of the Tehachapi Wind Resource Area.

5.5.2 STATE AND FEDERAL

States can control siting decisions either through specific decision-making bodies or by virtue of rules set for projects on state-controlled land. In addition, a state agency—such as a wildlife agency—might establish guidelines for siting wind projects. State guidelines can require maintaining certain sound levels or conducting environmental studies.

A few states have an energy siting board, which places the authority to review energy facilities with the state utility commission (i.e., a public service commission). The governor or legislature usually appoints representatives, and because they are more accountable to the public, they tend to be generally more familiar with this sector. The charge of these state commissions or boards often includes supplying reliable electric service at reasonable prices. Concerned individuals or project opponents have legal recourse to raise objections by formally challenging a commission decision.

The federal government participates in regulating wind energy projects through several different agencies, depending on the circumstances. Unless there is federal involvement, such as when developers propose a project on federally-managed land or there is a potential effect on areas of federal oversight, wind energy projects are not usually subject to the National Environmental Policy Act (NEPA). An agency can trigger the provisions of NEPA by undertaking a major federal action, such as allowing construction of a large energy project on or adjacent to federal lands (NEPA 1969).

The federal agencies that follow have mandates that may be related to wind energy:

- The **Federal Aviation Administration (FAA)** conducts aeronautical studies on all structures taller than 60 m for potential conflicts with navigable airspace and military radar, and ensures proper marking and lighting. Developers are required to submit an application for each individual turbine. From 2004 to 2006, the FAA approved almost 18,000 wind turbine proposals, nearly half in 2006 alone, and issued only eight determinations of hazard (Swancy 2006).
- The **Bureau of Land Management (BLM)** manages 105 million hectares of public land, mostly in the western United States. In 2005, the BLM finalized a programmatic environmental impact statement for wind energy development on BLM lands in the West. This statement includes best management practices for wind energy projects, sets standard requirements for projects, and allows for site-specific studies. As an alternative, wind developers can rely on the previous programmatic NEPA document and provide a development plan without having to do a full environmental impact statement (EIS) at each site, which can save valuable resources and time.
- The **U.S. Army Corps of Engineers (USACE)** issues permits for any development that will affect wetlands. Roads, project infrastructure, and foundations at some wind project sites have the potential to affect wetlands. Projects must also comply with the Endangered Species Act if any threatened or endangered species will be adversely affected.
- **The U.S. Fish & Wildlife Service (USFWS)** can pursue prosecution for violations of the Migratory Bird Treaty Act, which prohibits the killing or harming of almost all migratory birds. Some migratory birds, however, can be taken under a permit or license. The USFWS also enforces the Bald and Golden Eagle Protection Act, which gives additional protection to eagles. The USFWS exercises prosecutorial discretion under these statutes. To date, no wind energy companies have faced action under either law, but flagrant violations without mitigation could be subject to prosecution.
- The **Minerals Management Service (MMS)** oversees permitting for offshore ocean-based wind energy projects proposed for the outer continental shelf (OCS). MMS is developing the rules and issued a programmatic environmental impact statement for all alternative energy development on the OCS. New regulations are expected in 2008.
- The **U.S. Department of Agriculture Forest Service (USFS)** manages 78 million hectares of public land in national forests and grasslands. Projects sited on any Forest Service lands are subject to NEPA, and potentially to siting guidelines that the Forest Service is currently developing.

- The **U.S. Department of Defense (DOD)** has no formal review process for wind energy projects, although DOD does participate in the FAA studies. Wind energy companies planning a project near an Air Force base, however, generally work with base leadership to address and avoid conflicts.
- The **U.S. Department of Energy (DOE)** has taken the lead in creating an interagency project siting team. The team reviews how wind sites affect government assets such as radar installations, and decides how to plan for and mitigate those impacts.

5.6 ADDRESSING ENVIRONMENTAL AND SITING CHALLENGES

In order to install more than 10 GW of wind capacity per year by 2014, the United States will need to have a consistent way to review and approve projects. Examples below reflect what mature energy industries are doing to address concerns about wildlife and energy facility siting issues. The approaches described outline steps that could be adopted for a 20% Wind Scenario. The wind energy industry—in partnership with the government and nongovernmental organizations (NGOs)—will need to address environmental and siting issues.

5.6.1 EXPAND PUBLIC-PRIVATE PARTNERSHIPS

States, collaboratives, and the National Academy of Sciences (NAS) have identified gaps in the knowledge base about wind energy and its risks. This situation is not surprising for a relatively new energy technology. The knowledge gaps are framed in questions such as:

- How can large deployments of wind energy generation contribute to national climate change goals and significantly reduce GHG emissions?
- Can bats be deterred from turbines?
- How high do night-migrating songbirds fly over ridgelines?

Sometimes developers address these questions at specific sites, but broader research is urgently needed on a few of the most significant questions.

Several research collaboratives have been formed (see sidebar entitled "Examples of Existing Wind Energy Research Collaboratives") to ensure that the interests of various stakeholders are represented, that research questions are relevant, and that research results are widely disseminated. Collaboratives can help to avoid relying on industry-driven research, which critics often perceive as biased. Various combinations of technical experts and informed representatives from industry, relevant NGOs, and government agencies currently participate in ongoing collaboratives on wind energy. For example, BWEC is exploring the effectiveness of an acoustic deterrent device to warn bats away from the spinning blades of wind turbines. Although the risk to bats might be greater at some sites than at others, it is not necessarily feasible or appropriate for one company or one project to foot the entire bill for this research. A public–private partnership is often a more effective way to undertake and fund the research needed, and might also lead to more credible results.

Examples of Existing Wind Energy Research Collaboratives

Bats and Wind Energy Cooperative (BWEC)

After learning in 2003 that thousands of bats had been killed at a West Virginia site, the wind energy industry collaborated with Bat Conservation International, the USFWS, and the National Renewable Energy Laboratory (NREL) to form BWEC. This organization has developed a research program to explore ways to reduce fatalities. Its work currently centers on two areas: (1) understanding and quantifying what makes a site more risky for bats and (2) field-testing deterrent devices to warn bats away from wind turbine blades.

National Wind Coordinating Collaborative (NWCC)

NWCC is a forum for defining, discussing, and addressing wind–avian interaction issues, with a focus on public policy questions. Supported by funds from DOE, the NWCC Wildlife Workgroup (WWG) serves as an advisory group for national research on wind–avian issues. The group released a report, *Studying Wind Energy/Bird Interactions: A Guidance Document*, which is the first-ever comprehensive guide to metrics and methods for determining and monitoring potential impacts on birds at existing and proposed wind energy sites. Additionally, the WWG has facilitated six national research meetings. It is subdivided into a number of groups focused on specific tasks, such as development of a "mitigation toolbox."

Grassland/Shrub-Steppe Species Collaborative (GS3C)

The GS3C is a voluntary cooperative to identify what impacts, if any, wind energy has on grassland and shrub steppe avian species. Established in 2005 as the Grassland/Shrub Steppe Species Subgroup, the GS3C includes representatives from state and federal agencies, academic institutions, NGOs, and the wind industry.

As development levels ramp up, an overarching research consortium that would combine the work of these collaboratives could focus on addressing potential risks and ensuring that the most critical uncertainties are research priorities. As the more focused groups come together in a region, they could examine some of the habitat and biological sensitivity issues to understand which areas are most appropriate for development. With the public–private nature of the consortium, the conversation might shift from where development is inappropriate to where it is most promising. These groups, or a larger institute, could also identify priority conservation areas and work toward enhancing key habitat areas.

If the wind energy sector is to increase installations to more than 16 GW of capacity per year after 2018, research consortia could be created to take part in sustainable growth planning. A region, for example, might decide to open to development because new transmission lines are planned. In this case, a collaborative research body could determine what baseline wildlife and habitat studies are needed; organize and fund researchers to begin the work; and determine what mitigation, habitat conservation, or other activities might be appropriate for the area.

5.6.2 Expand Outreach and Education

Public acceptance of wind projects may increase if the local community directly shares in the benefits from a new wind energy development. In Europe, for example, tax law allows individuals to invest directly in wind projects. Those individuals might well view their turbines as a source of income and feel more positive about the siting of turbines nearby.

In the United States, examples of direct community impacts include:

- **Community wind:** Groups of individuals join together to develop and own a project. Although this can be risky because of the significant complexity and capital required to successfully build a wind project, the rewards are significant. A town or municipality sometimes purchases a turbine to generate power and lower public electricity bills. These groups might develop a smaller project in conjunction with a commercial development to leverage the economies of scale available for turbine purchases, construction, and operations and maintenance.
- **Property tax payments:** Wind projects are multimillion-dollar facilities that can make a significant contribution to a community's tax base. Projects are usually on leased private property, with the project owner paying any related property taxes.
- **Payment in lieu of taxes:** In places where property taxes are not required, project owners often contribute to a local community fund in lieu of taxes.

In other energy facility siting programs, communities might protect property values. Desired facilities are also sometimes collocated in the community as a form of incentive. Many such options exist and any combination might be part of a siting strategy. Wind energy developers can engage residents in a prospective host community to explain potential impacts, share information about the project, and learn about community concerns. This early involvement gives citizens an opportunity to ask questions and have their concerns addressed.

5.6.3 Coordinate Land-Use Planning

Successfully addressing numerous inconsistencies in permitting and regulation will require government and industry stakeholders to review the policies and procedures currently being implemented across multiple jurisdictions. In the long term, it may be necessary to create a sustainable growth planning effort as new areas of development open. A number of NGOs already have ecoregional plans that may yield a solid baseline of biological data. In 2006, states also completed wildlife action plans that identify high-priority actions needed to preserve and enhance their wildlife resources.

Numerous states and federal agencies have developed, or are in the process of developing, siting guidelines for wind power developments. Some states have created siting guidelines in conjunction with implementation of their state renewable portfolio standards (RPS). Other collaborative efforts to develop guidelines are moving forward through wildlife or energy agencies. Development of siting guidelines gives developers and agency officials a clear pathway to what may be required in certain jurisdictions, although time and cost considerations are involved.

5.7 Prospects for Offshore Wind Energy Projects in the United States and Insights from Europe

Europe's experience with offshore wind energy projects is instructive for how the United States might address environmental and siting challenges. European developments are supported by ambitious national goals for wind deployment, financial instruments and subsidies, and commitments to reduce GHGs. Direct comparisons and lessons learned would be instructive but need to be applied with appropriate cautions about different public policies.

A growing awareness of the large potential for electricity contributions from offshore wind energy has led to numerous proposals for siting offshore wind plants in European seas. Currently, 26 projects are installed in the North and the Baltic Seas in eight nations with a combined capacity of more than 1200 MW. A major scientific effort is in progress to support these projects. More than 280 research studies and assessments are examining environmental and human effects from installed offshore wind installations (SenterNovem 2005). Studies have also been conducted on birds, marine ecology, and animal physiology (Gerdes et al. 2005). Others have addressed the planning, construction, operations, maintenance, and decommissioning of turbines.

By contrast, the United States does not yet have any commercial-scale offshore wind power sites, and proposals for developing them are still limited. Preliminary environmental analyses relating to offshore installations are restricted to NEPA-related requirements for specific projects in federal waters. (Table 5-3 lists proposed projects and the documentation relating to the permitting and NEPA process.)

The state of knowledge and assessment of risks surrounding offshore wind energy are still emerging, which is characteristic of the early stages of any energy technology. To date, Denmark has conducted the most extensive before-after-control-impact study in the world. The most recent environmental monitoring program from this study, spanning more than five years, concluded that none of the potential ecological risks appear to have long-term or large-scale impacts (DEA 2006). Denmark intends to do further research, however, to assess the effects over time of multiple projects within the same region.

Offshore Wind Plant Siting and Seabed Rights in the United States

Various U.S. government agencies are responsible for evaluating and approving the siting, installation, and operation of wind power plants in the ocean. Until recently, offshore siting was notably more complex than land-based siting because of unclear and overlapping legal and jurisdictional authorities.

Before the passage of the Energy Policy Act (EPAct) of 2005, the USACE assumed permitting authority over proposed offshore wind energy developments. With EPAct 2005, Congress delegated authority to grant easements, leases, or rights-of-way in coastal waters to the MMS under the DOI.

Uncertainty about the extent of potential impacts of offshore wind projects—in addition to the lack of well-designed siting strategies— and the lack of long-term scientific information to fully evaluate the technology can contribute to delays in deployment (Musial and Ram, 2007).

Table 5-3. Status of offshore wind energy applications in state and federal waters

Type of Initiative[a]	Developer	Project Location	Number of Turbines Proposed	Federal Application Filed	Status as of June 2007
Project	Cape Wind Associates	Nantucket Sound	130	November 2001	Received permit approval for the met tower in 2002; USACE issued a draft EIS in November 2004; MMS issued a notice of intent (NOI) to prepare a new EIS in May 2006. Massachusetts issued a final environmental impact report (FEIR); draft environmental impact statement (DEIS) in progress by MMS.
Project	Long Island Power Authority and Florida Power & Light	Long Island Sound	40	July 2005	Joint application submitted to USACE April 2005; MMS issued an NOI to prepare an EIS in June 2006; project cancelled in October 2007.
Project	Wind Energy Systems Technologies	Galveston, TX	50–60	N/A (Texas state waters)	Signed lease with Texas General Land Office in 2005. Meteorological tower installed to begin collecting data in 2007.
Project	Bluewater Wind LLC	Delaware	70	TBD	Won competition May 22, 2007, with Delmarva Power & Light
Project	Hull Municipal	Boston Harbor	4	N/A (Massachusetts state waters)	Collecting data. Received funding from Massachusetts Technology Collaborative to support permitting and siting analyses.
Announced	Patriot Renewables LLC	Buzzards Bay, MA	90–120	N/A (Massachusetts state waters)	Applied for state approval with Massachusetts Environmental Affairs, May 2006. Conducting feasibility studies.
Announced	Southern Company	Off the coast of Savannah, GA	3–5	No current plans	Two-year collaborative study with Georgia Institute of Technology (Georgia Tech) concluded that conditions are favorable but current cost and regulatory situation precludes development.

[a] In this table, a "Project" is a planned commercial development or demonstration where complete state or federal applications have been submitted to appropriate permitting agencies. "Announced" refers to proposals at the feasibility study and data collection stage, with no commercial plans as yet and no permit applications completed.

To date, members of the European wind industry and other stakeholders have largely mitigated risks related to wind energy or decided that the local siting risks are less of a concern than other factors, such as air emissions and the larger global risks of climate change. Precautionary principles apply during the adoption of facility siting and design, as well as risk management principles. Because risks are highly site-specific, well-planned siting strategies are critical to future offshore wind developments. Successful strategies in Europe have recognized the need to engage local populations in siting decisions and development planning. This builds community support for wind facilities by addressing local and site-specific concerns, including:

- Fish and benthic communities
- Undersea sound and marine mammals
- Electromagnetic fields and fish behaviors
- Human intrusion on seascape environments
- Competing commercial and recreational uses of the ocean
- Other socioeconomic effects, including tourism and property values.

As the United States establishes a regulatory process and siting strategies for offshore wind projects, much can be learned from Europe's decades of experience with offshore wind. If the United States supports a major increase of offshore wind deployments over the next two decades, it will need to develop an ambitious and well-managed environmental research and siting program and lay the groundwork for collaborative approaches that engage the public and interested stakeholders.

5.8 Findings and Conclusions

To scale up wind energy development responsibly, benefits and risks should be considered in context with other energy options. Remaining uncertainties associated with overall risks, cost-effective opportunities for risk mitigation, strategic siting approaches, enlarged community involvement, and more effective planning and permitting regimes can also be considered. Figure 5-5 outlines activities that may be needed over the near and longer terms. Some of the activities would begin now and continue through 2030; more details are given in the subsections that follow. Given the significant ramp-up of wind installations by 2018 in the 20% Wind Scenario, these actions would need to occur within the next decade, in time to anticipate and plan for siting strategies and potential environmental effects.

Figure 5-5. Actions to support 20% wind energy by 2030

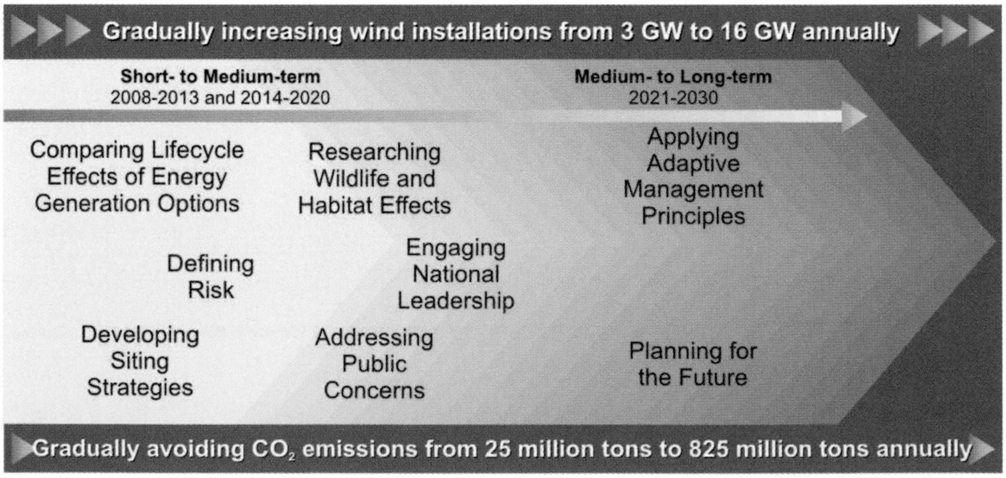

Near- and Mid-Term Actions

Comparing lifecycle effects of energy generation options: The knowledge base for comparing wind energy with other energy options—according to their climate change implications—is still uneven and incomplete. Such knowledge could prove helpful to wind energy developers; electric utilities; and national, state, and local regulators in evaluating wind energy developments. In fact, EPAct 2005 included authorization language for an NAS study of the comparative risk and benefits of current and prospective electricity supply options; the study has not begun.

Researching wildlife and habitat effects: The current research program on wind energy is largely driven by the problems that have arisen at specific sites, such as bird mortality in California and bat mortality in West Virginia. Additional research on wildlife and habitat fragmentation, which takes a collaborative approach and involves interested parties, affected communities, and subject matter experts, would be informative and should be placed within the context of other energy risks.

Defining risks: A systematic risk research program that addresses the full range of human, ecological, and socioeconomic effects from wind project siting is needed. Such a study would establish a systematic knowledge base to inform research priorities and decision makers. A comprehensive survey of risk issues that might arise at different sites has yet to be designed and undertaken, although several state agencies—such as the California Energy Commission—are developing these priorities. Along with these risk research programs, the associated cost and time implications must be demarcated.

Engaging national leadership: Evolving national and state policies and corporate programs seek to minimize human-caused emissions of greenhouse gases. Wind energy is an important part of the portfolio of energy technologies that can contribute to this goal. Many positive impacts are projected from wind energy comprising a larger share of the U.S. electricity grid, but these data must be quantified and made publicly available. National leadership could facilitate rapid progress toward 20% wind energy.

Develop siting strategies: The risks associated with wind energy deployment are heavily site-specific, and public responses will vary among potential sites. Siting strategies are needed to identify sites that are highly favored for wind energy developments, but also to avoid potential ecological risks and minimize community conflict. The American Wind Energy Association (AWEA) is currently developing a siting handbook, which may be valuable as a first step in addressing this need. Further work could continue to enhance collaborative siting processes that engage states, NGOs, host community officials, and various other stakeholders.

Addressing public concerns: Building public support is essential if wind energy is to supply 20% of the nation's electricity by 2030. Although substantial national experience exists with siting different types of energy facilities, that experience has not yet been incorporated into wind siting strategies. The roots of public perceptions of and concerns about wind energy are not well understood.

Long-Term Actions

Applying adaptive management principles: As with other technologies, wind energy will continue to pose new uncertainties as existing ones are reduced. The knowledge base is certain to evolve as new sites are developed and the scale of wind development expands in the United States, in Europe, and in other parts of the world. Adaptive management concepts and approaches, which have been applied to the development of numerous other technologies, should also be considered for incorporation into wind energy development.

5.9 References and Suggested Further Reading

ALA (American Lung Association). 2005. *State of the Air 2005*. New York: ALA. http://lungaction.org/reports/sota05exec_summ.html.

AMA (American Medical Association) Council on Scientific Affairs. 2004. *Mercury and Fish Consumption: Medical and Public Health Issues*. Report No. A-04. Chicago, IL: AMA. http://www.ama-assn.org/ama/pub/category/15842.html.

AMA. 2006. "AMA Adopts New Policies on Mercury Pollution, Hormone Compounds and Smoke-Free Meetings." *American Medical Association*, November 13. http://www.ama-assn.org/ama/pub/category/17086.html.

Arnett, E., W. Erickson, J. Horn, and J. Kerns. 2005. *Relationships between Bats and Wind Turbines in Pennsylvania and West Virginia: An Assessment of Fatality Search Protocols, Patterns of Fatality, and Behavioral Interactions with Wind Turbines.* Bat and Wind Energy Cooperative. http://www.batcon.org/wind/BWEC2004Reportsummary.pdf.

AWEA (American Wind Energy Association). 2007. *Wind Energy Basics*. Washington, DC: AWEA. http://www.awea.org/newsroom/pdf/Wind_Energy_Basics.pdf.

Baum, E., J. Chaisson, B. Miller, J. Nielsen, M. Decker, D. Berry, and C. Putnam. 2003. *The Last Straw: Water Use by Power Plants in the Arid West*. Boulder, CO and Boston, MA: The Land and Water Fund of the Rockies and Clean Air Task Force. http://www.catf.us/publications/reports/The_Last_Straw.pdf.

BLM. "BLM Facts." U.S. Bureau of Land Management. www.blm.gov/nhp/facts/index.htm.

BWEA (British Wind Energy Association). 2007. *Noise from Wind Turbines: The Facts*. London: BWEA. http://www.britishwindenergy.co.uk/pdf/noise.pdf.

Cryan, P. 2006. "Overview of What We Know About the Bat/Wind Interaction as of November of 2004." Presented at the National Wind Coordinating Collaborative (NWCC) Wildlife Research Meeting VI, November 14, San Antonio, TX. http://www.nationalwind.org/events/wildlife/2006-3/presentations/bats/cryan.pdf.

DEA (Danish Energy Authority). 2006. *Danish Offshore Wind – Key Environmental Issues*. ISBN: 87-7844-625-2. Stockholm, Sweden: DONG Energy. http://www.ens.dk/graphics/Publikationer/Havvindmoeller/danish_offshore_wind.pdf.

DOE (U.S. Department of Energy). 2006. *The Wind/Water Nexus*. DOE/GO-102006-2218. Golden, CO: National Renewable Energy Laboratory (NREL). http://www.eere.energy.gov/windandhydro/windpoweringamerica/pdfs/wpa/wpa_factsheet_water.pdf.

DOE. 2007. “Federal Interagency Wind Siting Collaboration.” DOE Office of Energy Efficiency and Renewable Energy, Wind & Hydropower Technologies Program. http://www1.eere.energy.gov/windandhydro/federalwindsiting/about_collaboration.html.

DOI (U.S. Department of the Interior). 2004. *State and Indian Regulatory Program Permitting: 2004.* Washington, DC: DOI Office of Surface Mining. http://www.osmre.gov/progpermit04.htm

EIA. 2006. *Emissions of Greenhouse Gases in the United States 2005*. Washington, DC: Department of Energy. Report No. DOE/EIA-0573. www.eia.doe.gov/oiaf/1605/ggrpt/index.html.

EPA (U.S. Environmental Protection Agency). 2007. Mercury: Health Effects. Washington, DC: EPA. http://www.epa.gov/mercury/effects.htm.

Erickson, W., G. Johnson, and D. Young. 2002. “Summary of Anthropogenic Causes of Bird Mortality.” Presented at Third International Partners in Flight Conference, March 20–24, Asilomar Conference Grounds, CA.http://www.dialight.com/FAQs/pdf/Bird%20Strike%20Study.pdf.

FAA. 2007. (U.S. Department of Transportation, Federal Aviation Administration), Advisory Circular, AC 70/7460-1K, Obstruction Marking and Lighting. Washington, DC. https://oeaaa.faa.gov/oeaaa/external/content/AC70_7460_1K.pdf

Flavin, C., and J. Sawin. 2006. *American Energy: The Renewable Path to Energy Security*. Washington, DC: Worldwatch Institute Center for American Progress. http://www.americanprogress.org/issues/2006/09/american_energy.html/AmericanEnergy.pdf.

GAO (Government Accountability Office). 2005. *Wind Power: Impacts on Wildlife and Government Responsibilities for Regulating Development and Protecting Wildlife*. Report No. GAO-05-906. Washington, DC: GAO. http://www.gao.gov/new.items/d05906.pdf

Gerdes, G., A. Jansen, K. Rehfeldt, and S. Teske. 2005. *Offshore Wind: Implementing a New Powerhouse for Europe; Grid Connection, Environmental Impact Assessment & Political Framework*. Amsterdam: Greenpeace International. http://www.greenpeace.org/raw/content/international/press/reports/offshore-wind-implementing-a.pdf

Global Strategy Group. 2005. *Survey on American Attitudes on the Environment – Key Findings*. New Haven, CT: Yale University School of Forestry & Environmental Studies. http://www.loe.org/images/070316/yalepole.doc

Government Accountability Office. 2005. *Wind Power: Impacts on Wildlife and Government Responsibilities for Regulating Development and Protecting Wildlife*. Report No. GAO-05-906. Washington, DC: Government Accountability Office. www.gao.gov/new.items/d05906.pdf

Hoen, B. 2006. *Impacts of Windmill Visibility on Property Values in Madison County, New York*. Unpublished M.S. thesis. Annandale on Hudson, NY: Bard College, Bard Center for Environmental Policy. http://www.aceny.org/pdfs/misc/effects_windmill_vis_on_prop_values_hoen2006.pdf.

Holling, C.S. 1978. *Adaptive Environmental Assessment and Management*. New York: John Wiley & Sons.

Hutson, S.S., N. Barber, J. Kenny, K. Linsey, D. Lumia, and M. Maupin. 2005. *Estimated Use of Water in the United States in 2000*. Circular No. 1268. Reston, VA: U.S. Geological Survey. http://pubs.usgs.gov/circ/2004/circ1268/index.html.

IPCC (Intergovernmental Panel on Climate Change). 2007. *Climate Change 2007: Impacts, Adaptation and Vulnerability: Contribution of Working Group II Contribution to the Intergovernmental Panel on Climate Change [IPCC] Fourth Assessment Report*. Report presented at 8th session of Working Group II of the IPCC, April 2007, Brussels, Belgium. http://www.usgcrp.gov/usgcrp/links/ipcc.htm

IATP 2006. Water Use by Ethanol Plants: Potential Challenges. Institute for Agriculture and Trade Policy (IATP). October 2006. http://www.agobservatory.org/library.cfm?refid=89449

Jetz, W., D. Wilcove, and A. Dobson. 2007. "Projected Impacts of Climate and Land-Use Change on the Global Diversity of Birds." *PLoS Biology*, 5(6): 1211–1219. http://www.pubmedcentral.nih.gov/picrender.fcgi?artid=1885834&blobtype=pdf.

Lee, K.N. 1993. *Compass and Gyroscope: Integrating Science and Politics for the Environment.* Washington, DC: Island Press.

Logan, J. and J. Venezia. 2007. *Coal-To-Liquids, Climate Change, and Energy Security*. World Resources Institute. http://www.wri.org/climate/topic_content.cfm?cid=3993.

Morgan, G., et al. 2007. *Best Practice Approaches for Characterizing, Communicating, and Incorporating Scientific Uncertainty in Climate Decision Making, Advance Copy*. Washington, DC: The National Academies Press (NAP).

Musial, W. and B. Ram. 2007. *Large Scale Offshore Wind Deployments: Barriers and Opportunities*, NREL, Golden, CO: Draft.

NACO (National Association of Counties). 2003. *A Brief Overview of County Government*. Washington, DC: NACO. http://www.naco.org/

NEPA 1969. The National Environmental Policy Act of 1969, as amended, 42 U.S.C. 4321–4347. http://ceq.eh.doe.gov/nepa/regs/nepa/nepaeqia.htm

NESCAUM (Northeast States for Coordinated Air Use Management). 2003. *Mercury Emissions from Coal-Fired Power Plants: The Case for Regulatory Action*. Boston, MA: NESCAUM.http://bronze.nescaum.org/airtopics/mercury/rpt031104mercury.pdf.

NRC (National Research Council). 2004. *Endangered and Threatened Fishes in the Klamath River Basin: Causes of Decline and Strategies for Recovery*. Washington, DC: NAP. http://books.nap.edu/openbook.php?isbn=0309090970.

NRC. 2007. *Environmental Impacts of Wind-Energy Projects*. Washington, DC: NAP. http://dels.nas.edu/dels/reportDetail.php?link_id=4185

NWCC. 1999. *Studying Wind Energy/Bird Interactions: A Guidance Document. Metrics and Methods for Determining Or Monitoring Potential Impacts On Birds At Exisitng and Proposed Wind Energy Sites.* Washington, DC: NWCC.

NWCC. 2004. *Wind Turbine Interactions with Birds and Bats: A Summary of Research Results and Remaining Questions*. Washington, DC: NWCC. http://www.nationalwind.org/publications/wildlife/wildlife_factsheet.pdf.

NWCC. 2006. "Collaborative Study of the Effects of Wind Power on Prairie-Chicken Demography and Population Genetics." Press release from National Wind Coordinating Collaborative, May 9. http://www.nationalwind.org/workgroups/wildlife/060509-press_release.pdf.

Pasqueletti, M.J. 2005. "Visual Impacts." Presented to National Wind Coordinating Collaborative Technical Considerations in Siting Wind Developments: Research Meeting, December 1–2, Washington, DC. http://www.nationalwind.org/events/siting/presentations/pasqualetti-visual_impacts.pdf.

Piwko, R., B. Xinggang, K. Clark, G. Jordan, N. Miller, and J. Zimberlin. 2005. *The Effects of Integrating Wind Power on Transmission System Planning, Reliability and Operations.* Schenectady, NY: GE Energy. http://www.nyserda.org/publications/wind_integration_report.pdf.

Price, J. and P. Glick. 2002. *The Birdwatchers Guide to Global Warming*. Reston, VA: National Wildlife Federation. www.abcbirds.org/climatechange/birdwatchersguide.pdf.

RBA Research. 2002. *Lambrigg Residents Survey*. London: BWEA. http://www.bwea.com/ref/lambrigg.html.

Regional Greenhouse Gas Initiative. "About RGGI." Regional Greenhouse Gas Initiative. www.rggi.org/about.htm.SenterNovem. 2005. *Concerted Action for Wind Energy Deployment (COD): Principal Findings 2003-2005*. Utrecth, Netherlands: SenterNovem . http://www.offshorewindenergy.org/cod/COD-Final_Rept.pdf.

Robel, R., Harrington, J., Hagen, C., Pitman, J., and Reker, R.. 2004. Effect of energy development and human activity of the use of sand sagebrush habitat by Lesser Prairie-chickens in southwest Kansas. Transactions of the North American Wildlife and Natural Resources Conference 68. http://kec.kansas.gov/wptf/robertrobel.pdf

Stern, P.C., and H.V. Fineberg. 1996. *Understanding Risk: Informing Decisions in a Democratic Society*. Washington, DC: NAP. http://books.nap.edu/openbook.php?isbn=030905396X.

Sterzinger, G., B. Fredric, and D. Kostiuk. 2003. *The Effect of Wind Development on Local Property Values: Analytical Report*. Washington, DC: Renewable Energy Policy Project (REPP). http://www.repp.org/articles/static/1/binaries/wind_online_final.pdf.

Strickland, D., and D. Johnson. 2006. "Overview of What We Know About Avian/Wind Interaction." Presented at the National Wind Coordinating Collaborative Wildlife Workgroup Research Meeting VI, November 14, San Antonio, TX. http://www.nationalwind.org/events/wildlife/2006-3/presentations/birds/strickland.pdf.

Swancy, H. 2006. "Wind Power and Aviation." Presented at the AWEA 2006 Fall Symposium, December 6–8, Phoenix, AZ.

Thomas, C.D., A. Cameron, R. Green, M. Bakkenes, L. Beaumont, Y. Collingham, B. Erasmus, et al. 2004. Extinction Risk from Climate Change. *Nature*, 427: 145–148. http://www.gbltrends.com/doc/nature02121.pdf.

USDA. "Forest Service: About Us." U.S. Department of Agriculture Forest Service. www.fs.fed.us/aboutus.

USFWS (U.S. Fish & Wildlife Service). Migratory Birds & Habitat Programs. "Migratory Bird Treaty Act." U.S. Fish & Wildlife Service. http://www.fws.gov/pacific/migratorybirds/mbta.htm.

Walters, C.A. 1986. *Adaptive Management of Renewable Resources*. New York: MacMillan.

White, S., and G. Kulsinski. 1998. *Net Energy Payback and CO_2 Emissions from Wind Generated Electricity in the Midwest*. Report No. UWFDM-1092. Madison, WI: Fusion Technology Institute, University of Wisconsin. http://fti.neep.wisc.edu/pdf/fdm1092.pdf.

Chapter 6. Wind Power Markets

Wind power suppliers and consumers span a broad range. Currently, wind power serves primarily large-scale utility markets, and smaller scale community-based projects are playing an increasing role in some regions. In addition, the eastern and Gulf Coast states are considering offshore proposals.

If 20% wind energy by 2030 were to be reached, supply and demand markets would need to expand to deliver wind energy to end-use customers throughout the United States. This chapter presents a brief overview of U.S. electricity markets, major wind power supply chain segments, market drivers, and their potential impacts on U.S. wind power expansion.

6.1 U.S. Market Evolution Background

The U.S. Department of Energy (DOE) projects U.S. electricity demand to increase by 39% from 2005 to 2030 (EIA 2007). Taking into account projected plant retirements and the implementation of energy efficiency and demand reduction programs, meeting this increased demand could require new electricity generation to increase by more than 50% over that period. Wind power is a viable option for meeting a substantial portion of this growing demand for electricity.

During the past seven years, the total number of wind installations worldwide has grown at an average annual rate of 27%. Recent growth of the wind power market in the United States has been driven by a dramatic reduction in the cost of wind energy, public interest in renewable energy, state renewable energy standards, federal production tax credits (PTCs), and volatile natural gas prices. Historically, however, periodic expiration and subsequent extensions of federal PTCs have resulted in intervals of no growth followed by explosive growth, as shown in Figure 6-1.

The U.S. wind power industry has experienced two major transformations in its history. In 1940, more than 100,000 wind turbines—many of them Jacobs Windmasters—were in operation across the Midwest, producing electricity for isolated farms and ranches. Their use declined, however, as electrification connected rural U.S. regions to electricity grids in the 1940s and 1950s. The oil price shocks of the 1970s stimulated new interest in renewable energy and led to the establishment of the Public Utility Regulatory Policies Act (PURPA) of 1978. By requiring utility companies to buy electricity from independent power producers (including wind companies), PURPA provided the foundation for the emergence of a second wind energy market in a few states in the 1980s. A key catalyst for wind's further development was California's investment tax credit and supportive state policies that jump started the bulk power wind industry in the early 1980s. The addition of federal tax credits also contributed to industry expansion. Several firms pioneered

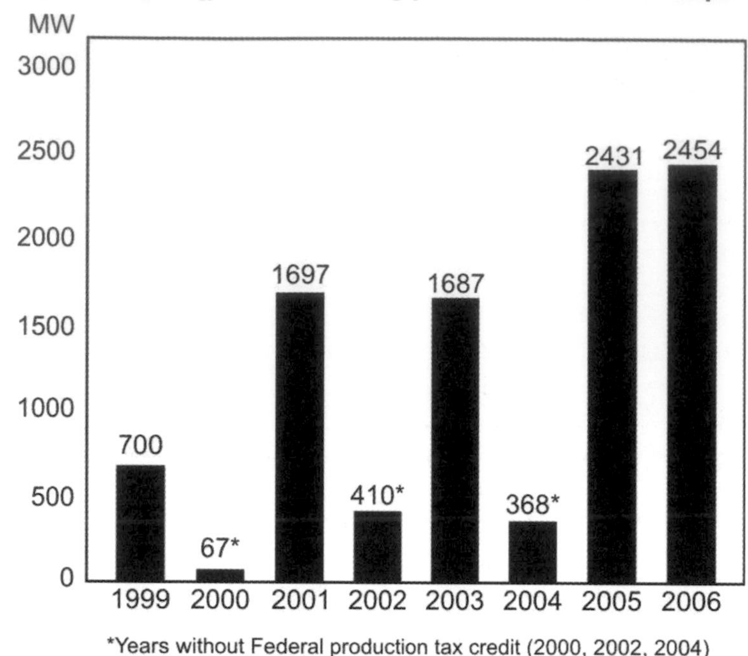

Figure 6-1. U.S. wind energy capacity growth (shown in megawatts [MW]) slowed during years when the PTC expired

*Years without Federal production tax credit (2000, 2002, 2004)

modern wind turbine technology during this period, and by 1990 more than 6,000 turbines were operating in the state.

The significantly broader and larger wind electricity supply today originated in the late 1990s. This most recent expansion resulted from a technical revolution that is influencing electricity markets in dozens of countries around the globe. Public and private research and technological innovation have rapidly improved wind resource assessment and siting, wind turbine aerodynamics and component design, and power electronics. Turbine sizes have increased steadily, leading to improvements in wind generation economics. Wind plant reliability has also improved—today, manufacturers routinely guarantee the availability of their turbines at 97% or higher. Although the wind resource is variable, wind turbines are highly reliable and operate whenever winds are sufficient to generate electricity. The current U.S. wind energy market is robust and expanding at unexpected rates.

6.2 U.S. Electricity Market

Electricity in the United States is supplied mainly by the more than 3,000 utilities across the country, some of which are owned by shareholders, others by the customers they serve. State public utility commissions and the Federal Energy Regulatory Commission (FERC) oversee these utilities and specific electricity markets. Utilities and commissions work within a regulatory framework based on federal and state legislation and jurisdiction-specific regulations that vary throughout the country. As a result of these regulatory differences, the roles of utilities and commissions also differ, creating a variety of market structures at the local and regional levels. To bring wind energy to customers nationwide, wind project developers must accommodate these local and regional market features.

6.2.1 Electric Utilities

Approximately 200 investor-owned utilities (IOUs), 70 large municipal and federal or state systems, and 50 rural generation and transmission cooperatives supply power for more than 3,000 local distribution companies across the country. The largest of these utilities typically own power plants and generate much of the power they supply. They purchase the rest of the electricity needed to serve their customers from other utilities or from nonutility generators through power purchase agreements (PPAs).

Utilities serve a variety of customers with differing needs and priorities, both retail and wholesale. Retail customers are divided into three categories: residential, commercial, and industrial. Residential customers use energy in a single dwelling for personal service. Commercial customers often have multiple dwellings, offices, or business enterprises located in a multifunction building. Industrial customers are typically large manufacturing or assembly plants that have hundreds of workers and multiple electricity applications. Special forms of commercial and industrial customers include the federal, state, and local public sectors.

Retail electricity service to end-use customers is regulated by state commissions in many states and jurisdictions. Some states have implemented restructuring or deregulation of their electricity markets, increasing competition among electricity providers and retailers. In states where competitive entities are vying to supply electric generation and to serve retail customers, wind developers have the opportunity to build projects and deliver energy directly to customers. In states that have not restructured, wind developers can sell into wholesale markets or sell to the incumbent utilities under a PPA. Some utilities are pursuing options for owning and operating their own wind projects.

At the national level, FERC policies have been implemented to foster competitive wholesale electricity markets and spur innovation and efficiency improvements. FERC continues to review and modify, as appropriate, its policies concerning competition in wholesale power markets. FERC policies cover transmission lines, treated as a common carrier, meaning that it requires transmission providers to allow nondiscriminatory access to their wires. The large wholesale markets enable a more effective exchange of services and compensation for all electricity generators, including wind power generators, helping them compete for larger shares of generation markets.

To regulate their utilities, roughly half of states in the country have integrated resource planning (IRP) policies in place. An IRP policy requires utilities to evaluate opportunities to serve loads through energy efficiency and demand reduction programs on the same basis they use to plan new generation. In addition, utilities must compare supply alternatives—including fossil and non-fossil resources—on a risk-adjusted basis. Some decisions made under the IRP process consider local customer preference, which can influence decisions made by commissions in selecting generation options. As a result, the IRP process has been an important factor in establishing wind power markets.

6.2.2 Federal Agencies

In aggregate, the federal government is the largest single consumer of electricity in the world. Federal agency electricity consumption in 2005 was more than 55,000

gigawatt-hours (GWh), which would equate to approximately 18 gigawatts (GW) of wind capacity at a 35% capacity factor.
Federal agencies were encouraged to meet an executive order goal of 2.5% of site electricity from new renewable energy sources by the end of 2005. Agencies exceeded the goal with a final tally of about 3,800 GWh (6.9%) of electricity consumed coming from renewable sources (DOE 2006). There was a dramatic increase in 2004 and 2005, largely because of renewable energy certificate (REC) purchases by the Air Force, the General Services Administration, and the Environmental Protection Agency (EPA). Overall, 96% of federal renewable energy—outside the Department of Defense—was purchased with RECs.

The Energy Policy Act (EPAct) of 2005 also guides federal agency energy use. It requires the agencies to incorporate renewable energy into their electricity supply mix at an escalating rate beginning at 3.0% in 2007 and increasing up to 7.5% by 2013, to the extent economically feasible and technically practicable. Wind energy could play a significant role in meeting this goal, particularly through projects sited on federal lands, and both EPAct 2005 and the executive order goal will help advance wind power use across federal facilities.

6.2.3 Power Marketing Administrations

Starting in the 1930s, the federal government created Power Marketing Administrations (PMAs) to market electricity generated by government-owned hydropower projects. The PMAs include the Bonneville Power Administration (BPA), the Western Area Power Administration (Western), the Southwestern Power Administration (SWPA), and the Southeastern Power Administration (SEPA). Though not technically a PMA, the Tennessee Valley Authority (TVA) has a similar purpose. Each of these entities operates as a utility, supplies power to other utilities, and often owns extensive transmission networks that are important to generators, including the wind industry. Western and BPA, in particular, have extensive transmission grids in regions with significant wind potential. Generally, the PMAs and the TVA are mandated by Congress to set rates at the lowest possible levels consistent with sound business principles. The PMAs provide access to available transmission capacity on their systems under FERC-approved transmission tariffs.

6.2.4 Compliance, Voluntary, and Emissions Markets

Under a scenario of significant wind energy expansion, multiple revenue streams and diverse markets for wind generation output will be increasingly important. Compliance and voluntary markets, which have the potential to create separate and complementary revenue streams for supporting wind energy generation, can reduce risks. Emerging emissions reduction markets might also provide revenue streams.

Policy-Driven Markets

Compliance markets, or markets where there are standards for renewable energy contributions, play an important role in supporting the development of wind energy resources. Today, 25 states plus the District of Columbia have established renewable portfolio standards (RPS) requirements, which proscribe the amount of renewable energy that must be produced within the state. These compliance markets have been growing rapidly in recent years and hold the potential to substantially expand wind energy capacity. Current state RPS policies call for about 55 GW of new renewable energy capacity by 2020, and a number of states are considering increasing their targets.

Voluntary or Green Power Markets

Voluntary markets for renewable energy also play a key role in supporting new wind energy development. Today, more than 500,000 electricity customers across the nation are purchasing green power products through regulated utility companies, from green power marketers in a competitive market setting, or in the form of RECs.

These voluntary purchasers support about 2 GW of new renewable energy capacity, mostly wind. Sales have recently grown at annual rates exceeding 60%. Large nonresidential customers—including businesses; universities; and federal, state, and local governments—are driving much of the growth, and this trend is likely to continue.

Voluntary REC markets can also be important because they might be able to support wind energy projects in regions that have good wind regimes but no compliance markets (e.g., RPS). Because RECs are sold separately from commodity electricity, they can be used to support wind energy facilities in regions with the best resources. Some factors do limit the effectiveness of RECs, though, including the lack of a national REC tracking system, the lack of a national REC trading system, and the difficulty of using RECs in project financing.

Air Quality Markets

Throughout the past several decades, approaches for controlling pollution from fossil-based power generators have moved from traditional command and control strategies to market-oriented trading regimes that allow the most cost-effective emission reduction techniques to be applied first. Sulfur dioxide (SO_2) emissions were the first to be controlled with cap and trade programs, and now nitrogen oxides (NO_x) and mercury (Hg) programs have been added. Others, such as carbon dioxide (CO_2) programs, are currently under serious consideration. Markets must have accurate price information to operate efficiently, and these programs help to incorporate the external costs of pollutants from carbon-based fuels into power prices.

6.3 Wind Power Applications

There are four basic wind applications:

- Utility-scale wind power plants, both land-based and offshore
- Community-owned projects, which often produce power for local consumption and sell bulk power under contracts
- Institutional and business applications

- Off-grid home installations and behind-the-meter farm/ranch/home systems.

The size and number of turbines vary in each of these applications. Utility-scale wind power plants typically use turbines larger than 1,000 kW to produce large amounts of wholesale power, accounting for more than 90% of all wind power generated in the United States. A 1,000 kW turbine can supply electricity for about 300 homes. Off-grid and behind-the-meter projects usually employ turbines smaller than 100 kilowatts (kW).

Wind projects range from less than 400 watts (W) to more than 400 megawatts (MW), with much larger projects expected in the future. The utility-scale technology that started in California in the early 1980s revolved around 50- to 100 kW machines, while the standard size of today's more efficient and reliable turbines ranges from 1,500 kW to 2,500 kW.

6.3.1 Large-Scale Wind Power Plants

Wind power plants consist of a number of individual wind turbines that are generally operated through a common control center. The number can range from a few, to dozens, to hundreds of energy-producing turbines.

Wind projects that are 2,000 megawatts or larger have been proposed. Such large-scale wind projects will bring about new challenges and benefits, requiring (and large enough to justify) dedicated large-scale transmission infrastructure to carry power long distances on land or shorter distances offshore to urban demand centers.

Accelerated growth of wind power in the United States would almost certainly require developing a number of very large-scale projects, considering:

- **Siting constraints on traditional projects:** Installing large numbers of turbines in remote regions minimizes landowner objections to dense turbine siting in populated areas.
- **Geographic distribution of the wind resource:** Most high-quality land-based wind resources in the nation are in mountain and plains states. The 20% Wind Scenario would require significant amounts of these resources to be captured.
- **Development pace and scale of development:** A few very large projects can add as much wind generation capacity as hundreds of traditional 100 MW projects and can be developed and built much more quickly.
- **Restrictions on land-based deployment:** Some energy-constrained coastal areas will depend on offshore wind resources that will require large-scale project development to reduce overall infrastructure costs.

6.3.2 Offshore Wind

Coastal areas, especially in California and the northeastern United States, pay higher than average prices for electricity, so offshore wind developers have an added incentive—in the form of high market prices—to enter these markets. There are uncertainties with permitting requirements in federal waters. However, the Minerals Management Service (MMS) is in the process of developing proposed rules, along with a programmatic environmental impacts statement. The MMS program is expected to be in place toward the end of 2008. Still, technical, market and policy uncertainties are limiting the deployment of offshore wind turbines alone (see chapters 3 and 5 for more discussion of offshore wind).

In addition, the cost of offshore wind projects is higher than land-based turbines by about 40%, according to a study conducted by Black & Veatch, an engineering company based in Overland, Kansas (Black & Veatch, 2007). This higher cost can be attributed to the added complexity of siting wind turbines in a marine (and potentially harsher) environment, higher foundation and infrastructure costs, and higher operations and maintenance (O&M) costs because of accessibility issues and O&M associated with offshore locations and the marine environment.

In the next 10 years, the U.S. offshore wind market could play a more significant role in bringing new power generation online in selected regions of the country where electricity prices are higher than average, population density restricts power plant installations, shallow water sites are available, state governments have passed aggressive RPS requirements, and coastal communities support this energy option.

6.3.3 Community Wind

Community stakeholders have started to evaluate wind development as a way to diversify and revitalize rural economies. Schools, universities, farmers, Native American tribes, small businesses, rural electric cooperatives, municipal utilities, and religious centers have installed their own wind projects. Although community wind projects can be of any size, they are usually commercial in scale, with capacities greater than 500 kW, and are connected on either side of the meter. Community wind includes both on-site wind turbines used to offset customer's loads and wholesale wind generation sold to a third party.

Community wind is likely to advance wind power market growth because it has the following advantages:

- **Strengthens communities:** Locally-owned and -controlled wind development substantially broadens local tax bases and generates new income for farmers, landowners, and entire communities.
- **Galvanizes support:** Local ownership and increased local impacts broaden support for wind energy, engage rural and economic development interests, and build a larger constituency with a direct stake in the industry's success. Local investments and local impacts produce local advocates.

6.3.4 Small Wind

Small wind (sometimes called "distributed wind energy") refers to wind turbines that are generally smaller than 100 kW. Residences or businesses can install small wind turbines on-site to meet their local electricity demands, often selling excess electricity sold back to the grid on distribution lines. On-grid behind-the-meter applications, where turbines are connected to distribution lines and supply electricity to partially meet local loads, comprise the primary market for small wind. On-grid installations are currently supported by a variety of state and utility financial incentives, which reduce up-front capital costs to the consumer. Small wind can also include small units for off-grid applications, such as remote homes and livestock watering facilities as well as wind–diesel hybrid systems that are deployed in remote village settings, such as, Alaska.

Small wind has lower wind speed requirements, so more locations can accommodate and harvest wind. The U.S. small wind manufacturing industry dominates today's world markets, and deploying distributed wind energy in rural or remote parts of the United States can help to build acceptance of future wind power plants. As markets continue to expand and manufacturers increase their volume, the result will be lower cost turbines. An additional benefit, although small wind systems have higher per-kilowatt costs than utility-scale systems, they compete with retail instead of wholesale electricity rates, which are also higher.

Community Wind in Minnesota

Minnesota took major steps to encourage the development of renewables by requiring the state's largest utility, Xcel Energy, to acquire a growing amount of wind energy. The target was 425 MW in 1994, 825 MW by 1999, and 1,125 MW by 2003. This created a reliable wind energy market in the state which, in turn, helped wind energy find its way into many areas of Minnesota's economy, including construction, O&M, and engineering. It also forged the path for development of permitting rules that other states and counties use as models for writing their own regulations.

Community wind began in the United States in Minnesota in 1997, when local advocates worked with the legislature to create the Minnesota Renewable Energy Production Incentive (REPI). Local ownership was a priority for those who created this incentive, which paid \$0.01 to \$0.015/kWh for the first 10 years of production for projects smaller than 2 MW. In the beginning, local wind developers had to individually negotiate with utilities for interconnection and PPAs. It was not until a special community wind tariff—establishing a set power purchase rate of \$0.033/kWh and standard procedures for interconnection for wind projects below 2 MW—was created in 2001 as part of Xcel Energy's merger settlement, that community wind projects really became feasible. The initial Minnesota REPI allocation was then quickly subscribed, and a second round was fully subscribed within 6 months. Pairing of these complementary policies allowed the community wind market to really take off.

Small wind energy market challenges include turbine availability (product gaps exist for 5-, 15-, and greater than 100 kW turbines); economics and lack of financial incentives across all market segments; turbine reliability; utility interconnections; and zoning and permitting.

6.3.5 Native American Wind Projects

Native American reservations constitute a special community with emerging interests in wind power development. Wind-generating potential on tribal lands, which is conservatively estimated at more than 1.5 GW, could make an important contribution toward the 20% Wind Scenario. At least 39 Native American reservations with significant wind power potential (Class 4 and higher) are located in remote areas that could support development. Self-governed Native American tribes also have a unique legal relationship with the U.S. federal government and are afforded increased opportunities under EPAct 2005.

6.4 Stakeholder Involvement and Public Engagement

As wind energy development proceeds in the United States, site selection and development will require well-designed and effective stakeholder engagement. The preceding sections outlined the markets and supply segments that can contribute to the 20% Wind Scenario. The types of stakeholders and their perceptions of wind energy are likely to vary markedly from one location to another. An important part of any stakeholder initiative is to identify the full range of interested parties and decision makers, such as public utility commissions and their staffers, utilities and regional transportation organizations and their customers, state and federal legislators, and financiers. Understanding stakeholder interests and how to effectively communicate with these various groups is central to the pursuit of 20% wind energy by 2030.

Experience with past wind and other energy facility development in the United States has brought home the critical importance of stakeholder involvement. The energy community now generally recognizes that effectively engaging stakeholders in siting-related decisions requires attention to a number of key factors:

- State and local siting guidelines and procedures are needed to establish a known and deliberate siting process in which local concerns and siting issues are fully considered. Developers must also be able to plan for and manage a predetermined and predictable process.
- The developer, state and local officials, and the host communities should collaborate on designing stakeholder outreach
- A comprehensive list of stakeholders—including those who will be targeted in the engagement efforts—should be compiled early in the process.
- Concerns and requirements of various stakeholders should be assessed. Needs should be identified and defined through interviews with stakeholders.
- The stakeholder-engagement process should begin before the site is assessed and selected so that baseline information can be established. Stakeholders should continue to be actively engaged throughout facility development and operation, with an emphasis on two-way communications.
- A neutral third party should carefully evaluate effectiveness of the engagement process along the way, to ensure that any initiatives incorporate new stakeholders that might appear and new concerns that might arise. This will also allow deficiencies in engagement and communications to be forthrightly addressed.

Finally, no element in an engagement and communications effort is more important than building trust among the developers, state and local officials, and members of the host community. Although this is a much more difficult task than is generally understood, experience has shown that openness, serious consideration of local concerns, and a participatory process all contribute substantially to successful outcomes.

6.5 CONCLUSIONS

Within the 20% Wind Scenario, multiple revenue streams and multiple markets for wind generation output would be increasingly important. Standards for renewable energy contributions as well as voluntary markets have the potential to create separate and complementary revenue streams for supporting wind energy generation while reducing risks. Today, 25 states have established RPS requirements. Compliance markets, which have been growing rapidly in recent years, can make substantial contributions to the expansion of wind energy capacity. Emerging emissions markets can also be a source of revenue streams.

To create the catalyst necessary to support aggressive wind energy growth, many different market drivers must converge; and if the significant increase in wind power development under the 20% Wind Scenario is to be realized, many stakeholders will need to embrace a robust wind future. Stakeholder interests are as diverse as stakeholder types; a long-term commitment to understanding and working with stakeholders will be critical for deploying significant levels of wind power. All segments of the market must be taken into account when planning for the wide adoption of wind-generated electricity. Market forces need to be targeted and utilized efficiently to leverage stakeholder interests if 20% of U.S. electricity from wind is to be realized.

6.6 REFERENCES

DOE (U.S. Department of Energy). 2006. *Annual Report to Congress on Federal Government Energy Management and Conservation Programs Fiscal Year 2005*. Washington, DC: DOE. http://www1.eere.energy.gov/femp/pdfs/annrep05.pdf.

EIA (Energy Information Administration). 2007. *Annual Energy Outlook 2007 with Projections to 2030*. Washington, DC: EIA. Report No. DOE/EIA-0383. http://www.eia.doe.gov/oiaf/archive/aeo07/index.html

Black & Veatch. 2007. 20 *% Wind Energy Penetration in the United States: A Technical Analysis of the Energy Resource*. Walnut Creek, CA.

Appendix A. 20% Wind Scenario Impacts

A.1 Introduction

This appendix describes the analytic tool and assumptions that were used to identify some key components of the impacts, and technical challenges of providing 20% of the nation's electricity from wind in 2030. The 20% level was chosen exogenously as the central assumption of the evaluation. The relative cost difference between a scenario including 20% wind-generated electricity and a scenario in which no additional wind technology is installed after 2006 is the primary metric. All modeling assumptions contribute to this incremental cost of wind energy. Thus, changes to the assumptions increase or decrease the incremental cost of the 20% Wind Scenario over the scenario that does not include wind energy. No sensitivities exploring changes to the assumptions were performed for this analysis. Modeling assumptions are described in this appendix (See Table A-1) and Appendix B.

The National Renewable Energy Laboratory's (NREL) Wind Deployment System (WinDS) model was employed to simulate generation capacity expansion of the U.S. electricity sector through 2030. This model used a wind energy generation rate that would result in the production of 20% of projected electricity demand from wind by 2030. Carbon emission reductions in this 20% Wind Scenario have also been derived from the WinDS model outputs. Water savings associated with significant wind energy generation has been externally calculated as well. The assumptions used for these analyses were developed from a variety of sources and experiences that span the wind and electricity generation industries; model-specific details of these assumptions are presented in Appendix B.

The 20% Wind Scenario requires U.S. wind power capacity to grow from the current 16–17 gigawatts (GW) to more than 300 GW over the next 23 years. This ambitious growth could be reached in many different ways, with varying challenges, benefits, costs, and levels of success. This report examines one particular scenario for achieving this dramatic growth and contrasts it to another scenario called No New Wind, which assumes no wind growth after 2006 for analytic simplicity.

Considerations in the 20% Wind Scenario

- Wind resources of varying quality exist across the United States and offshore.
- Although land-based resources are less expensive to capture, they are sometimes far from demand centers.
- Typically, wind power must be integrated into the electric grid with other generation sources.
- Technology and power market innovations would make it easier to handle a variable energy resource such as wind.
- New transmission lines would be required to connect new wind power sources to demand centers.
- Transmission costs add to the cost of delivered wind energy costs, but today's U.S. grid requires significant upgrading and expansion under almost any scenario.
- Wind installations will require significant amounts of land, although actual tower footprints are relatively small.
- Domestic manufacturing capacity might not be sufficient to accommodate near-term rapid growth in U.S. wind generation capacity; the gap may be filled by other countries.

The authors recognize that U.S. wind capacity today is growing rapidly, although from a very small base, and that wind energy technology will be a part of any future electricity generation scenario for the United States. At the same time, there is still a great deal of uncertainty about what level of contribution wind could or is likely to make. In its *Annual Energy Outlook 2007 with Projections to 2030* (AEO), the Energy Information Administration (EIA) forecasts that an additional 7 GW—beyond the 2006 installed capacity of 11.6 GW—will be installed by 2030 (EIA 2007).[12] Other organizations are projecting higher capacity additions, and given today's uncertainties, developing a "most likely" forecast would be difficult. The analysis presented here sidesteps these uncertainties and contrasts the impacts of producing 20% of the nation's electricity from wind with No New Wind. This yields a parameterized estimate of some of the impacts associated with increased reliance on wind energy generation.

The analysis was also simplified by assuming that the contributions to U.S. electricity supplies from other renewable sources of energy would remain at 2006 levels in both scenarios. In addition, no sensitivity analyses have been done to identify how the results would differ if assumptions were changed.

Broadly stated, this 20% Wind Scenario is designed to optimize costs while recognizing certain constraints and considerations (see sidebar above). Specifically, the scenario describes the mix of wind resources that would have to be captured, the geographic distribution of the wind power installations, estimated land needs, and required utility and transmission infrastructure changes associated with 20% wind in 2030. It is not a definitive identification of the exact locations of wind turbines and transmission lines.

The scenario reflects several assumptions about generation technology cost and performance as well as electric grid system operation and expansion. For example, wind technology development is projected to continue based on a history of performance improvements. The national transmission system is assumed to evolve in ways favorable to wind energy development by shifting toward large regional markets. In addition, future environmental study and permit requirements are not expected to add significant costs to wind technology.

The 20% Wind Scenario was constructed by specifying annual wind energy generation in each year from 2007 to 2030, based on a trajectory proposed in a previous NREL study (Laxson, Hand, and Blair 2006). The investigators forced the WinDS model to reach the 20% level for wind-generated electricity by 2030 and evaluated aggressive near-term growth rates. Next, they examined sustainable levels of wind capacity installations that would maintain electricity generation levels at 20% and accommodate the repowering of aging wind turbine equipment in wind installations beyond 2030. The 20% wind by 2030 trajectory from the NREL study was implemented in WinDS by calculating the percentage of annual energy production from wind, an increase of approximately 1% per year. Figure A-1 illustrates the energy generation trajectory proposed in the study, and the corresponding annual wind capacity installations that the WinDS model projects will meet these energy generation percentages.

[12] AEO data from 2007 were used in this report. AEO released new data in March of 2008, which were not incorporated into this report. While the new EIA data could change specific numbers in the report, it would not change the overall message of the report.

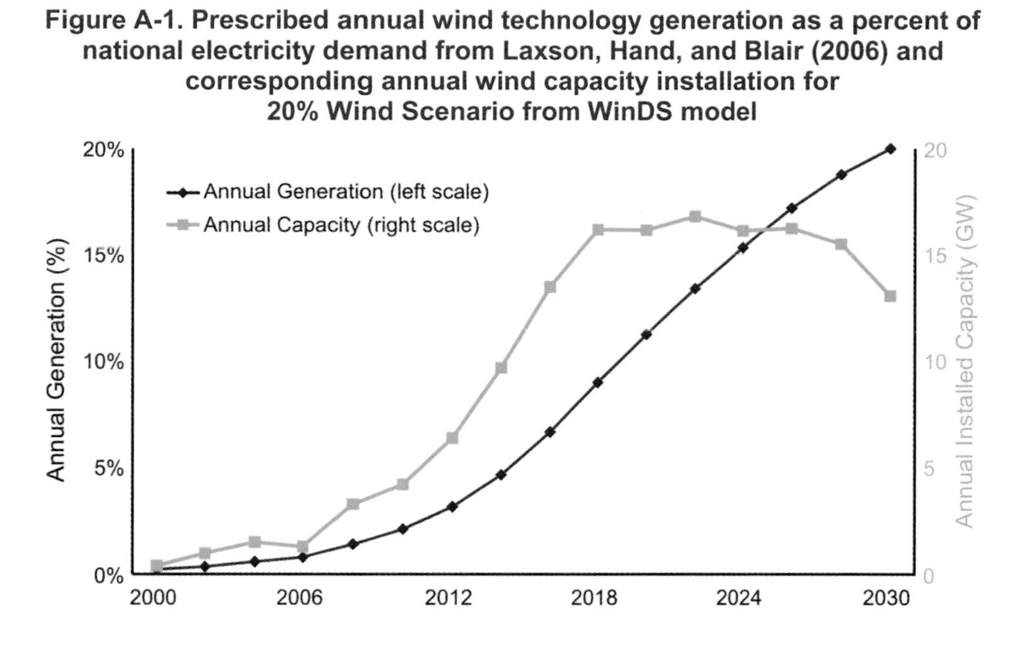

Figure A-1. Prescribed annual wind technology generation as a percent of national electricity demand from Laxson, Hand, and Blair (2006) and corresponding annual wind capacity installation for 20% Wind Scenario from WinDS model

The combined cost, technology, and operational assumptions in the WinDS model show that reaching an annual installation rate of about 16 GW/year by 2018 could result in generation capacity capable of supplying 20% of the nation's electricity demand by 2030. This annual installation rate is affected by the quality of wind resources selected for development as well as future wind turbine performance. The declining annual installed capacity after 2024 is an artifact of the prescribed energy generation from the NREL study (Laxson, Hand, and Blair 2006), in which technology improvements and wind resource variations were not considered. The NREL study (Laxson, Hand, and Blair 2006) provides an upper level of about 20 GW/yr, because turbine performance is unchanged over time. Based on the wind resource data and the projected wind technology improvements presented in this report, sustaining a level of annual installations at approximately 16 GW/yr beyond 2030 would accommodate the repowering of aging wind turbine equipment and increased electricity demand, so that the nation's energy demand would continue to be met at the 20% wind level.

The 20% Wind Scenario does not include policy incentives such as a production tax credit (PTC) or carbon regulations, although such policies may make this growth trajectory more likely. It is implicitly assumed that a stable policy environment that recognizes wind's impacts could lead to growth rates that would result in the 20% Wind Scenario.

Some of the consequences of a 20% Wind Scenario in 2030, including carbon emission reductions and natural gas demand reduction, were calculated based on the results of the WinDS model. To estimate the impacts associated with incorporating electricity from wind into the grid at this level, a comparison has been made with a scenario in which no additional wind power would be installed after 2006. The differences between the two cases are attributed to the incorporation of wind power.

From a planning and operational perspective, integrating wind generation into the U.S. electricity grid at the 20% level appears to be technically feasible without significantly and unrealistically constraining the WinDS model (e.g., assuming no new transmission will be built). In addition to modeling the expansion of the electricity grid to transmit power from wind-rich geographic areas to demand (load) centers, the model treats wind resource variability on time scales ranging from multiyear capacity planning to minute-to-minute ancillary service requirements. (WinDS does not perform minute-to-minute ancillary service calculations, but it uses statistics to approximate these requirements; see Appendix B for a detailed discussion of the treatment of wind variability.) The 20% Wind Scenario presented here includes future reductions in wind technology costs and increased performance, coupled with transmission system expansion that is favorable to wind energy. These assumptions affect only the direct cost to the electricity sector associated with this level of wind energy expansion. These cost and performance assumptions differ from those used by EIA; the assumptions are based on 2006 market data developed by Black & Veatch for all generation technologies. Explicit cost and performance projections are used rather than learning algorithms generally used by EIA. See Appendix B for more information.

A.2 Methodology

The WinDS model was used to identify some key components of the impacts of producing 20% of the nation's electricity from wind energy by 2030. WinDS is a geographic information system (GIS) and linear programming model of electricity capacity expansion for the U.S. wholesale market.[13] The model operates over multiple regions and time periods. Generation capacity expansion is selected to achieve a cost-optimal generation mix to meet 20% wind generation over a 20-year planning horizon for each 2-year period from 2000 to 2030.

The assumptions used for the WinDS model were obtained from a number of sources, including technical experts (see Appendix D), the WinDS base case (Denholm and Short 2006), AEO (EIA 2007), and a study performed by Black & Veatch (2007). These assumptions include projections of future costs and performance for all generation technologies, transmission system expansion costs, wind resources as a function of geographic location within the continental United States, and projected growth rates for wind generation. Appendix B describes these assumptions in detail.

A.2.1 Energy Generation Technologies

Wind-generation technologies are contained in the WinDS model, along with conventional technologies such as coal plants (pulverized coal and integrated gasification combined cycle [IGCC]), nuclear plants, and natural-gas-fired combustion turbine and combined cycle plants. The model does not include technologies installed "behind the meter," such as cogeneration or other distributed generation systems, nor does it include energy efficiency or demand response technologies. Table A-1 summarizes the modeling assumptions.

Wind technology options include land-based and offshore technologies. Wind resource classes 3 through 7 (at 50 meters [m] above ground level) are specified for 358 wind supply regions across the continental United States. Each wind supply region in WinDS includes a mix of these wind resource classes. Offshore wind resources are associated with coastal and Great Lakes regions. Resource maps reference those produced by the Wind Powering America (WPA) initiative or by individual state programs, and include environmental and land use exclusions. In

[13] The model, developed by NREL's Strategic Energy Analysis Center (SEAC), is designed to address the principal market issues related to the penetration of wind energy technologies into the electric sector. For additional information and documentation, see http://www.nrel.gov/analysis/winds/.

Table A-1. Assumptions used for scenario analysis

	Scenario Assumptions
Renewable Energy Technologies (other than wind)	• Contributions to U.S. electricity supply from renewable energy (other than wind) are held constant at 2006 levels through 2030
Land-Based Wind Technology Cost	• $1,730/kW in 2005 and 2010, decreasing 10% by 2030 • Regional costs vary with population density, with an additional 20% in New England
Shallow Offshore Wind Technology Cost	• $2,520/kW in 2005, decreasing 12.5% by 2030
Wind Technology Performance	• Capacity factor improvements about 15% on average over all wind classes between 2005 and 2030
Existing Transmission	• 10% of existing transmission capacity available to wind plants at point of interconnection
New Transmission	• Transmission will be expanded • $1,600/megawatt-mile (MW-mile) • 50% of cost covered by wind project • Regional cost variations prescribed as follows: 40% higher in New England and New York, 30% higher in Pennsylvania-New Jersey-Maryland (PJM) East interconnection, 20% higher in PJM West, 20% higher in California
Wheeling Charges	• No wheeling charges between balancing areas
Conventional Generation Technology Cost and Performance	• Natural gas plant cost ($780/kW in 2005) and performance flat through 2030 • Coal plant capital cost ($2,120/kW in 2005) increases about 5% through 2015 and then remains flat through 2030 • Coal plant performance improvement of about 5% between 2005 and 2030 • Nuclear plant capital cost ($3,260/kW in 2005) decreases 28% between 2005 and 2030 • Nuclear plant performance stays flat through 2030
Fuel Prices	• Natural gas prices follow AEO high fuel price forecast • Coal prices follow AEO reference fuel price forecast • Uranium fuel price is constant

addition to the geographic display of wind resources, seasonal and diurnal variations in capacity factor (CF) are computed based on wind resource data. Appendix B contains more information about the wind resource data used for this study.

Experts at Black & Veatch Corporation developed wind technology cost and performance projections based on their experience and market knowledge, discussions with wind industry professionals, and review of cost and performance trends (Black & Veatch 2007). Wind technology costs in 2005 are assumed to be $1,730/kW[14] (kilowatt; in real US$2006), which reflects recent cost increases attributed to current exchange rates between the euro and the dollar, increased commodity prices, a constrained supply of wind turbines, and construction

[14] All dollar values in appendices A and B are in $US2006. These capital costs include construction financing, which adds approximately 5% to the "overnight" capital cost given in Appendix B. The WinDS model applies financing costs in each solution period that requires overnight capital costs as input.

A

financing. Wind technology costs are projected to decrease to $1,550/kW by 2030 (in US$2006 including construction financing). The cost of offshore wind energy technology is projected to decrease 12.5% over the same period, from $2,500/kW in 2005 to $2,200/kW in 2030 (real US$2006 including construction financing).

Specialists at Black & Veatch developed wind technology performance projections, in the form of capacity factors, by extrapolating historical performance data from 2000 to 2005. Appendix B gives more details on the cost and performance estimates for current and future years for both land-based and offshore wind technologies.

Black & Veatch experts also developed conventional generation technology cost and performance projections based on reported engineering, procurement, and construction costs for currently proposed plants through 2015. Fossil plant costs were assumed to remain flat beyond 2015 with modest performance improvements for coal plants. Cost and performance projections for nuclear plants assume continued technology development. Appendix B presents these cost and performance assumptions.

A.2.2 Transmission and Integration

Wind energy can be used to meet local loads (i.e., loads in the same wind supply region), as well as those in other geographic locations. Local loads can be met either by transmitting on the existing grid where capacity is available (10% of the capacity of each existing line is assumed to be available for wind) or by building a short, dedicated transmission line directly to the local load. Wind energy can also be transmitted to another locale, either on the existing grid when capacity is available or on a new transmission line. If the transmission line crosses between two balancing areas, there must be enough capacity on the line in each seasonal or diurnal time frame to take advantage of wind's full nameplate capacity, as well as the energy associated with other generators transmitting power over that line. If no capacity remains on that transmission line, the model assumes that a dedicated transmission line will be constructed for wind.

When integrating wind resources into the grid, the model considers both planning and operating reserve margins for all North American Electric Reliability Corporation (NERC) regions. For both types of reserve margin, WinDS accounts for the variability that occurs when wind generation is used from disparate wind sites whose output is not fully synchronized. The wind plant's capacity value is a function of the CF, seasonal and diurnal wind variations, and correlation with existing wind capacity installations. In this way the variability of the wind resource is assessed in combination with conventional generation within each NERC region.

The transmission system is assumed to expand under large, regional operation and planning entities, which incorporate polices that favor wind energy. Operating grid systems on large, regional bases, such as through the Midwest Independent Transmission System Operator (Midwest ISO), mitigates the variability of wind power. The WinDS model calculates reserve and planning margins at the NERC regional level, which is representative of these large operating structures. A wind energy penetration limit of 25% has been assumed at the interconnect level. Also, based on the participant funding principle adopted by Midwest ISO, the cost of new transmission is assumed to be split equally between the originating project, be it wind or conventional generation, and the ratepayers within the region. This tariff structure is assumed to apply nationwide. The exception is for new transmission lines that are built at a project in one interconnection region to meet loads in another interconnection region. In this case, the transmission cost is borne entirely by the project. Consistent with the large, regional planning process, this scenario assumes that there are no wheeling charges between balancing areas. Finally, the 20% Wind Scenario assumes that 10% of existing grid capacity is available for wind energy.

A.2.3 Quantification of Impacts

Projected electricity demand estimates, as well as financing and economic assumptions, were obtained from the AEO reference case (EIA 2007). Total direct costs of all generation technologies were estimated over a 20-year planning horizon in each two-year solution period. The sum of these direct costs represents the total cost to the electricity sector for generation choices through 2030, including costs for capital investment, operations and maintenance (O&M), new transmission, and fuel.

To calculate the impacts of 20% wind energy, the authors of this report constructed the aforementioned No New Wind Scenario. This scenario assumes that the conventional generation mix expands to meet electricity demand with currently enacted policies. Table A-1 outlines the major assumptions in the scenario and supporting analyses. The difference between the two cases, 20% Wind and No New Wind, represents the impact of wind energy.

A.3 Wind Capacity Supply Curves

In economic analysis, a supply curve is used to determine the quantity of a product that is available at various prices. For this report, wind generation potential is plotted against its calculated levelized cost (LC) of electricity in ascending order. See Appendix B for more information. For example, the potential (in gigawatts) from high-speed wind resources has been plotted against its levelized cost (dollars per megawatt-hour [MWh]); lower-speed wind projects have higher costs and represent the next step up on the supply curve. Cost and potential were estimated for each region based on a GIS optimization strategy developed by NREL. The regions were aggregated such that an overall supply curve for national wind potential could be developed. The following supply curves compare the quantities and costs for wind resources and show which products can be brought to market at the lowest cost (resources on the left side of Figure A-2 "Supply Curve for wind energy: current bus-bar energy costs"). See Appendix B for wind resource estimates.

The national supply curve for bus-bar energy costs—for the wind plant alone, excluding transmission costs—is shown in Figure A-2. The figure illustrates that more than 8,000 GW of wind energy is available in the United States at $85/MWh or less. This is a huge amount of capacity, equivalent to roughly eight times the existing nameplate generating capacity in the country, which is estimated at 983 GW (EIA 2007). This price, however, excludes the cost of transmission or integration. The supply curve uses today's cost and performance figures, which are projected to improve with future technology development.

The supply curve shows the simple relationship between wind power class and cost, as the higher classes are the lowest cost (and least abundant resources); Classes 3 and 4 are much more prevalent. At today's costs, offshore wind is not cost-competitive with land-based wind technologies. Finally, the national resource potential for land-based wind technologies exceeds the existing nameplate generating capacity in the country by a factor of eight. However, that does not mean capturing that full potential is economically, technically, or politically viable.

Figure A-2. Supply curve for wind energy—current bus-bar energy costs

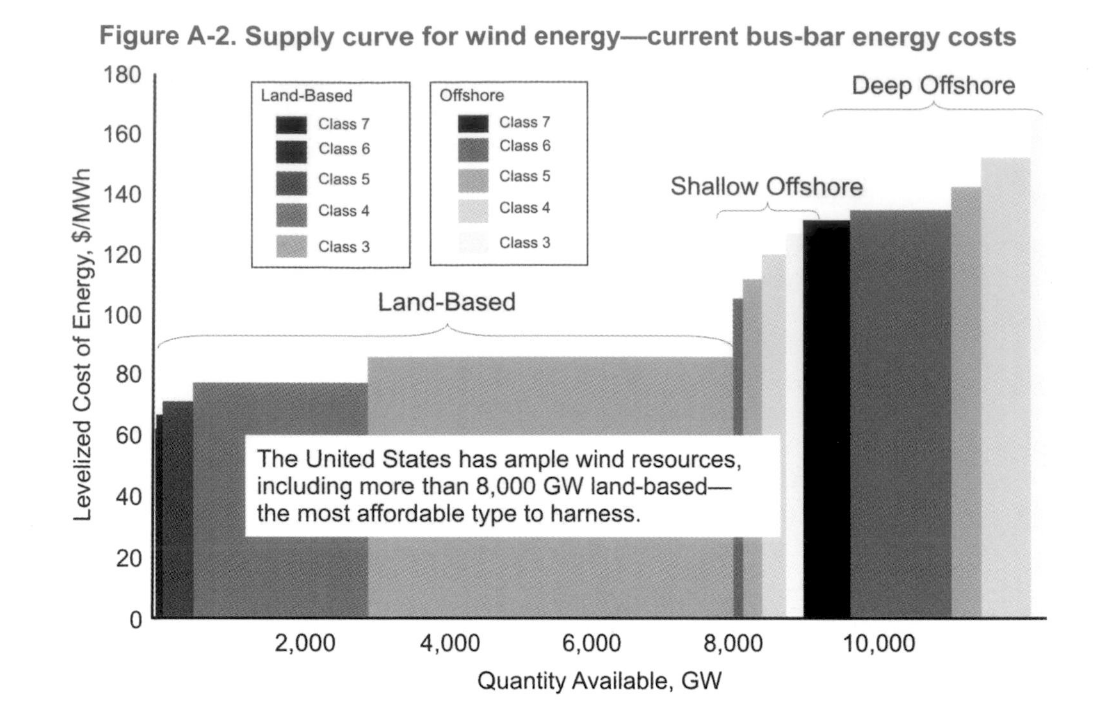

The national supply curve in Figure A-3 shows the costs of connecting to the existing transmission system, given that 10% of capacity is available for new wind generation. This supply curve also shows the cost of connecting directly to load centers that are in the same balancing area as the wind resource, given that a maximum of 100% of that load can be served by wind. This curve is produced as an input to the WinDS model. Please see Appendix B or (Black & Veatch 2007) for more information.

Figure A-3. Supply curve for wind energy: energy costs including connection to 10% of existing transmission grid capacity

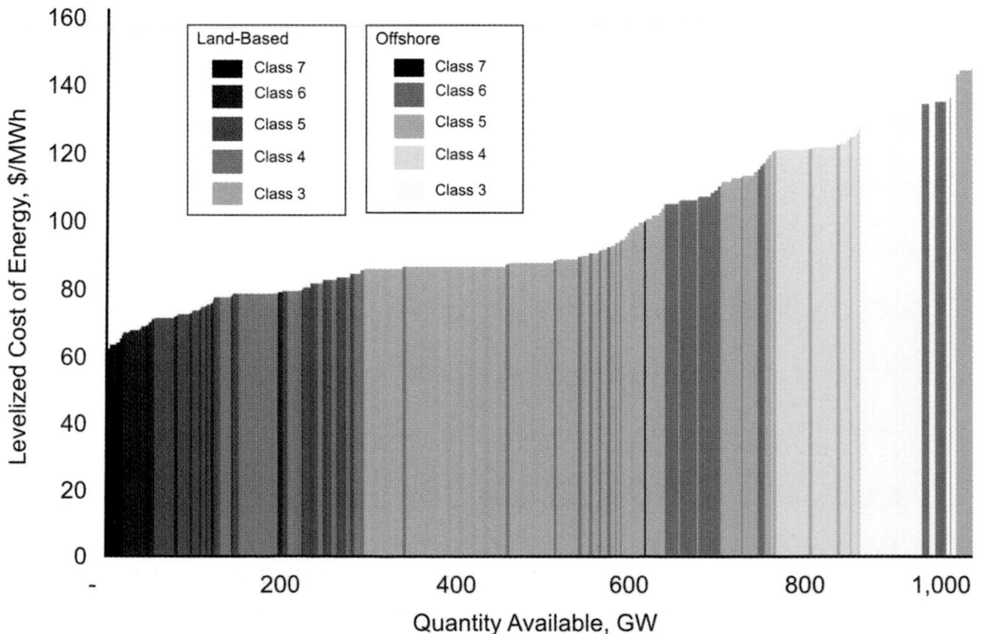

Figure A-3 shows only the supply curve for wind projects that can enter the existing transmission system (or that can power nearby loads), and does not include wind projects that would require new transmission to deliver power to markets distant from the generation system. The supply curve, however, shows more than 1,000 GW of wind energy— approximately 600 GW of land-based and roughly 400 GW of offshore capacity. Developing all of this resource is not economical and would require significant modifications in the transmission system, but under certain conditions it could produce enough energy to greatly exceed 20% of the nation's electricity supply in the future. The supply curve further illustrates that more than 600 GW of wind are available at or below $100/MWh at current bus-bar energy costs and performance indicators. These supply curves do not factor in transmission or integration costs or technology improvements.

A.4 Impacts

Based on the assumptions used to create the 20% Wind Scenario, providing 20% of the nation's projected electricity demand by 2030 would require the installation of 293.4 GW of wind technology (in addition to the 11.4 GW currently installed) for a cumulative installed capacity of 304.8 GW, generating nearly 1,200 terawatt-hours (TWh) annually. Offshore wind technology would account for about 18% (54 GW) of total wind capacity by 2030. Figure A-4 shows the cumulative installed capacity of land-based and offshore wind technologies required to generate 20% of projected electricity demand by 2030.

Figure A-4. Cumulative installed wind power capacity required to produce 20% of projected electricity by 2030

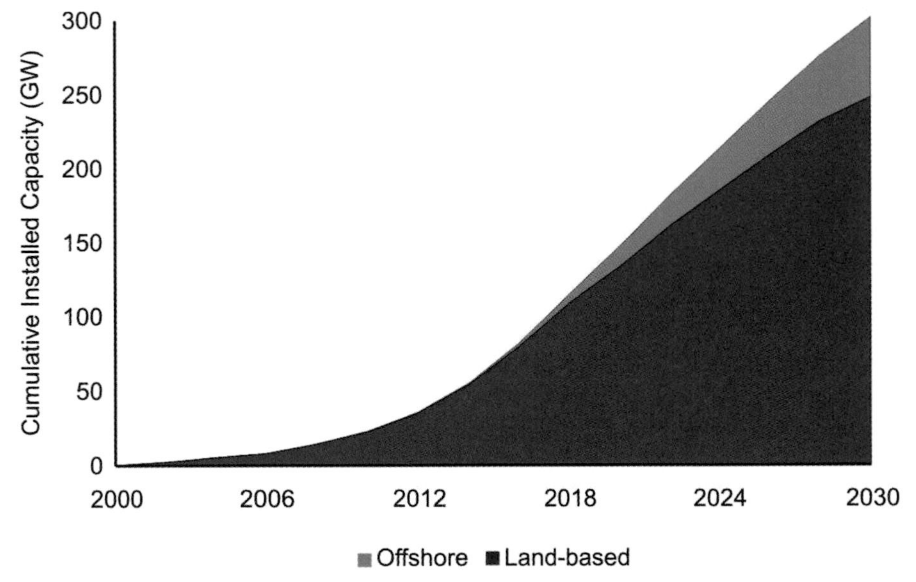

A.4.1 Generation Mix

This section presents impacts on the remaining generation mix and on the emissions of carbon from producing 20% of the nation's electricity from wind in 2030. The geographic distribution of wind turbines and the transmission expansion required to accommodate them are also addressed. Sophisticated routines in the WinDS model use existing transmission or build new transmission while incorporating associated wind integration costs. This scenario shows that with wind technology advancement associated reductions in costs and changes in the grid system, producing 20% wind energy in the nation's portfolio by 2030 could be technically feasible.

The generation mix produced by the WinDS model based on the requirement that 20% electricity generation will come from wind is pictured in Figure A-5. This scenario does not assume that carbon regulation policies are in place and reflects the assumptions listed in Table A-1 as well as others. The resulting generation mix, excluding wind, is made up of the most cost-effective conventional technologies in place today. Wind energy grows as a percentage of the nation's generation mix, and coal-generated electricity remains the major generation technology in 2030. Nuclear power generation declines slightly as a fraction of the total generation mix. Natural gas technologies make a greater contribution to the total mix through 2016 and then decline to a level similar to today's level by 2030. Changes in assumptions would produce a different mix of conventional generation technologies.

Figure A-5. 20% Wind Scenario electricity generation mix 2000–2030

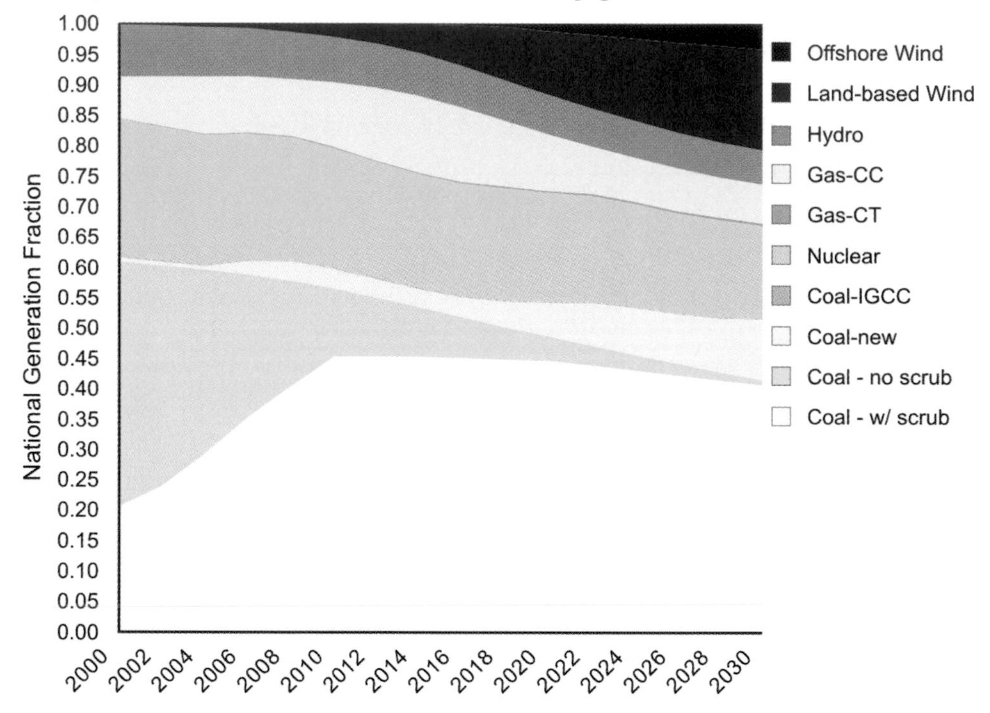

Figure A-6 illustrates the comparison in net generation in 2030 between conventional energy and wind energy generation, when applied to the 20% Wind and the No New Wind Scenarios. The 20% Wind Scenario, of course, would result in dramatically higher levels of wind energy generation. This figure also shows a significant reduction of energy generated from combined cycle natural gas plants (Gas-CC) as well as reduced energy from new pulverized coal plants (Coal-New). Figure A-7 compares generating capacity by 2030 between the 20% Wind Scenario and the No New Wind Scenario. Again, the contribution from wind is the primary difference. The 20% Wind Scenario requires less Coal-New and Gas-CC capacity.

The 20% Wind Scenario does require additional gas combustion turbine capacity (Gas-CT) to maintain grid reliability when wind resources vary. As shown in Figure A-6, relatively little electricity is generated from these plants in both scenarios.

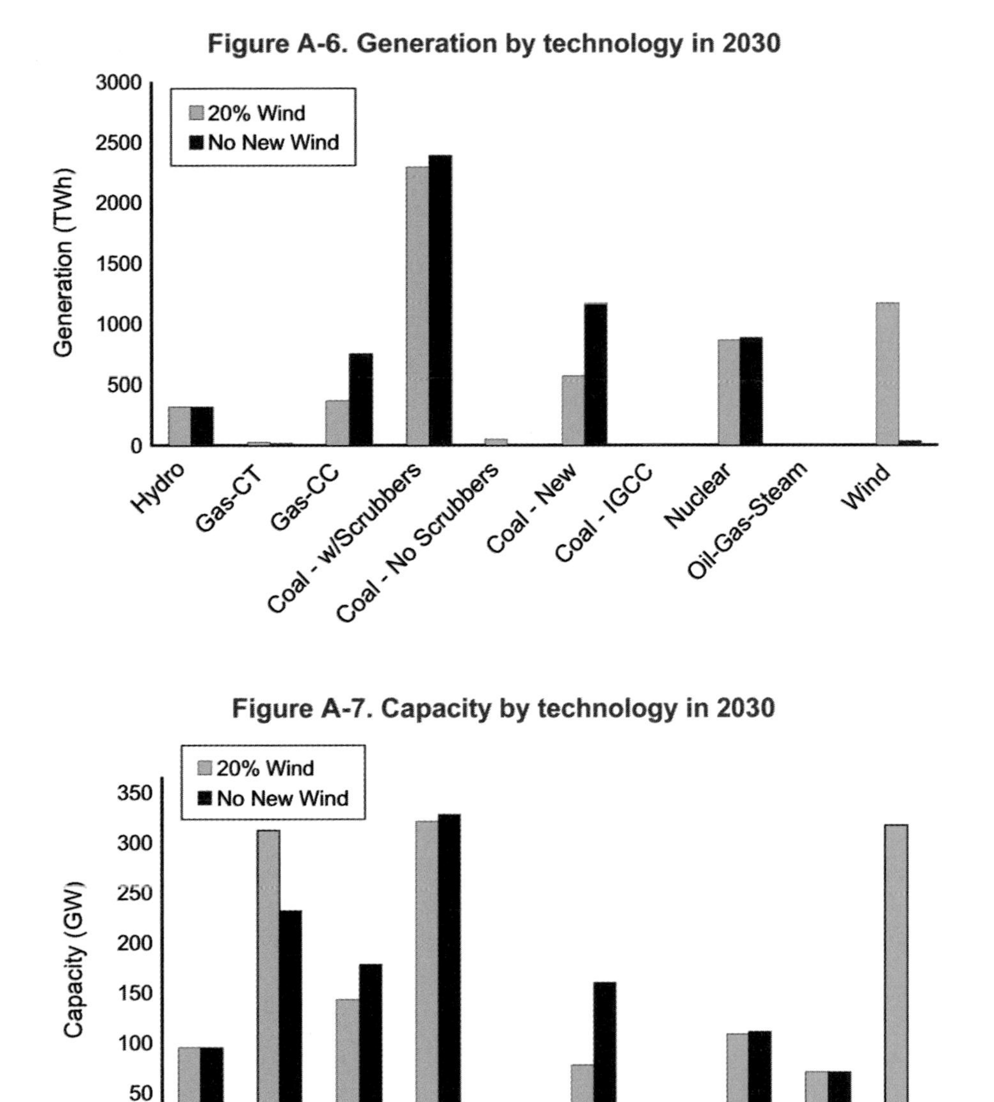

Figure A-6. Generation by technology in 2030

Figure A-7. Capacity by technology in 2030

Several important assumptions could affect the resulting mix of conventional generation in output from the WinDS model, including the following:

- **Fuel price forecasts:** The WinDS model uses regional gas and coal fuel price projections from the AEO (EIA 2007*)*. The reference-case coal fuel projections were implemented, but the natural gas price forecast from the high-price case was deemed more probable. Other gas and coal future price projections could be used, and modifying these prices would affect generation from gas, coal, and other sources.
- **Fuel price elasticity:** For this analysis, the WinDS model does not include fuel price elasticity. This could be important in scenarios that differ significantly from the scenario assumed in the AEO (EIA 2007). For example, assuming wind generation at 20% of U.S.

electricity, the demand for gas and coal would decrease, resulting in a lower price for both (thereby conversely driving up demand) and settling on a cost value lower than that currently used in the model.

- **Carbon regulation:** The imposition of a carbon constraint would also change this generation mix significantly, increasing future Coal-IGCC and Nuclear capacity, reducing future Coal-New and Gas-CC capacity, and leading to significantly more plant retirements and less use of existing coal plants.

A.4.2 Carbon Emission Reduction

Comparing the 20% Wind Scenario with the No New Wind Scenario provides one way of estimating the potential carbon emissions reductions that could be attributed to wind energy. This scenario assumes that the conventional generation mix is allowed to expand while optimizing total costs without any carbon regulation policy. Figure A-8 illustrates the cumulative carbon emissions reduction of more than 2,100 million metric tons of carbon equivalent (MMTCE) attributed to producing 20% of the nation's electricity from wind during the significant wind energy expansion period, 2005 to 2030. Extrapolating cumulative carbon emissions avoidance over the 20-year wind plant life through 2050 results in avoided emissions of more than 4,000 MMTCE, and avoided carbon emission in 2030 alone of 225 MMTCE.

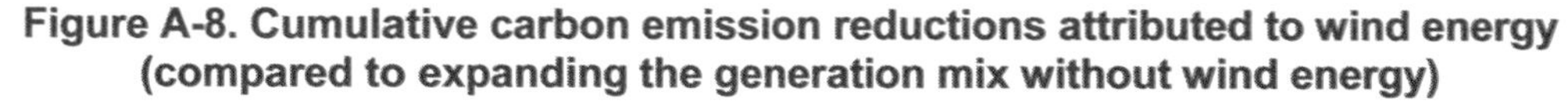

Figure A-8. Cumulative carbon emission reductions attributed to wind energy (compared to expanding the generation mix without wind energy)

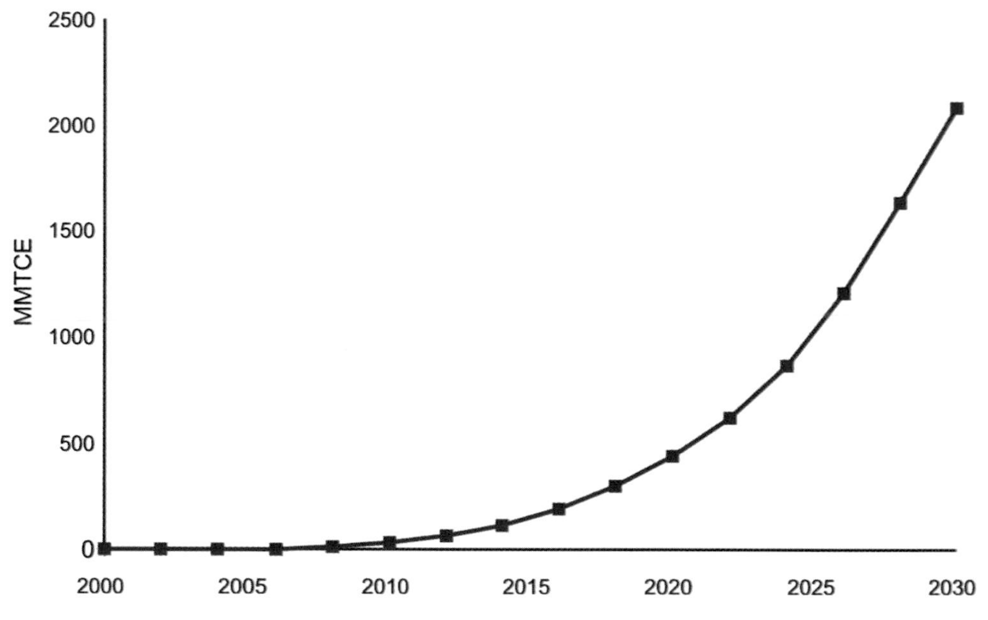

A.4.3 Reduced Natural Gas Demand

Figure A-9 demonstrates the decrease in coal and gas fuel use for the 20% Wind Scenario relative to the No New Wind Scenario. This graph indicates that a reduction in coal use across all coal technologies and a reduction in natural gas use comprise a significant portion of the total amount that would be used without additional wind installations. Incorporating enough wind generation technology to produce 20% of the nation's electricity demand by 2030 could reduce the electricity sector's natural gas requirements by about 50% and its coal requirements by about 18%. This shift translates into a reduced national demand for natural gas of 11%.

Figure A-9. Fuel usage and savings resulting from 20% Wind Scenario

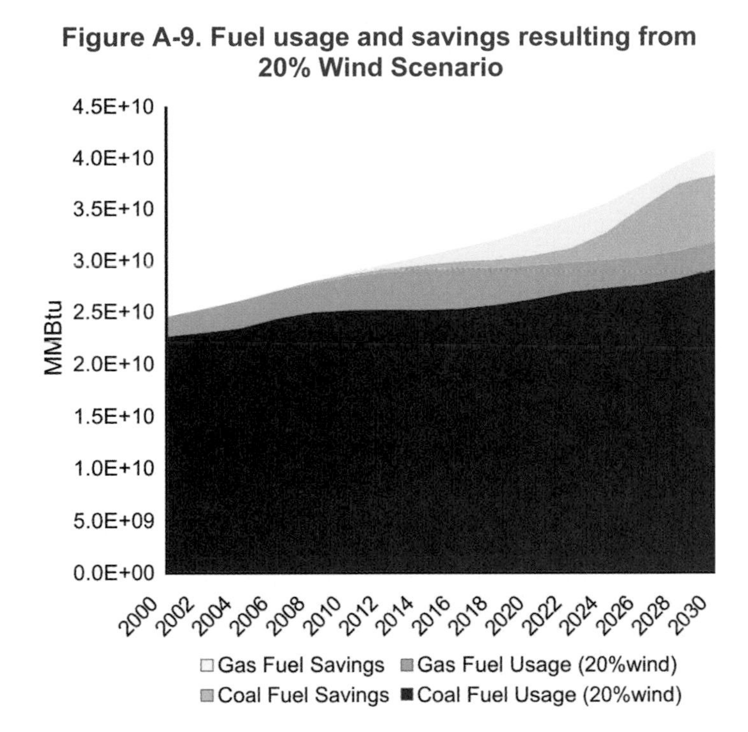

Wind power offers the country important resource diversification benefits, including the prospect for moderating natural gas demand. In 2006, gas-fired generation accounted for nearly 20% of the nation's electricity generation capacity. Because of the way electricity markets operate, the price of gas-fired generation determines the price of electricity. Wellhead natural gas prices, which hovered near $2/MMBtu in the 1990s, have risen to more than $6/MMBtu, and most forecasts expect prices to remain high relative to historical standards. Past efforts to forecast natural gas prices have not been very successful (e.g., Wiser and Bolinger 2004 and Bolinger and Wiser 2006).

A.4.4 Land Use

Under the 20% Wind Scenario, wind turbines required to supply 20% of the nation's electricity (over 300 GW) would be broadly distributed across the United States; at least 100 MW would be installed in 43 of the 48 contiguous states. Hawaii and Alaska have not been represented in this study, but both states are expected to install more than 100 MW of wind capacity. The WinDS model uses the best available assessment of local wind resources to expand wind technology capacity. Limitations of wind resource input data, which could significantly affect the wind technology capacity installed in a given state, are discussed in Appendix B. In addition to wind resources, other factors related to the model logic can influence the amount of wind capacity installed in a given state. For instance, current long-term power purchase agreements are not implemented in WinDS. The model assumes that local load is met by the generation technologies in a given region.

The lack of wind capacity installed in Ohio is assumed to be primarily a result of the amount of existing conventional energy resources that supply the state, reducing the need for additional generating capacity, regardless of the fact that Ohio's wind resources are sufficient to support wind technology development. Additionally Ohio's wind resources are concentrated in the western part of the state. The transmission cost assumptions are higher in Ohio than in neighboring Indiana and

Michigan, which makes Ohio's wind resource appear less cost-effective in comparison. Some states such as Louisiana, Mississippi, and Alabama have lower quality wind resources than Ohio, but under the right economic circumstances some wind energy development could occur in those states. The WinDS model optimizes the installation of wind energy capacity within each of the three large interconnection areas in the United States. The model shows that broad geographic distribution of wind energy capacity serves to meet the broadly distributed national electricity load. Figures A-10 to A-13 illustrate capacity expansion of wind energy representing the years 2012, 2018, 2024, and 2030 (approximately 3%, 9%, 15%, and 20% electricity generation, respectively). The specific assumptions used in this model significantly affect each state's projected wind capacity. See Table A-1 and Appendix B for more information on the assumptions. In reality, these levels will vary significantly as electricity markets evolve and state policies promote or restrict wind energy production.

The black outline in each state in Figures A-10 to A-13 represents land area required for a wind farm, corresponding to the capacity shown on the green scale. These figures use standard exclusion practices, which are detailed in Appendix B. The total land area of the United States required for 305 GW of wind energy, assuming a turbine density of 5 MW per square kilometer (km^2), would be smaller than 61,000 km^2 (50,000 km^2 for land-based projects and 11,000 km^2 for offshore projects). Only about 2% to 5% of the wind farm area, which is represented by the brown square within each black outline, is occupied by towers, roads, and other infrastructure components, and the balance of the area remains available for its original use (such as farming or ranching).

Figure A-10. Projected wind capacity installations in 2012

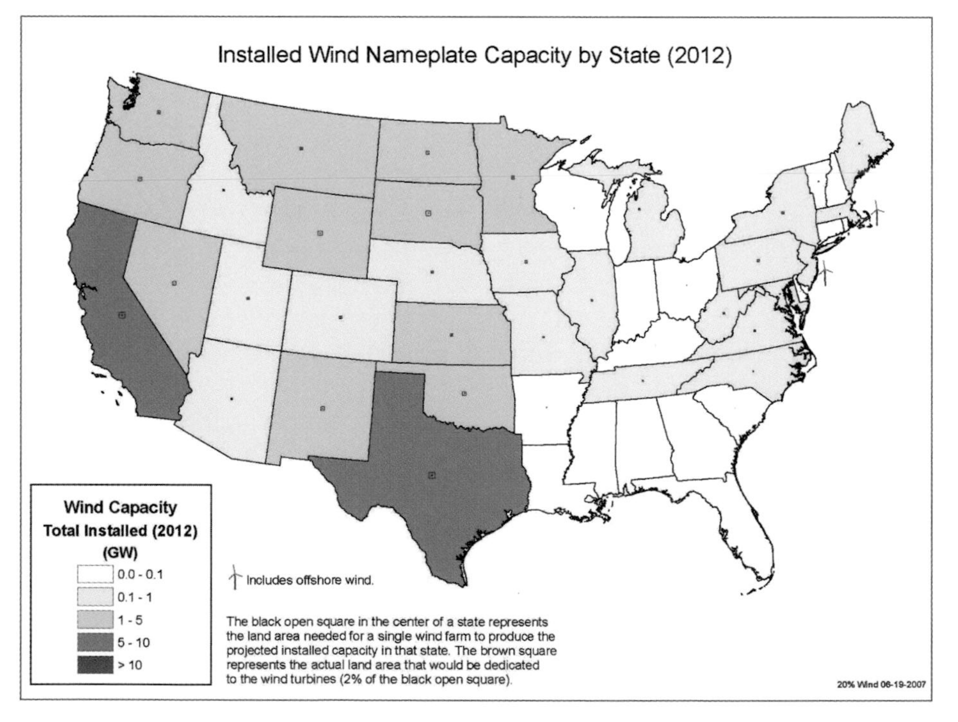

Figure A-11. Projected wind capacity installations in 2018

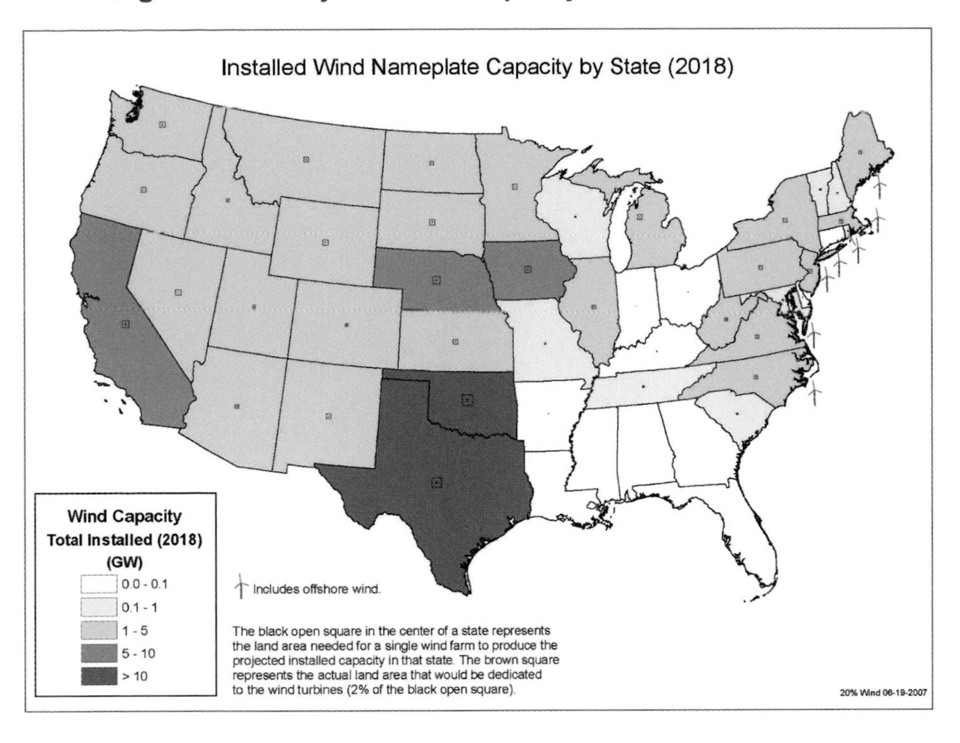

Figure A-12. Projected wind capacity installations in 2024

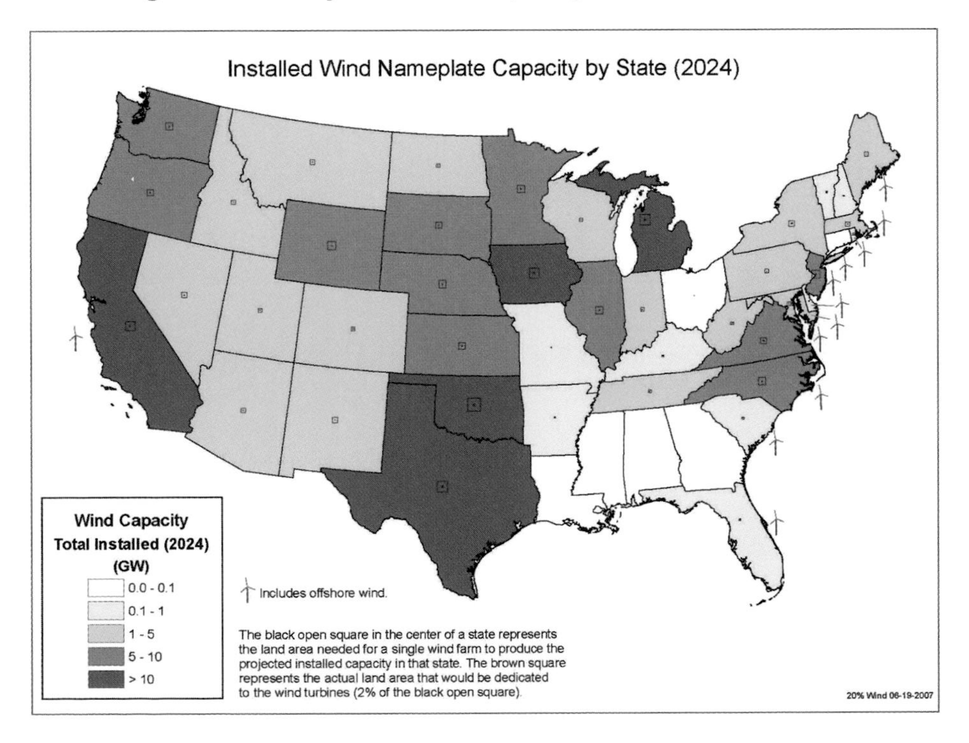

A

Figure A-13. Projected wind capacity installations in 2030

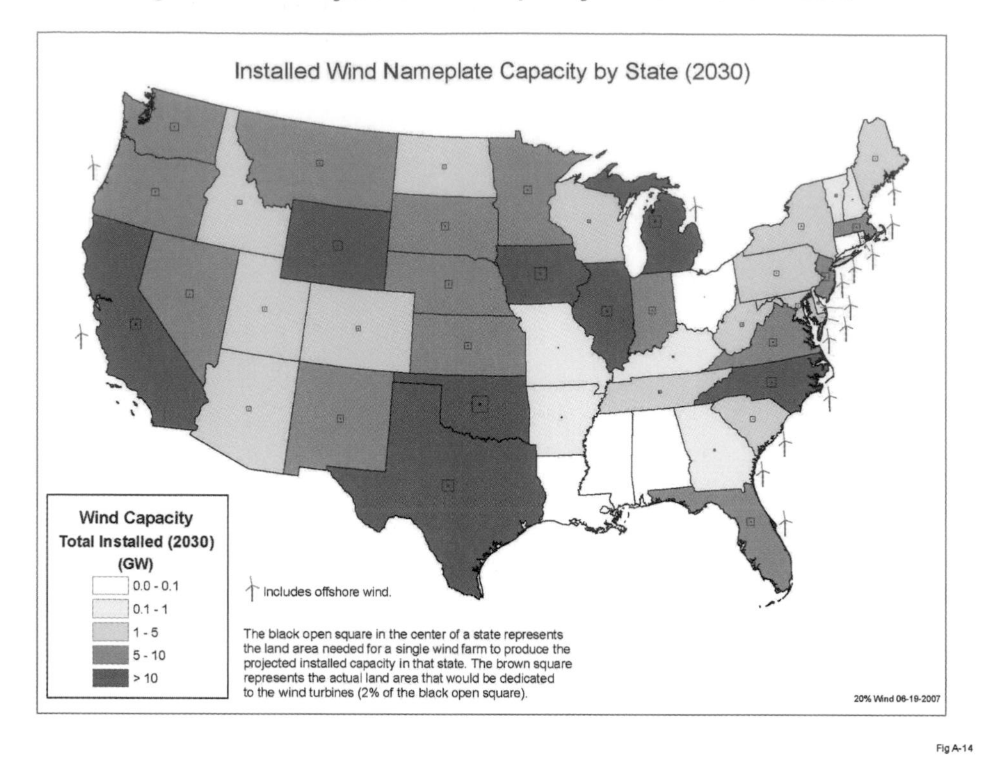

Fig A-14

A.4.5 Transmission

To meet the nation's growing demand for electricity, significant transmission expansion will be required. Meeting the 20% Wind Scenario requires transmission expansion to accommodate such a geographically dispersed resource. Three types of transmission systems included in the WinDS model could be used to transport wind power around the country:

- **Existing grid:** The model assumes that 10% of the existing grid could be used for new wind capacity, either by improving the grid or by drawing on existing unused capacity.
- **New lines:** The WinDS model can evaluate the use of straight-line transmission lines in the 358 wind regions. The model assumes that appropriate planning will allow new transmission lines to be constructed as additional capacity is needed.
- **In-region transmission:** In any of the 358 wind regions in the United States, the model can assess transmission lines directly from the wind site to loads within the same region.

Figures A-14 to A-17 illustrate the expansion of the transmission system required under the 20% Wind Scenario for the years 2012, 2018, 2024, and 2030 (approximately 3%, 9%, 15%, and 20% wind-electricity generation, respectively).

The 20% Wind Scenario assumes that transmission planning and grid operations occur on several levels—planning at the national level, reserve margin constraint planning at the NERC level, and load growth planning and operations at the balancing area (BA) level. For visual clarity, these figures display wind capacity only at the balancing area level.

Figure A-14. Transport of wind energy over existing and new transmission lines projected for 2012

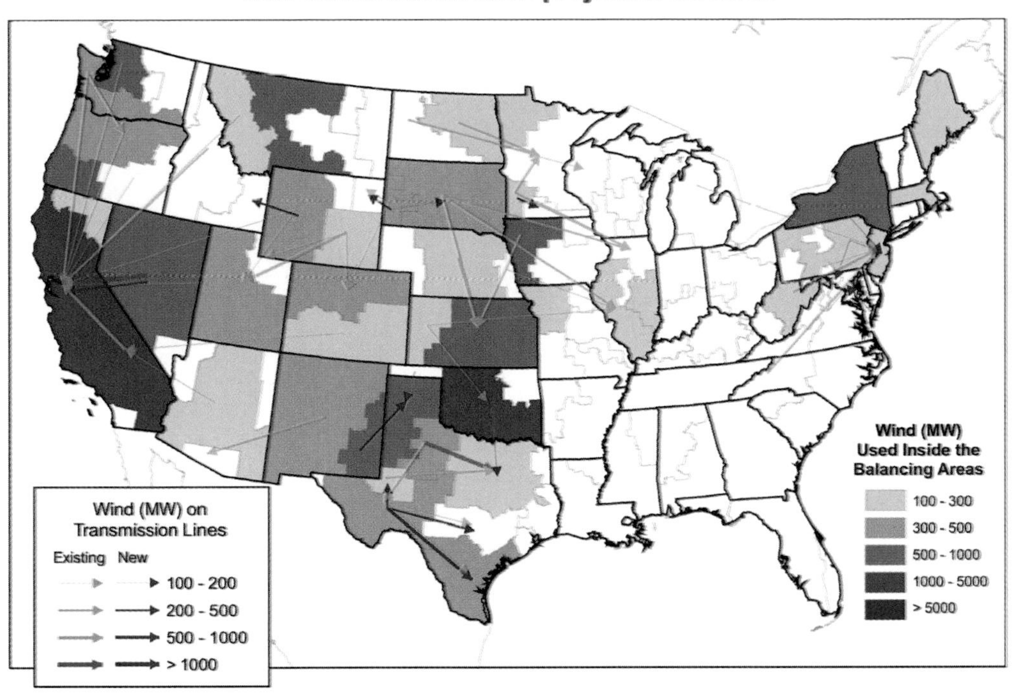

Total Between Balancing Areas Transfer >= 100 MW (all power classes, land-based and offshore) in 2012.
Wind power can be used locally within a Balancing Area (BA), represented by purple shading, or transferred out of the area on new or existing transmission lines, represented by red or blue arrows. Arrows originate and terminate at the centroid of the BA for visualization purposes; they do not represent physical locations of transmission lines.

A

Figure A-15. Transport of wind energy over existing and new transmission lines projected for 2018

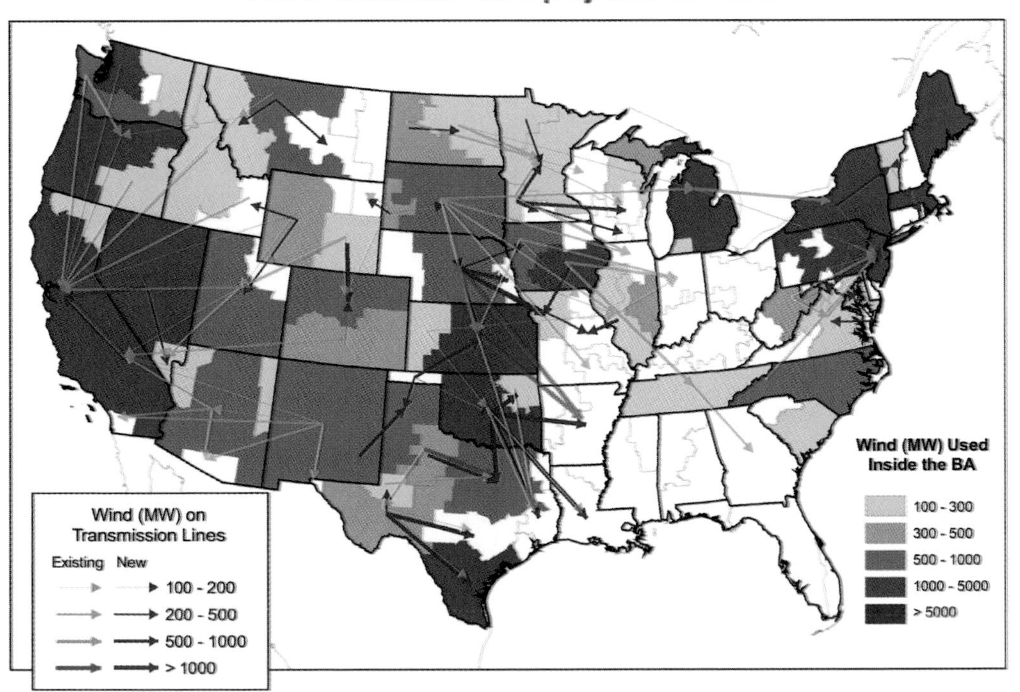

Total Between Balancing Areas Transfer >= 100 MW (all power classes, land-based and offshore) in 2018.
Wind power can be used locally within a Balancing Area (BA), represented by purple shading, or transferred out of the area on new or existing transmission lines, represented by red or blue arrows. Arrows originate and terminate at the centroid of the BA for visualization purposes; they do not represent physical locations of transmission lines.

Figure A-16. Transport of wind energy over existing and new transmission lines projected for 2024

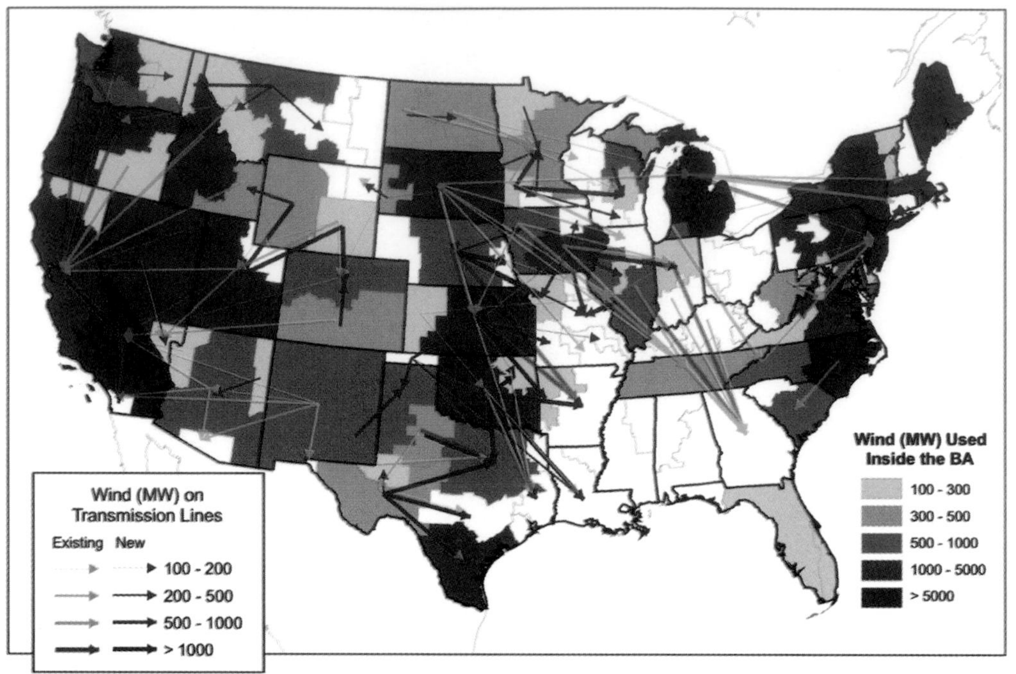

Total Between Balancing Areas Transfer >= 100 MW (all power classes, land-based and offshore) in 2024.
Wind power can be used locally within a Balancing Area (BA), represented by purple shading, or transferred out of the area on new or existing transmission lines, represented by red or blue arrows. Arrows originate and terminate at the centroid of the BA for visualization purposes; they do not represent physical locations of transmission lines.

A

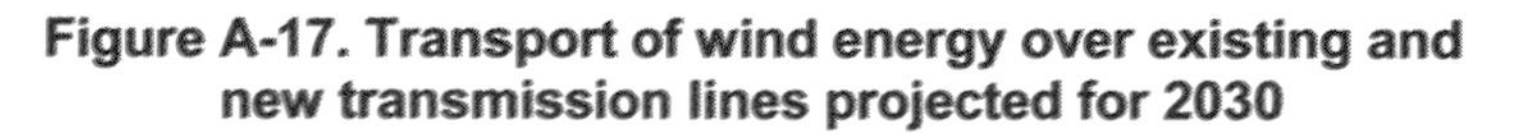

Figure A-17. Transport of wind energy over existing and new transmission lines projected for 2030

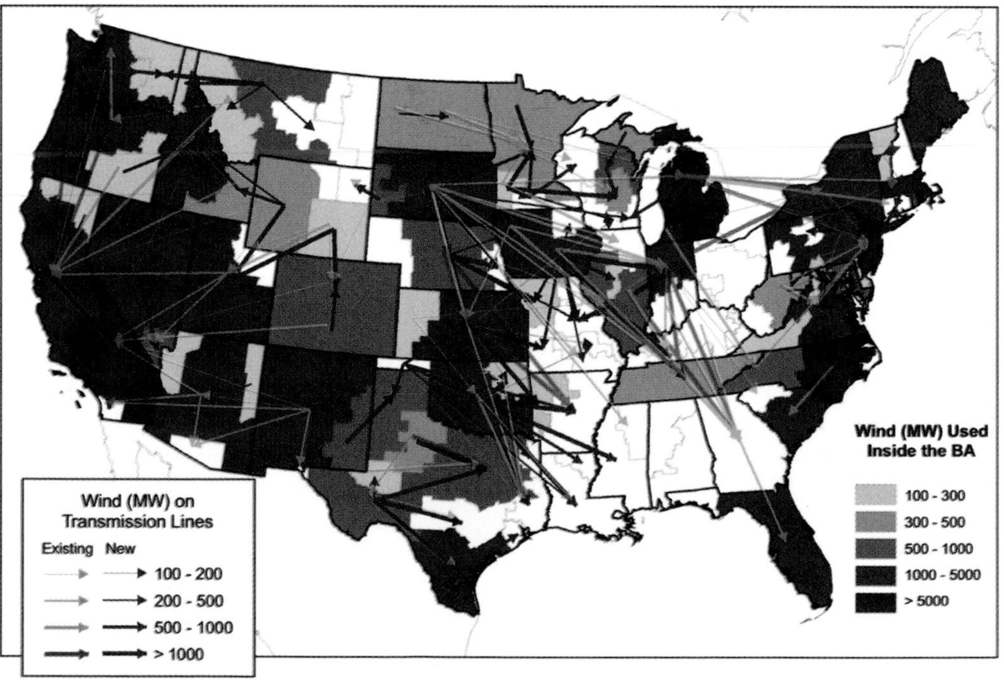

Total Between Balancing Areas Transfer >= 100 MW (all power classes, land-based and offshore) in 2030.
Wind power can be used locally within a Balancing Area (BA), represented by purple shading, or transferred out of the area on new or existing transmission lines, represented by red or blue arrows. Arrows originate and terminate at the centroid of the BA for visualization purposes; they do not represent physical locations of transmission lines.

The balancing areas, shaded in purple, depict the amount of locally installed wind, which is assumed to meet local load levels. Generally, the first wind system installed either uses the existing grid or is accompanied by a short transmission line built to supply local loads. In later years, as the existing grid capacity is filled, additional transmission lines are built. New transmission lines built to support load in a balancing area with wind resources within that same area are not pictured in these figures; only transmission lines that cross balancing area boundaries are illustrated.

In each figure, the blue arrows represent wind energy transported on existing transmission lines between balancing areas. The red arrows represent new transmission lines constructed to transport wind energy between balancing areas. The arrows originate and terminate at the centroid of a balancing area and do not represent the physical location of demand centers or wind resources. The location and relative number of red or blue arrows depend on the relative cost of using existing transmission lines or building new lines.

Table A-2 summarizes the projected installed wind capacity in 2030 by transmission type, number of megawatt-miles of transmission, and the resulting average distance traveled by each megawatt. Transmission options are based on a variety of factors; the cost of using existing transmission compared with new transmission can shift the relative amounts significantly. Appendix B contains a more complete discussion of transmission options used in the WinDS model.

Table A-2. Distribution of wind capacity on existing and new transmission lines

Transmission Type	2030 Wind Capacity	2030 MW-Miles	Average Distance Traveled for Each MW
Existing Transmission Lines	71 GW	20 million MW-miles	278 miles
New Capacity Lines within a WinDS region	67 GW	N/A	N/A (estimated at 50 miles)
New Capacity Lines that Cross One or More WinDS Region Boundaries	166 GW	30 million MW-miles	180 Miles

A.5 Direct Electricity Sector Cost

WinDS has been used to estimate the direct costs of meeting 20% of the nation's electricity requirements with wind power in accordance with the 20% Wind Scenario (see Appendix B for detailed calculations of each cost element). Direct costs to the electricity sector for each scenario include the capital costs of wind and conventional energy equipment, as well as transmission, O&M, and fuel costs. External analyses based on the WinDS model have estimated water consumption reductions. By comparing this scenario with a reference case that involves No New Wind generation after 2006, the potential costs of future wind development were estimated as the incremental change between these two scenarios.

Capital and transmission expansion costs are calculated for generation capacity added through 2030. Other costs presented in this section assume a 20-year project life for wind technology installed after 2010. Thus, the incremental differences in

A

fuel consumption, carbon emissions, and water consumption between the two scenarios in 2030 are reduced proportionally for wind systems that achieve a 20-year operational history between 2030 and 2050.

Direct costs to the electricity sector for each scenario include the capital costs of wind and conventional energy equipment, as well as transmission, O&M, and fuel costs. Table A-3 and Figure A-18 illustrate costs for the 20% Wind Scenario as well as the No New Wind Scenario. These costs represent the effect of investment decisions made over 20 years. The primary difference between the two scenarios is the higher capital investment for the 20% Wind Scenario, which is offset somewhat by additional fuel costs for the No New Wind Scenario. Both scenarios show a significant investment—exceeding $2 trillion—in generation capacity expansion through 2030. The capital costs include all financing costs applied to WinDS model investment selection, as described in Appendix B. The discounted capital costs, excluding financing, are $717 billion for the 20% Wind Scenario and $580 billion for the No New Wind Scenario.

Table A-3. Direct electricity sector costs for 20% Wind Scenario and No New Wind Scenario (US$2006)

	Present Value Direct Costs for 20% Wind Scenario* (billion US$2006)	Present Value Direct Costs for No New Wind after 2006* (billion US$2006)
Wind Technology O&M Costs	$51	$3
Wind Technology Capital Costs	$236	$0
Transmission Costs	$23	$2
Fuel Costs	$813	$968
Conventional Generation O&M Costs	$464	$488
Conventional Generation Capital Costs	$822	$905

* 7% real discount rate is used, per Office of Management and Budget (OMB) guidance; the time period of analysis is 2007-2030. WinDS modeling is used through 2030 and extrapolations of fuel usage and O&M requirements are used for 2030-2050.

A

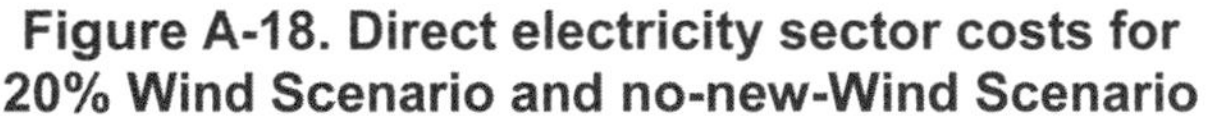

Figure A-18. Direct electricity sector costs for 20% Wind Scenario and no-new-Wind Scenario

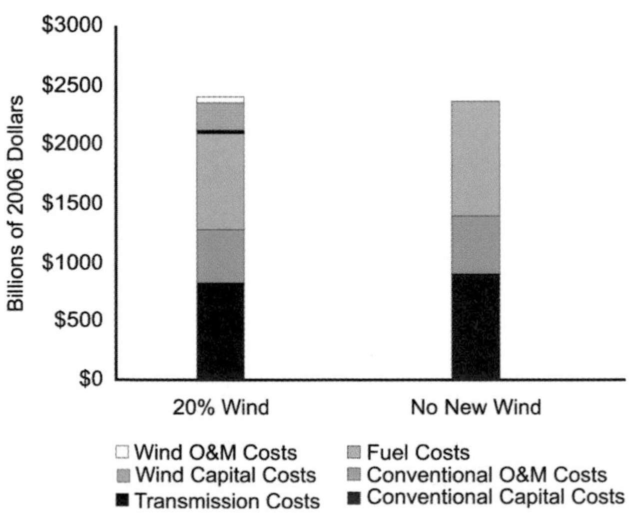

The WinDS model assumes that conventional generation systems, including coal and nuclear plants, are sited near load centers (except for California, which restricts the installation of coal and nuclear plants). Wind resources, on the other hand, tend to be geographically distant from load centers, requiring transmission lines to move electricity to the load. Estimated costs of transmission expansion in the No New Wind Scenario, then, are much lower than those for the 20% Wind Scenario, which might be overly conservative. Assuming that conventional plants are built near load centers is a simplifying assumption for modeling purposes, but may not reflect real siting issues that the coal and nuclear industries face today.

The WinDS model also estimates construction of a portion of a duplicate transmission line to maintain system reliability while expanding transmission capacity, but the model does not explicitly model system reliability conditions and resulting transmission upgrades.

Table A-4 summarizes the key findings of this analysis, focusing on direct electricity sector costs and ignoring the benefits of wind generation in reducing carbon emissions, or reducing water consumption. All costs are shown in US$2006, and the difference between the present values of the two cost streams is the total cost difference; in effect, WinDS calculates the incremental cost of achieving 20% wind (considering costs of capital, O&M, transmission and integration, and decommissioning) relative to the No New Wind Scenario.

Table A-4. Incremental direct cost of achieving 20% wind, excluding certain benefits (US$2006)

Present Value Direct Costs (billion US $2006) [a]	**Average Incremental LC of Wind** ($/MWh-Wind)[b]	**Average Incremental Levelized Rate Impact** ($/MWh-Total)	**Impact on Average Household Customer** ($/month)[c]
43 billion	$8.6/MWh	$0.6/MWh	$0.5/month

[a] Per Office of Management and Budget (OMB) guidance, a 7% real discount rate is used. The time period of analysis is 2007–2030. WinDS modeling is used through 2030 and extrapolations of fuel usage and O&M requirements are used for 2030–2050.

[b] The levelized cost per kilowatt-hour of wind produced is found by solving the following formula: $\sum$ wind generation * LC $/(1+d)^t$ = PV of costs in 20% Wind Scenario–PV of costs in No New Wind Scenario.

[c] Assumes 11,000 kWh/year average consumption.

The result of this analysis suggests rather modest incremental electricity-sector costs.[15] The direct incremental cost of 20% wind is estimated to be $43 billion in net present value terms, increasing electricity rates by only $0.6/MWh on average over the 2007–2050 analysis period, and raising average residential monthly electricity bills by just $0.5/MWh over that same time period. The average incremental LC imposed by each megawatt-hour of wind is estimated at $8.6/MWh. Because WinDS considers not just bus-bar energy costs, but also transmission costs and the cost of integrating the variable output pattern of wind into electricity grids, the analysis presented here suggests that the potential direct costs of achieving 20% wind, relative to meeting load with conventional technologies, need not be overwhelming.

[15] These costs reflect the model inputs and could vary significantly with different fossil fuel price assumptions, carbon taxes or caps, or additional breakthroughs in renewable technologies.

A.5.1 Water Consumption Savings

In the energy sector, water is used primarily for cooling in steam plants, but it is also used in boilers and in air pollution reduction processes. Several technologies are used to condense steam (EPRI 2002; Feeley et al. 2005):

- **Recirculating steam plant cooling:** Water is reused to cool steam in a closed-loop system using a cooling tower or cooling pond.
- **Once-through cooling:** Water from a lake, a river, or the ocean is used to condense steam, and the water is then returned to its source, but at a higher temperature.
- **Dry cooling:** Air cools steam, using far less water than the first two "wet" cooling technologies. Although dry cooling is not widely used, it can be the cooling technology of choice where water supplies are limited.

Two types of water use are generally considered:

- **Water withdrawal:** Water is removed from the ground or diverted from a surface source for use.
- **Water consumption:** Water is withdrawn from a source but not directly returned to the source because it is evaporated, transpired, incorporated into products and crops, or consumed by people or livestock.

In this analysis, water consumption projections were made by applying water consumption rates (gallons per megawatt-hour generated) to projected megawatt-hours of generation for each type of power plant. These calculations were made on a yearly basis for the 20% Wind Scenario and No New Wind Scenario. Water savings from deploying large amounts of wind-generated electricity are calculated as the difference in water consumption between the two scenarios. Water consumption rates were developed from several data sources, the most important of which are:

- **EIA Form 767 for 2002:** This database includes water consumption rates for each steam power plant, including the steam portion of combined cycle plants. Because these data often contain unrealistic values for water consumption (e.g., no water consumption or very large amounts of water consumption per megawatt-hour), observations with extremely high and low values have been removed before computing average consumption rates for each type of power plant (EIA 2002).
- **EPRI's water and sustainability study:** This report contains typical water consumption values for steam and combined cycle power plants (EPRI 2002).
- **A Clean Air Task Force/Western Resource Advocates study:** This report supplements the EPRI estimates with other sources of data (Baum et al. 2003).

Because of the quality and availability of data from these sources, the authors have assumed in this study that existing and new power plants have the same water consumption rates. Although once-through cooling plants withdraw more water per megawatt-hour than recirculating plants, the mix of power plant cooling types (once-through and recirculating) was assumed to stay the same over the study period. No systematic regional variations in water consumption for coal-fired steam plants were found, and the number of realistic observations for the other technologies was too small to permit a useful geographic disaggregation. Therefore, only national average water consumption rates were used. Table A-5 illustrates water consumption rates used in the analysis, and Figure A-19 shows annual water savings resulting from the deployment of wind resources.

Table A-5. Water consumption rates for power plants

Generation Type	Water Consumption Rate: Gallons per MWh	Source (see list of references for full citation)
Coal-Fired Steam	541	EIA Form 767 for 2002
Gas-Fired Combined Cycle	180	EPRI; Clean Air Task Force & Western Resource Advocates
Nuclear	609	EIA Form 767 for 2002
Oil- or Gas-Fired Steam	662	EIA Form 767 for 2002
Combustion Turbine	0-100	See note below
Wind	0	Clean Air Task Force & Western Resource Advocates

Note: Data on water consumption rates for combustion turbines are sparse. Estimated consumption rates range from 0 to about 100 gall/MWh. For example, the U.S. Department of Energy *Environmental Assessment for the Installation and Operation of Combustion Turbine Generators at Los Alamos National Laboratory* (December 2002) estimated that water use by planned combustion turbines would be 0 (p. 17). A California Energy Commission study (2005) indicated that water consumption for combustion turbines is less than 100 gal/MWh. We analyzed total water savings, assuming combustion turbine water consumption is 0 gal/MWh and 100 gal/MWh and found that the difference in total water savings in any year was only 0.3% or less. Therefore, water savings are not sensitive to assumptions about water consumption rates for combustion turbines.

Figure A-19. Annual water consumption savings due to deployment of wind energy

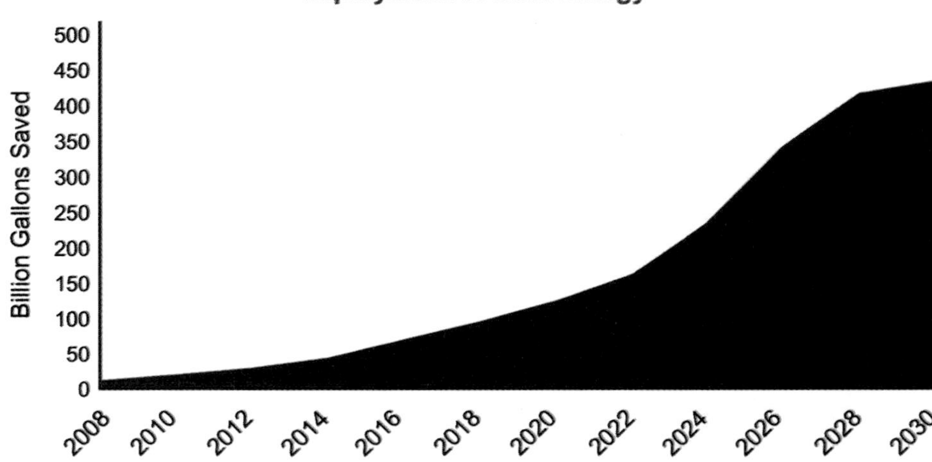

Displacing large amounts of fossil-fueled power generation with wind energy reduces water consumption. Based on the authors' estimates, if the current conventional generation mix is expanded to meet electricity needs, approximately 51 trillion gallons of water will be consumed for electricity production from 2007 to

A

2030. If wind energy deployment gradually increases to 20% of the nation's electricity over the same time period, however, 47 trillion gallons of water will be consumed. This is a saving of 4 trillion gallons; an 8% reduction in water consumption. Of the 4 trillion gallons of water saved nationally, 29% will be in the West, 41% will be in the Midwest/Great Plains, 14% will be in the Northeast, and 16% will be in the Southeast (see Table A-6). Extrapolating the savings beyond 2030 to account for the 20-year investment benefit from installing wind energy yields cumulative water consumption savings of 6 trillion gallons by 2050.

Table A-6. U.S. states, by region

Region	States
West	Alaska, Arizona, California, Colorado, Hawaii, Idaho, Montana, Nevada, New Mexico, Oregon, Washington, Wyoming, Utah
Midwest/Great Plains	Illinois, Indiana, Iowa, Kansas, Michigan, Minnesota, Missouri, Nebraska, North Dakota, Ohio, Oklahoma, South Dakota, Texas, Wisconsin
Northeast	Connecticut, Delaware, Maine, Maryland, Massachusetts, New Hampshire, New Jersey, New York, Pennsylvania, Rhode Island, Vermont
Southeast	Alabama, Arkansas, Florida, Georgia, Louisiana, Kentucky, North Carolina, South Carolina, Tennessee, Mississippi, Virginia, West Virginia

A

A.6 Other Effects

Appendix C describes the jobs and economic impacts directly associated with the manufacturing, construction, and operational sectors of the wind industry.

Other benefits associated with wind energy include an improved environment and better health resulting from reduced particulate or other chemical emissions such as acid rain or mercury, and market benefits including diversification of the electricity sector. These benefits, and others, have not been quantified in this study.

A.7 References & Suggested Further Reading

Awerbuch, S. 1993. "The Surprising Role of Risk in Utility Integrated Resource Planning." *The Electricity Journal* 6(3): 20–33.

Awerbuch, S. 2003. "Determining the Real Cost: Why Renewable Power is More Cost-Competitive than Previously Believed." *Renewable Energy World* 6(2), March-April 2003.

Baum, E., J. Chaisson, B. Miller, J. Nielsen, M. Decker, D. Berry, and C. Putnam. 2003. *The Last Straw: Water Use by Power Plants in the Arid West*. Boulder, CO and Boston, MA: The Land and Water Fund of the Rockies and Clean Air Task Force. http://www.catf.us/publications/reports/The_Last_Straw.pdf.

Black & Veatch Corporation. 2007. Wind Supply Curves. September. Overland, KS: Black & Veatch.

Bokenkamp, K., H. LaFlash, V. Singh, and D.B. Wang. 2005. "Hedging Carbon Risk: Protecting Customers and Shareholders from the Financial Risk Associated with Carbon Dioxide Emissions." *The Electricity Journal* 18(6): 11–24.

Bolinger, M., and R. Wiser. 2005. *Balancing Cost and Risk: The Treatment of Renewable Energy in Western Utility Resource Plans*. LBNL-58450. Berkeley, CA: Lawrence Berkeley National Laboratory (Berkeley Lab).

Bolinger, M., and R. Wiser. 2006. *Comparison of AEO 2007 Natural Gas Price Forecast to NYMEX Futures Prices*. LBNL-62056. Berkeley, CA: Berkeley Lab..

Bolinger, M., R. Wiser, and W. Golove. 2006. "Accounting for Fuel Price Risk When Comparing Renewable to Gas-Fired Generation: The Role of Forward Natural Gas Prices." *Energy Policy* 34(6): 706–720.

CEC (California Energy Commission). 2005. Page 43 in *A Preliminary Environmental Profile of California's Imported Electricity.* CEC-700-2005-017. Sacramento: CEC. http://www.energy.ca.gov/2005_energypolicy/documents/2005-06-27+28_workshop/presentations/2005-06-27_CECSTAFF+ASPEN_PRESENTATION.PDF

Cavanagh, R., A. Gupta, D. Lashof, and M. Tatsutani. 1993. "Utilities and CO_2 Emissions: Who Bears the Risks of Future Regulation?" *The Electricity Journal*, 6(2): 64–75.

Denholm, P., and W. Short. 2006. *Documentation of WinDS Base Case.* Version AEO 2006 (1). Golden, CO: National Renewable Energy Laboratory (NREL). http://www.nrel.gov/analysis/winds/pdfs/winds_data.pdf.

DOE (U.S. Department of Energy), NNSA (National Nuclear Security Administration), and Los Alamos Site Office (US). 2002. *Environmental Assessment for the Installation and Operation of Combustion Turbine Generators at Los Alamos National Laboratory.* DOE/EA-1430. Los Alamos, NM: DOE, NNSA, and Los Alamos Site Office.

EIA (Energy Information Administration). 2002. *A Steam-Electric Plant Operation and Design Report, Schedule 6, Cooling System Information.* Form EIA-767. Washington, DC: EIA. http://www.eia.doe.gov/cneaf/electricity/forms/eia767/eia767.pdf

EIA (Energy Information Administration). 2007. *Annual Energy Outlook 2007 with Projections to 2030.* Washington, DC: EIA. Report No. DOE/EIA-0383. http://www.eia.doe.gov/oiaf/archive/aeo07/index.html.

EPRI (Electric Power Research Institute). 2002. *Water and Sustainability (Volume 3): U.S. Water Consumption for Power Production – The Next Half Century.* Report 1006786. Prepared by Bevilacqua-Knight, Inc. Palo Alto, CA: EPRI.

Feeley, T. III, L. Green, J. Murphy, J. Hoffman, and B. Carney. 2005. *Department of Energy/Office of Fossil Energy's Power Plant Water Management R&D Program.* Washington, DC: U.S. Department of Energy (DOE), National Energy Technology Laboratory (NETL), Science Applications International Corporation (SAIC).

A

Hoff, T.E. 1997. *Integrating Renewable Energy Technologies in the Electric Supply Industry: A Risk Management Approach.* NREL/SR-520-23089. Golden, CO: NREL. Hutson, S., N. Barber, J. Kenny, K. Linsey, D. Lumia, and M. Maupin. 2004. *Estimated Use of Water in the United States in 2000.* Circular 1268. Denver, CO: U.S. Geological Survey. http://pubs.usgs.gov/circ/2004/circ1268/.

Johnston, L., E. Hausman, A. Sommer, B. Biewald, T. Woolf, D. Schlissel, A. Roschelle, and D. White. 2006. *Climate Change and Power: Carbon Dioxide Emissions Costs and Electricity Resource Planning*. Cambridge, MA: Synapse Energy Economics, Inc.

Laxson, A., M. Hand, and N. Blair. 2006. *High Wind Penetration Impact on U.S. Wind Manufacturing Capacity and Critical Resources.* NREL/TP-500-40482. Golden, CO: NREL. http://www.nrel.gov/docs/fy07osti/40482.pdf.

NREL. 2006. *Projected Benefits of Federal Energy Efficiency and Renewable Energy Programs-FY2007 Budget Request*. NREL/TP-320-39684. Golden, Colorado: NREL. http://www1.eere.energy.gov/ba/pdfs/39684_00.pdf.

Rabe, B. 2002. *Greenhouse & Statehouse: The Evolving State Government Role in Climate Change*. Arlington, Virginia: Pew Center on Global Climate Change. http://www.pewclimate.org/docUploads/states_greenhouse.pdf.

Repetto, R., and J. Henderson. 2003. *Environmental Exposures in the US Electric Utility Industry.* New Haven, CT: Yale School of Forestry & Environmental Studies. http://environment.yale.edu/doc/969/environmental_exposures_in_the_us_electric/.

Wiser, R., and M. Bolinger. 2004. *An Overview of Alternative Fossil Fuel Price and Carbon Regulation Scenarios*. LBNL-56403. Berkeley, CA: Berkeley Lab. eetd.lbl.gov/EA/EMP/reports/56403.pdf.

Wiser, R., M. Bolinger, and M. St. Clair. 2005. Easing the Natural Gas Crisis: Reducing Natural Gas Prices through Increased Deployment of Renewable Energy and Energy Efficiency. LBNL-56756. Berkeley, CA: Berkeley Lab. http://www.lbl.gov/Science-Articles/Archive/sabl/2005/February/assets/Natural-Gas.pdf

A

Appendix B. Assumptions Used for Wind Deployment System Model

To define the 20% Wind Scenario, a number of modifications were made to the National Renewable Energy Laboratory's (NREL) Wind Deployment System (WinDS) base-case assumptions (which are described in the WinDS documentation; see Denholm and Short 2006). These changes include updating wind resource maps, accounting for seasonal and diurnal capacity factor (CF) variations, and including offshore wind resources from South Carolina to Texas. Black & Veatch developed the wind and conventional generation technology cost and performance projections in consultation with American Wind Energy Association (AWEA) industry experts. The assumptions about the large regional planning and operation structure of the transmission system were developed through collaboration with the experts who contributed to Chapter 4. The financial assumptions and the region definitions are unchanged from the WinDS base case. This appendix outlines the assumptions used in constructing the 20% Wind Scenario.

B.1 Financial Parameters

WinDS optimizes the electric power system "build" based on projected life-cycle costs, which include capital costs and cumulative discounted operating costs over a fixed evaluation period. The "overnight" capital costs supplied as inputs to the model are adjusted to reflect the actual total cost of construction, including tax effects, interest during construction, and financing mechanisms. Table B-1 summarizes the financial values used to produce net capital and operating costs. These assumptions are unchanged from the WinDS base case (Denholm and Short 2006) and correspond to assumptions made by the Energy Information Administration (EIA) in the *Annual Energy Outlook 2007 with Projections to 2030* (AEO); (EIA 2007a).

Table B-1. Baseline financial assumptions

Name	Value	Notes and Source
Inflation Rate	3%	Based on recent historical inflation rates
Real Discount Rate	8.5%	Equivalent to weighted cost of capital. Based on EIA assumptions (EIA 2006)
Marginal Income Tax Rate	40%	Combined federal/state corporate income tax rates
Evaluation Period	20 Years	Base Case Assumption
Depreciation Schedule Conventional Wind	15 Year 5 Year	MACRS (Modified Accelerated Cost Recovery Schedule) MACRS (Modified Accelerated Cost Recovery Schedule)
Nominal Interest Rate during Construction	10%	Base Case Assumption
Dollar Year	2004	All costs are expressed in year 2004 dollars.

B.2 Power System Characteristics

B.2.1 WinDS Regions

Four types of regions are included in the WinDS model (see Figure B-1):

- **Interconnect regions:** There are three major interconnects in the United States (all are electrically isolated): the Eastern Interconnect, Western Interconnect, and the ERCOT (Electric Reliability Council of Texas) Interconnect.
- **National Electric Reliability Council (NERC)[16] subregions:** WinDS uses 13 NERC regions, which are listed in Table B-2.
- **Balancing areas:** WinDS uses 136 balancing areas.
- **Wind resource regions:** There are 358 wind resource regions in WinDS.

Interconnect regions, NERC regions, and balancing areas are defined and operated by various regulatory agencies.

Figure B-1. WinDS regions

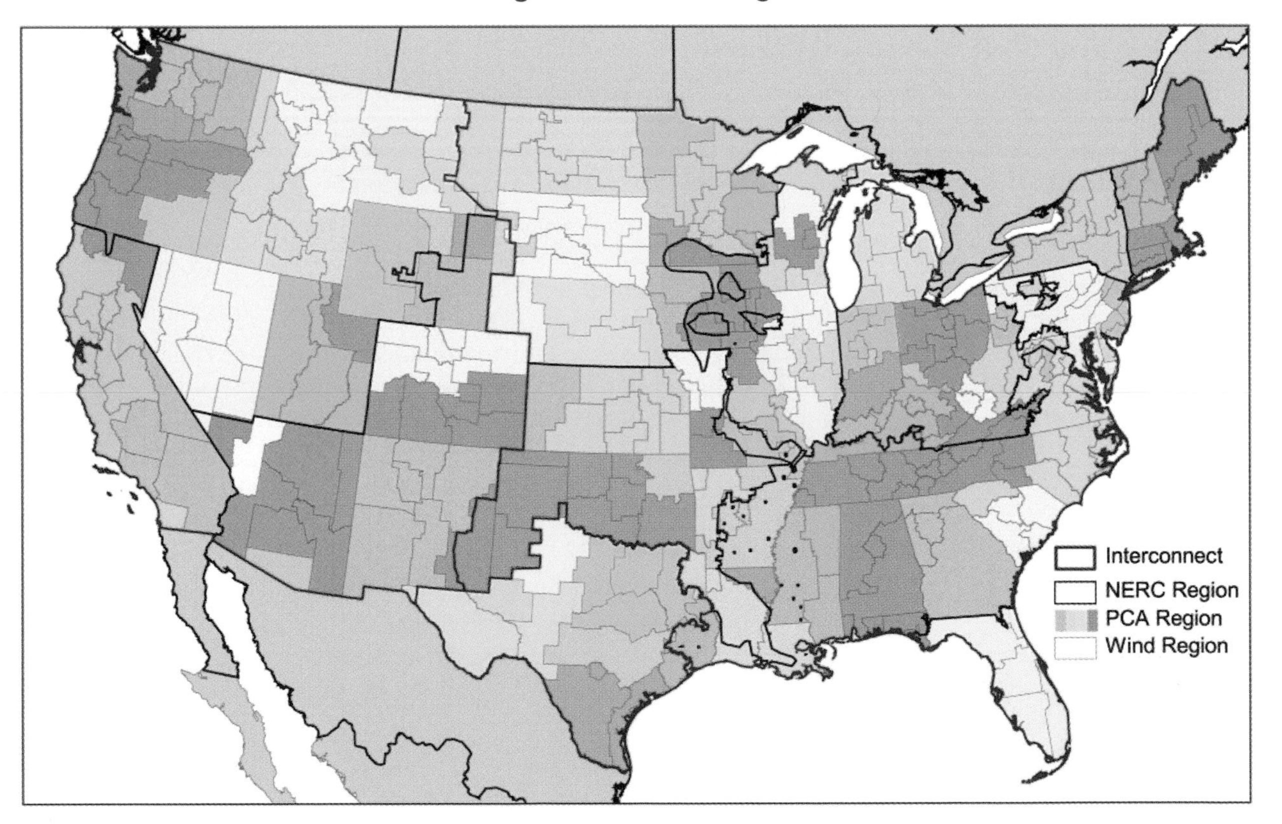

[16]For more information on NERC, see http://www.nerc.com/regional/.

Table B-2. NERC regions used in WinDS

NERC Region/ Subregion	Abbreviation	Region Name
1	ECAR	East Central Area Reliability Coordination Agreement
2	ERCOT	Electric Reliability Council of Texas
3	MAAC	Mid-Atlantic Area Council
4	MAIN	Mid-America Interconnected Network
5	MAPP	Mid-Continent Area Power Pool
6	NY	New York
7	NE	New England
8	FRCC	Florida Reliability Coordinating Council
9	SERC	Southeast Reliability Council
10	SPP	Southwest Power Pool
11	NWP	Northwest
12	RA	Rocky Mountain Area
13	CNV	California/Nevada

Note: NERC regions in WinDS are based on the pre-2006 regional definitions defined by the EIA (2000). In January 2006, NERC regions were redefined; however, the EIA has not incorporated these changes through publication of an AEO. Therefore, the WinDS will continue to use pre-2006 definitions until the EIA modifies its data. Similarly, some of the recent changes to balancing-area boundaries (now referred to as balancing authorities) are not yet reflected in WinDS (e.g., the formation of the Texas Regional Transmission Organization).

Wind resource regions were created specifically for the WinDS model. These regions were selected using the following rules and criteria:

- Incorporate buildup from counties (so the electricity load can be determined for each wind supply/demand region based on county population)
- Avoid crossing state boundaries (so that state-level policies can be modeled)
- Conform to balancing areas as much as possible (to better capture the competition between wind and other generators)
- Separate major windy areas from load centers (so that the distance from a wind resource to a load center can be well approximated)
- Conform to NERC region/subregion boundaries (so that the results are appropriate for use by integrating models that use the NERC regions and subregions).

Figure B-2 illustrates all wind regions and balancing areas in the United States.

Figure B-2. Wind region and Balancing Areas in WinDS base case

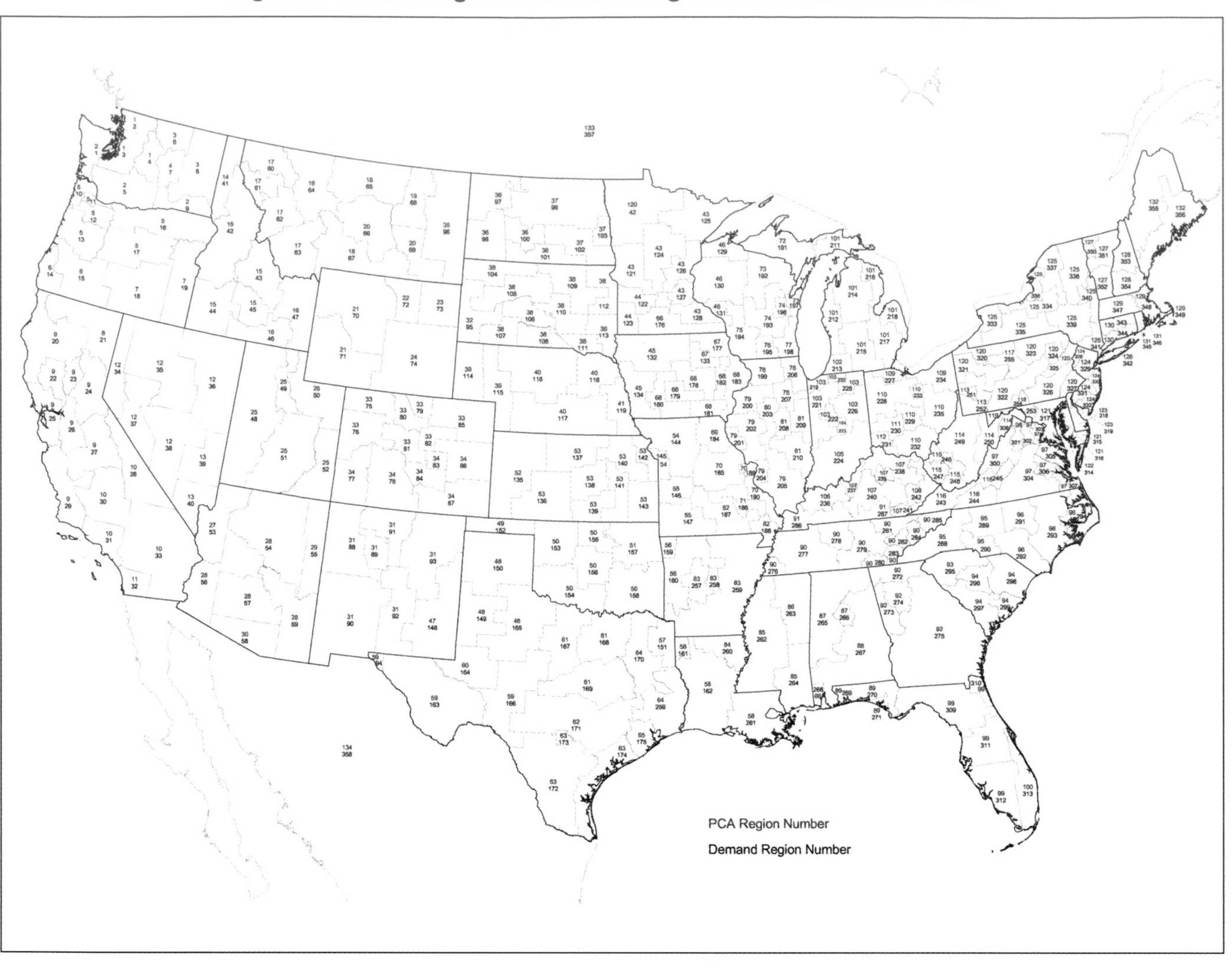

Several components of the WinDS model necessitate using four levels of geographic resolution. For example, electricity demand is modeled at the NERC region level, and wind-generator performance is modeled at the wind-resource region level.

B.2.2 Electric System Loads

Loads are defined by region and by time. WinDS meets the energy and power requirements for each of 136 balancing areas. Energy is met for each balancing area in each of 16 time slices, and within each year modeled. Table B-3 defines these slices.

The electricity load in 2000 for each balancing area and time slice is derived from an RDI/Platts database (Platts Energy Market Data; see http://www.platts.com). Figure B-3 illustrates the WinDS load duration curve (LDC) for the entire United States for the base year, showing the 16 load time slices. For reference, the actual U.S. coincident LDC—also derived from the Platts database—is depicted in the figure as well. The aggregated data for the United States that are shown in Figure B-3 are not used directly in WinDS because the energy requirement is met in each balancing area. This curve does, however, give a general idea of the WinDS

Table B-3. WinDS demand time-slice definitions

Slice Name	Number of Hours Per Year	Season	Time Period
H1	1,152	Summer	Weekends, plus 11:00 p.m. to 6:00 a.m. weekdays
H2	462	Summer	Weekdays, 7:00 a.m. to 1:00 p.m.
H3	264	Summer	Weekdays, 2:00 p.m. to 5:00 p.m.
H4	330	Summer	Weekdays, 6:00 p.m. to 10:00 p.m.
H5	792	Fall	Weekends, plus 11:00 p.m. to 6:00 a.m. weekdays
H6	315	Fall	Weekdays, 7:00 a.m. to 1:00 p.m.
H7	180	Fall	Weekdays, 2:00 p.m. to 5:00 p.m.
H8	225	Fall	Weekdays, 6:00 p.m. to 10:00 p.m.
H9	1,496	Winter	Weekends, plus 11:00 p.m. to 6:00 a.m. weekdays
H10	595	Winter	Weekdays, 7:00 a.m. to 1:00 p.m.
H11	340	Winter	Weekdays, 2:00 p.m. to 5:00 p.m.
H12	425	Winter	Weekdays, 6:00 p.m. to 10:00 p.m.
H13	1,144	Spring	Weekends, plus 11:00 p.m. to 6:00 a.m. weekdays
H14	455	Spring	Weekdays, 7:00 a.m. to 1:00 p.m.
H15	260	Spring	Weekdays, 2:00 p.m. to 5:00 p.m.
H16	325	Spring	Weekdays, 6:00 p.m. 10:00 p.m.

Figure B-3. National load duration curve for base year in WinDS

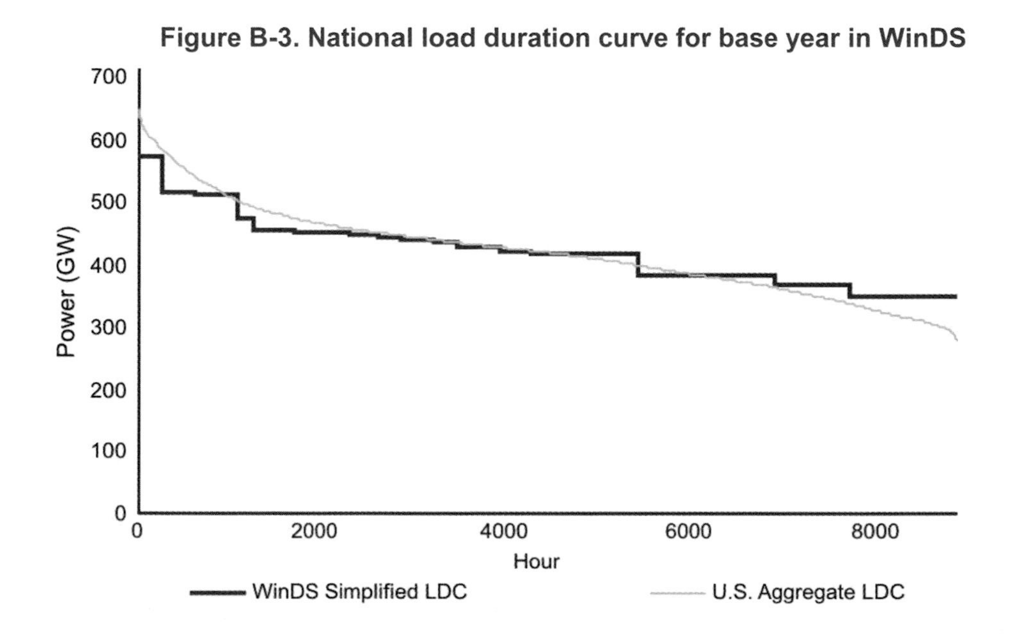

energy requirement. The LDC does not include the "super peak," which occurs in most systems for a few hours per year. These peak requirements are discussed in Section B.2.4.

B.2.3 Growth Rate

Load growth is defined at the NERC region level. Loads in all balancing areas within each NERC region are assumed to grow at the same rate to 2050. Table B-4 contains the 2000 load and annual growth rates for each NERC region.

Table B-4. Base load and load growth in the WinDS scenario

NERC Region/Sub-Region	Abbreviation	2000 Load TWh/year	Annual Load Growth
1	ECAR	370	1.010
2	ERCOT	205	1.016
3	MAAC	197	1.009
4	MAIN	184	1.010
5	MAPP	110	1.011
6	NY	109	1.006
7	NE	96	1.010
8	FL	141	1.022
9	SERC	589	1.015
10	SPP	132	1.013
11	NWP	176	1.017
12	RA	97	1.022
13	CNV	202	1.017

Source: EIA (2007b)

WinDS assumes that the growth rate in each time slice is also constant (i.e., the load shape remains the same over time).

B.2.4 Capacity Requirements

In each balancing area, WinDS requires that firm capacity be available to meet the demand in each time slice (see the national example of time-slice demand in Figure B-3). In addition, for every NERC and interconnect region, WinDS requires sufficient capacity to meet the peak instantaneous demand throughout the course of the year, plus a peak reserve margin. The instantaneous annual peak load is higher than the load in each of the 16 time slices, because the load in each time slice is the average load over the hours included in that time slice. The reserve margin requirement can be met by any generator type, although the generator must have the appropriate capacity value. In the case of wind power, the actual capacity value is a minority fraction of the nameplate capacity. Section B.6 discusses the treatment of resource variability within the model.

Although these capacity requirements are implemented regionally, Table B-5 illustrates their national impact.

Table B-5. National capacity requirements in the WinDS base case

Capacity Requirement	Total (GW)		Annual Growth Rate %
	2000	2050	
Average load in the summer peak time slice	571	1,249	1.6
Annual peak instantaneous load	702	1,531	1.6
Peak capacity value (not nameplate) to meet reserve margin	875	1,730	1.4

Table B-6 gives the peak reserve margin for each region. Reserve margin is ramped from its initial value in 2000 to the 2010 requirement, and maintained thereafter. It is assumed that energy growth and peak demand grow at the same rate, and that the load shape stays constant from one period to the next.

Table B-6. Peak reserve margin

NERC Region	Abbreviation	2010 Required Reserve Margin
1	ECAR	0.12
2	ERCOT	0.15
3	MAAC	0.15
4	MAIN	0.12
5	MAPP	0.12
6	NY	0.18
7	NE	0.15
8	FL	0.15
9	SERC	0.13
10	SPP	0.12
11	NWP	0.08
12	RA	0.14
13	CNV	0.13

Source: PA Consulting Group (2004)

B.3 Wind

B.3.1 Wind Resource Definition

Table B-7 defines wind power classes.

Table B-7. Classes of wind power density

Wind Power Class	Wind Power Density, W/m^2	Speed, m/s
3	300–400	6.4–7.0
4	400–500	7.0–7.5
5	500–600	7.5–8.0
6	600–800	8.0–8.8
7	>800	>8.8

Notes: W/m^2 = watts per square meter; m/s = meters per second. Wind speed measured at 50 m above ground level.
Source: Elliott and Schwartz (1993)

B

Wind power density and speed are not explicitly calculated in WinDS. Different classes of wind power are identified by resource level, CF, turbine cost, and so forth, which are discussed in the subsections that follow.

B.3.2 Wind Resource Data

The basic wind resource input for the WinDS model is the amount of available windy land area (in square kilometers [km^2]) by wind power class (Class 3 and higher). The amount of available windy land is derived from state wind resource maps and modified for environmental and land-use exclusions (as outlined in Tables B-8 and B-9). These maps are the most recent available from the Wind Powering America (WPA) initiative (EERE) and individual state programs. The maps depict estimates of the wind resource at 50 m above the ground.

The WinDS base case (Denholm and Short 2006) used only two data sources, the WPA maps validated by NREL and the *Wind Energy Resource Atlas of the United States* (PNL 1987). For this report, however, the WinDS model uses recent wind maps from individual state programs where available (instead of maps from the 1987 PNL atlas) and new WPA state maps.

Using the recent maps offers an advantage in that modern mapping techniques and recent measurement data are incorporated into the mapping process, resulting in a finer horizontal resolution (1 km or smaller size grid cells) of the wind resource. The disadvantage is that not all updated maps were created using the same technique. The difference in techniques leads to a "patchwork quilt" pattern in some regions. The differences also result in notable resource discontinuities at state borders. For this project, several 50 m state maps were adjusted to produce more interstate compatibility. Table B-8 summarizes the state sources and land-use exclusions for the land-based wind resource data used in WinDS, and Table B-9 presents the same information for offshore wind.

B

Most state maps were completed with direct support from WPA and cost-sharing from individual states and regional partners. Under the WPA initiative, state wind resource maps were produced as described here. The preliminary resource map was produced by AWS Truewind (AWST; Albany, New York). NREL validated this map in cooperation with private consultants who had access to proprietary data, special data, and knowledge of wind resources in each state, or both. The validation results were used to modify the preliminary map and to create a final wind map. NREL mapped three states—Illinois, North Dakota, and South Dakota—before AWST became involved. An important difference between the NREL and AWST maps is that the NREL mapping technique assumed low surface roughness (equivalent to short grasslands); AWST used digital land cover data sets for surface roughness values. Increases in surface roughness generally decreases the estimated 50 m wind resource, so the NREL maps might overestimate the wind resource in areas that do not have low surface roughness. The 50 m wind power classes for individual grid cells on the WPA maps were used to determine available windy land for the WinDS model.

Individual state programs have updated other (non-WPA) maps, which were created using a variety of mapping techniques. NREL has not, however, validated these

Table B-8. Data sources for land-based wind resource and environmental exclusions

Onshore Wind Resource Data Used in WinDS (10/23/2006)

Resource Data (50 m height):

State	Data Source*	State	Data Source*	State	Data Source*
Arizona	2003, N/AWST	Maine	2002, N/AWST	Ohio[a]	2004, N/AWST
Alabama	1987, PNL	Maryland	2003, N/AWST	Oklahoma[a]	2002, OTH
Arkansas	2006, N/AWST**	Massachusetts	2002, N/AWST	Oregon	2002, N/AWST
California	2003, N/AWST	Michigan[a]	2005, N/AWST	Pennsylvania[a]	2003, N/AWST
Colorado	2003, N/AWST	Minnesota	2006, OTH	Rhode Island	2002, N/AWST
Connecticut	2002, N/AWST	Mississippi	1987, PNL	South Carolina	2005, AWST
Delaware	2003, N/AWST	Missouri[a]	2004, N/AWST	South Dakota	2000 NREL
Florida	1987, PNL	Montana	2002, N/AWST	Tennessee	1987, PNL
Georgia	2006, AWST	Nebraska[a]	2005, N/AWST	Texas	2004, OTH/2000, NREL
Idaho	2002, N/AWST	Nevada	2003, N/AWST	Utah	2003, N/AWST
Illinois	2001, NREL	New Hampshire	2002, N/AWST	Vermont	2002, N/AWST
Indiana[a]	2004, N/AWST	New Jersey	2003, N/AWST	Virginia	2003, N/AWST
Iowa	1997, OTH	New Mexico	2003, N/AWST	Washington	2002, N/AWST
Kansas	2004, OTH	New York[a]	2004, AWST	West Virginia	2003, N/AWST
Kentucky	1987, PNL	North Carolina	2003, N/AWST	Wisconsin	2003, OTH
Louisiana	1987, PNL	North Dakota	2000 NREL	Wyoming	2002, N/AWST

* YrSource

Yr = Year produced (1987 to present); Source = PNL, NREL, N/AWST (NREL with AWS TrueWind), AWST (AWS TrueWind alone not validated by NREL) or OTH (data from other sources)

PNL data resolution is 1/4 degree of latitude by 1/3 degree of longitude, each cell has a terrain exposure percent (5% for ridgecrest to 90% for plains) to define base resource area in each cell. Ridgecrest areas have 10% of the area assigned to the next higher power class.

NREL data was generated with the WRAMS model, and does not account for surface roughness. Resolution is 1 km. Texas includes the Texas mesas study area updated by NREL using WRAMS.

N/AWST data was generated by AWS TrueWind and validated by NREL. Resolution is 400 m for the northwest states (WA, OR, ID, MT, and WY) and 200 m everywhere else. These data consider surface roughness in their estimates.

N/AWST** data was generated by AWS TrueWind, and will be validated by NREL. Data used is preliminary.

OTH data from other sources. The methods, resolution, and assumptions vary. These results have not been validated by NREL For most states, the data was taken at face value. However, some datasets were not available as 50 m power density. In those cases, assumptions were made to adjust the data to 50 m power density.

[a] In these states, the class 2, 3 and 4 wind power class estimates were adjusted upwards by 1/2 power class to better represent the likely wind resource at wind turbine height. For Nebraska, only the portion of the state east of 102 degrees longitude was adjusted.

Wind Resource Onshore Exclusions (last revised Jan 2004)

Criteria for Defining Available Windy Land (numbered in the order they are applied):

Environmental Criteria	*Data/Comments:*
2) 100% exclusion of National Park Service and Fish and Wildlife Service managed lands	USGS Federal and Indian Lands shapefile, Jan 2005
3) 100% exclusion of federal lands designated as park, wilderness, wilderness study area, national monument, national battlefield, recreation area, national conservation area, wildlife refuge, wildlife area, wild and scenic river or inventoried roadless area.	USGS Federal and Indian Lands shapefile, Jan 2005
4) 100% exclusion of state and private lands equivalent to criteria 2 and 3, where GIS data is available.	State/GAP land stewardship data management status 1, from Conservation Biology Institute Protected Lands database, 2004
8) 50% exclusion of remaining USDA Forest Service (FS) lands (incl. National Grasslands)***	USGS Federal and Indian Lands shapefile, Jan 2005
9) 50% exclusion of remaining Dept. of Defense lands***	USGS Federal and Indian Lands shapefile, Jan 2005
10) 50% exclusion of state forest land, where GIS data is available***	State/GAP land stewardship data management status 2, from Conservation Biology Institute Protected Lands database, 2004
Land Use Criteria	
5) 100% exclusion of airfields, urban, wetland and water areas.	USGS North America Land Use Land Cover (LULC), version 2.0, 1993; ESRI airports and airfields (2003)
11) 50% exclusion of non-ridgecrest forest***	Ridge-crest areas defined using a terrain definition script, overlaid with USGS LULC data screened for the forest categories.
Other Criteria	
1) Exclude areas of slope > 20%	Derived from elevation data used in the wind resource model.
6) 100% exclude 3 km surrounding criteria 2-5 (except water)	Merged datasets and buffer 3 km
7) Exclude resource areas that do not meet a density of 5 km^2 of class 3 or better resource within the surrounding 100 km^2 area.	Focalsum function of class 3+ areas (not applied to 1987 PNL resource data)

***50% exclusions are not cumulative. If an area is non-ridgecrest forest on FS land, it is just excluded at the 50% level one time.

B

Table B-9. Data sources for offshore wind resource and environmental exclusions

Offshore Wind Resource Data Used in WinDS (10/23/2006)					
Resource Data (50 m height):					
State	Data Source*	State	Data Source*	State	Data Source*
Alabama	2006, NREL3	Maine	2002, NREL1	North Carolina	2003, NREL1
California	2003, NREL1	Maryland	2003, NREL1	Ohio	2006, NREL2
Connecticut	2002, NREL1	Massachusetts	2003, NREL1	Oregon	2002, NREL1
Delaware	2003, NREL1	Michigan	2006, NREL2	Pennsylvania	2006, NREL2
Florida	2006, NREL3	Minnesota	2006, NREL2	Rhode Island	2002, NREL1
Georgia	2006, NREL3	Mississippi	2006, NREL3	South Carolina	2006, NREL3
Illinois	2006, NREL2	New Hampshire	2002, NREL1	Texas	2006, NREL3
Indiana	2006, NREL2	New Jersey	2003, NREL1	Virginia	2003, NREL1
Louisiana	2006, NREL3	New York	2003, NREL1	Washington	2002, NREL1
				Wisconsin	2006, NREL2
* YrSource					
Yr = Year produced (2002 to present); Source = NREL with different methods enumerated below					
NREL1: Validated near-shore data was supplemented with offshore resource data from earlier, preliminary runs which extended further from shore. In most cases, this still did not fill the modeling area of interest of 50 nm from shore. The resource estimates were extended linearly to obtain full coverage at 50 nm with little or no change in spatial pattern.					
NREL2: Similar to NREL1, but available resource data estimates and areas not covered by validated and preliminary data were evaluated by NREL meteorologist to establish a best estimate of resource distribution based on expert knowledge and available measured/modeled data sources.					
NREL3: No validated resource estimates existed to provide a baseline. NREL meteorologists generated an initial best estimate of resource distribution to be used in the model, based on expert knowledge and available measured/modeled data sources.					
Wind Resource Offshore Exclusions					
No exclusions were applied to the offshore resource data. It is characterized by power class and depth (0-30 m and >30m)					

maps, which do not necessarily show the 50 m wind power classes on the maps or the 50 m classes in geographic information system (GIS) format. For two states (Minnesota and Wisconsin) where the 50 m power classes for individual grid cells were unavailable, a methodology that applies basic assumptions to calculate wind power classes for each grid cell was used. This methodology calculates a combination of wind speed at the grid cells (direct or interpolated), extrapolates to adjust the wind speeds from map height(s) to 50 m, plots common wind speed frequency distribution, and takes air density into consideration. Next, environmental and land-use exclusions were applied to arrive at the final windy land area totals.

Updated wind resource maps were unavailable for six southeastern states—Alabama, Florida, Kentucky, Louisiana, Mississippi, and Tennessee. The underlying 50 m wind power class data from the maps contained in the 1987 atlas (PNL 1987) were used to calculate windy land area for these states. The horizontal resolution of the atlas maps is quite a bit larger (approximately 25 km grid cells) than that of the updated state maps, which feature 1 km or smaller grid cells. To compensate for the low resolution, landform classifications and environmental and land use exclusions were used to calculate the available windy land for these states.

As mentioned previously, several state maps were adjusted to produce more interstate compatibility. The Texas map was adjusted to include wind resources currently being developed on the mesas in western Texas. Because the mesas are relatively small terrain features, adequately depicting the available resources on these features is difficult. As a result, the Texas map underestimates the power class on the mesas where considerable wind energy development has taken place. In adjusting the maps, the power class values for the mesas were increased based on anemometer measurements, leading to a more realistic representation of the wind energy available. The maps for eight states—Oklahoma, Missouri, Nebraska (the eastern two-thirds of the state), Indiana, Michigan, Ohio, Pennsylvania, and New York—were adjusted because their 50 m wind power class maps underestimate the potential resource at modern turbine hub heights. The available resource increase results from the high wind speed shear that is present in these states. The available windy land in these states was increased based on the wind power density values of individual grid cells. Grid cells in classes 2, 3, and 4 that had 50 m power density values greater than the midpoint of the associated wind power class were adjusted to the next highest class. The these adjustments increased the estimated amount of land with class 3, 4, and 5 wind resources.

For each of the 358 WinDS regions, the total available land area corresponding to a particular wind resource power class was multiplied by an assumed turbine density of 5 megawatts per square kilometer (MW/km^2). This calculation yields the total wind-generation capacity available within each WinDS region for each wind power class.

The patchwork quilt effect that results from the varied resource input data affects the selection of wind energy capacity in the WinDS model. If a state's resource is underestimated, the WinDS model may select less wind energy capacity than is currently being developed in a given state. Similarly, if a state's resource is overestimated, the actual wind energy capacity could be significantly less than that calculated by the model.

All these resource maps were based on wind power estimates at 50 m above ground level. Today's wind turbines, however, have hub heights as high as 80 m to 100 m. As turbine technology improves and hub heights increase, wind resources could be significantly different. Many states that show poor wind capability for electricity generation at the 50 m level may have significantly improved wind speeds at heights of 80 m to 100 m. As an example, even though Missouri is currently developing several hundred megawatts of wind energy, WinDS does not specify significant wind energy capacity for the state.

B.3.3 WinDS Seasonal and Diurnal Capacity Factor Calculations

For each region and wind power class (classes 3 to 7), 16 time slices represent four seasons and four time periods (see Table B-3). The diurnal and seasonal variations of the wind are portrayed as the ratio of the average wind turbine output during the time slice with the annual average wind turbine output. Average CFs are calculated for each of the 358 WinDS regions for each power class.

Monthly and hourly wind variations were obtained from two databases:

- AWST text supplemental database files
- National Commission on Energy Policy/National Center for Atmospheric Research (NCEP/NCAR) global reanalysis mean values (Kalnay et al. 1996).

For states with AWST data, annual and monthly average wind speeds and power were selected from the fine map grid (400 m resolution in Washington, Oregon, Idaho, Montana, and Wyoming; 200 m resolution in all other states), and hourly wind speed profiles by season from the coarse map grid (10 km in Washington, Oregon, Idaho, Montana, and Wyoming; 2 km in all other states). States with AWST data are identified in Table B-8.

For monthly input data, only one 3 × 3 km cell for each region and power class was used. This cell was chosen because it has the lowest cost, based on the existing grid usage optimization that is normally done as an input to WinDS (Sabeff et al. 2004). The resulting monthly pattern is the average of the monthly values within the 3 × 3 km cell for all map points in the desired power class (plus or minus one class). For hourly input data, the closest grid point from the coarse grid for each 3 × 3 km cell was used. The hourly pattern is the average of hourly values for up to twenty 3 × 3 km cells for each region/power class combination. There are four patterns, one for each season. Seasons are three-month periods (March–May, June–August, September–November, and December–February).

For states without AWST data and for certain offshore regions, NCEP/NCAR reanalysis data were used. Reanalysis uses a dynamic data assimilation model to create worldwide data sets of wind, temperature, and other variables on a 208 km resolution grid, four times daily, throughout the depth of the atmosphere. Average values of wind speed, wind power, and air density were used, by month and by day (four times daily), over a 46-year period of record. Reanalysis wind characteristics from 120 m above ground level have been found to have the best correlation with measured wind data and wind maps. Reanalysis data, however, is suitable for use only over fairly level terrain at lower elevations. Fortunately, AWST data is available for most states that are not suitable for reanalysis.

For regions that use reanalysis, the reanalysis grid point closest to the geographic center of the region was chosen. For some offshore locations, the center of the offshore region was computed and the closest reanalysis grid point was used.

Using the AWST and NCEP/NCAR databases, input data sources were used to populate matrices of average wind speed, wind power, and air density by month and hour of day (24 hours × 12 months). The 24 × 12 array of wind speed, wind power, and air density was then divided into desired seasonal and diurnal time slices (see Table B-3). For each time slice, the power output of the General Electric International (GE) 1.5 MW wind turbine as a function of air density was estimated, and a histogram of wind speed probability as a function of wind speed and Weibull k factor was calculated.

The data was then combined to calculate the wind turbine CF for each time slice. In the AWST data, wind power is available only by month, so the Weibull k factor was calculated only once for each season. All times of day use the same Weibull k for calculating CF. Finally, a weighted average of CFs from the four time slices was used to revise nighttime values into a "nights and weekends" capacity factor. Time-slice CFs were then normalized by the total annual CF, resulting in values representing the ratio of power produced in the current time slice to annual average power produced. This is the desired input into the WinDS model.

This process creates a desired array of CF ratios only for regions and wind power classes with data. With reanalysis, each region has data from only one power class. A final data processing step is to populate the entire array of 358 regions × 5 power classes with results. If a power class is missing, data from the next-lower power class are chosen. If there are no available data from a lower power class, the next-higher power class is chosen. For reanalysis regions, all five power classes are given the same array of CF ratios.

B.3.4 Wind Technology Cost and Performance

Black & Veatch analysts (in consultation with AWEA industry experts) developed wind technology cost and performance projections for this report (Black & Veatch, forthcoming 2008). Costs for turbines, towers, foundations, installation, profit, and interconnection fees are included. Capital costs are based on an average installed capital cost of $1,775 per kilowatt (kW) in 2007. After adjusting for inflation and removing the construction financing charge, this reduces to $1,650/kW for 2006. Additional costs reflecting terrain slope and regional population density are described later in this subsection.

Technology development is projected to reduce future capital costs by 10%.Black & Veatch used historical capacity factor data to create a logarithmic best-fit line, which is then applied to each wind power class to project future performance improvements.[17] Black & Veatch's experience indicate that variable and fixed operations and maintenance (O&M) costs represent an average of recent project costs. Approximately 50% of variable O&M cost is the turbine warranty. These costs are expected to decline as turbine reliability improves and the scale of wind turbines increases. Other variable O&M expenses are tied to labor rates, royalties, and other costs that are expected to be stable. Fixed O&M costs, including insurance, property taxes, site maintenance, and legal fees, are projected to stay the same because they are not affected by technology improvements. Table B-10 lists cost and performance projections for land-based wind systems (Black & Veatch 2007).

Table B-11 lists cost and performance projections prepared by Black & Veatch for shallow offshore wind technology (in water shallower than 30 m). Capital costs for 2005 were based on publicly available cost data for European offshore wind farms. Capital costs are assumed to decline 12.5% as a result of technology development and a maturing market. The capacity factor projection, which is based on the logarithmic best-fit lines generated for land-based turbines, we increased 15% to account for larger rotor diameters and reduced wind turbulence over the ocean. By 2030 this adjustment factor is reduced to 5% as land-based development allows larger turbines to be used in turbulent environments. O&M costs are assumed to be three times those of land-based turbines (Musial and Butterfield 2004) with a learning rate commensurate to that projected by the U.S. Department of Energy (DOE; NREL 2006).

A number of adjustments, including financing, interest during construction, terrain slope, population density, and rapid growth were applied to the capital cost. Although financing has not been treated explicitly, it is assumed to be captured by the weighted cost of capital (real discount rate) of 8.5%.

A slope penalty that increases one-fourth of the capital cost by 2.5% per degree of terrain slope was used to represent expected costs associated with installations on mesas or ridge crests. Costs associated with installation represent 25% of the capital cost. Wiser and Bolinger (2007) present regional variations in installed capital cost for projects constructed in 2006. Applying a multiplier related to population density within each of the WinDS regions results in regional variations similar to the observed data. An additional 20% must be applied to the base capital cost in New

[17]Capacity factors for 2000 and 2005 fit to actual data. For the higher wind power classes (6 and 7), however, limited data are available for operating plants, so capacity factors were extrapolated from the linear relationships between wind classes.

B

Table B-10. Land-based wind technology cost and performance projections
(US$2006)

Wind Resource Power Class at 50 m	Year Installed	Capacity Factor (%)	Cost ($/kW)	Fixed O&M ($/kW-yr)	Variable O&M ($/MWh)
3	2005	32	1,650	11.5	7.0
3	2010	35	1,650	11.5	5.5
3	2015	36	1,610	11.5	5.0
3	2020	38	1,570	11.5	4.6
3	2025	38	1,530	11.5	4.5
3	2030	38	1,480	11.5	4.4
4	2005	36	1,650	11.5	7.0
4	2010	39	1,650	11.5	5.5
4	2015	41	1,610	11.5	5.0
4	2020	42	1,570	11.5	4.6
4	2025	43	1,530	11.5	4.5
4	2030	43	1,480	11.5	4.4
5	2005	40	1,650	11.5	7.0
5	2010	43	1,650	11.5	5.5
5	2015	44	1,610	11.5	5.0
5	2020	45	1,570	11.5	4.6
5	2025	46	1,530	11.5	4.5
5	2030	46	1,480	11.5	4.4
6	2005	44	1,650	11.5	7.0
6	2010	46	1,650	11.5	5.5
6	2015	47	1,610	11.5	5.0
6	2020	48	1,570	11.5	4.6
6	2025	49	1,530	11.5	4.5
6	2030	49	1,480	11.5	4.4
7	2005	47	1,650	11.5	7.0
7	2010	50	1,650	11.5	5.5
7	2015	51	1,610	11.5	5.0
7	2020	52	1,570	11.5	4.6
7	2025	52	1,530	11.5	4.5
7	2030	53	1,480	11.5	4.4

Note: MWh = megawatt-hour
Source: Black & Veatch (2007)

England to reflect observed capital cost variations. Slope and population density penalties have been applied to the capital cost listed in Tables B-10 and B-11 within the model to represent topographical and regional variations across the United States.

If the demand for new wind capacity significantly exceeds the amount supplied in the previous year, WinDS assumes that the price paid per unit of wind capacity can rise above the capital costs of Tables B-10 and B-11 as well as the multiplier factors.. In particular, installing more than 20% new wind generation over the preceding year, will increase capital costs by 1% for each 1% growth above 20% per year (EIA 2004).

Table B-11. Shallow offshore wind technology cost and performance projections (US$2006)

Wind Resource Power Class at 50 m	Year Installed	Capacity Factor (%)	Capital Cost ($/kW)	Fixed O&M ($/kW-yr)	Variable O&M ($/MWh)
3	2005	34	2,400	15	21
3	2010	37	2,300	15	18
3	2015	38	2,200	15	16
3	2020	39	2,150	15	14
3	2025	40	2,130	15	13
3	2030	40	2,100	15	11
4	2005	38	2,400	15	21
4	2010	41	2,300	15	18
4	2015	43	2,200	15	16
4	2020	44	2,150	15	14
4	2025	45	2,130	15	13
4	2030	45	2,100	15	11
5	2005	42	2,400	15	21
5	2010	45	2,300	15	18
5	2015	46	2,200	15	16
5	2020	47	2,150	15	14
5	2025	48	2,130	15	13
5	2030	48	2,100	15	11
6	2005	46	2,400	15	21
6	2010	48	2,300	15	18
6	2015	50	2,200	15	16
6	2020	51	2,150	15	14
6	2025	51	2,130	15	13
6	2030	51	2,100	15	11
7	2005	50	2,400	15	21
7	2010	52	2,300	15	18
7	2015	54	2,200	15	16
7	2020	55	2,150	15	14
7	2025	55	2,130	15	13
7	2030	55	2,100	15	11

Source: Black & Veatch (2007)

B.4 Conventional Generation

U.S. conventional energy generation included in the WinDS model, and most likely to be built in the United States, has been included in EIA's data reports (2007). Table B-12 illustrates expected construction time and schedules for conventional energy technologies.

WinDS considers outage rates when determining the net capacity available for energy (as described in Section 2), and also when determining the capacity value of each technology. Planned outages are assumed to occur in all seasons except summer. Table B-12 shows outage rates for each conventional technology.

B

Table B-12. General assumptions for conventional generation technologies

Technology Modeled	Capability for new builds in WinDS	Construction Time (years) (1)	Construction Schedule (2) Fraction of Cost in Each Year						Forced Outage Rate (%) (3)	Planned Outage Rate (%) (3)	Emissions Rates (4) (lbs/MMBTU fuel input)				Lifetime (years)
			1	2	3	4	5	6			SO_2	NO_x	Hg	CO_2	
Conventional Hydropower - Hydraulic Turbine	No	NA	-	-	-	-	-	-	2.0%	5.0%	0	0	0	0	100
Natural Gas Combustion Turbine	Yes	3	0.8	0.1	0.1	-	-	-	10.7%	6.4%	0.0006	0.08	0	33.2877	30
Combined Cycle Natural Gas Turbine	Yes	3	0.5	0.4	0.1	-	-	-	5.0%	7.0%	0.0006	0.02	0	33.2877	30
Conventional Pulverized Coal Steam Plant (No SO_2 Scrubber)	No-Scrubbers may be added to meet SO_2 constraints. Existing plants may also switch to low-sulfur coal.	6	0.1	0.2	0.2	0.2	0.2	0.1	7.9%	9.8%	0.2355	0.448	4.6E-06	55.77131	60
Conventional Pulverized Coal Steam Plant (With SO_2 scrubber)	No-see above	6	0.1	0.2	0.2	0.2	0.2	0.1	7.9%	9.8%	1.57	0.448	4.6E-06	55.77131	60
Advanced Supercritical Coal Steam Plant (with SO_2 and No_x Controls)	Yes	4	0.4	0.3	0.2	0.1	-	-	7.9%	9.8%	0.157	0.02	4.6E-06	55.77131	60
Integrated Coal Gasification Combined Cycle Turbine	Yes	4	0.4	0.3	0.2	0.1	-	-	7.9%	9.8%	0.0184	0.02	4.6E-06	55.77131	60
Oil/Gas Steam Turbine	No - Assumes Gas-CT or Gas-CC will be built instead.	NA	-	-	-	-	-	-	7.9%	9.8%	0.026	0.1	0	33.2877	50
Nuclear	Yes	6	0.1	0.2	0.2	0.2	0.2	0.1	5.0%	5.0%	0	0	0	0	30
Geothermal	No	NA	-	-	-	-	-	-	5.0%	5.0%	0	0	0	0	20
Biomass (as Thermal Steam Generator)	No	NA	-	-	-	-	-	-	5.0%	5.0%	0	0	0	0	45
Concentrating Solar Power with Storage	Yes	3	0.5	0.4	0.1	-	-	-	35.0%	5.0%	0.00015	0.02	0	8.321926	30
Municipal Solid Waste / Landfill Gas	No	NA	-	-	-	-	-	-	5.0%	5.0%	0	0	0	0	30

Emission rates are estimated in Table B-12 for SO_2, NO_x, mercury, and CO_2 and provides input-specific emission rates (in pounds per million British thermal units) for plants that use combustible fuel. Output emission rates (in pounds per megawatt-hour) are calculated by multiplying input emission rate by heat rate.

B.4.1 Conventional Generation Cost and Performance

Table B-13 also gives capital cost values, heat rates (efficiency), and fixed and variable O&M costs for conventional technologies that might be added to the electric system. Cost and performance values for natural gas, nuclear, and coal technologies are based on recent project costs according to Black & Veatch experience. Pulverized coal plants continue to operate in WinDS, and SO_2 scrubbers can be added to unscrubbed coal plants for $200/kW. Oil, gas, steam, and unscrubbed coal plants cannot be added to the electric system, but those currently in operation are maintained until retired. WinDS sites conventional generation technology where it is least expensive (generally adjacent to load centers) and does not require new transmission. California is the exception because its legislative requirements prohibit siting new coal plants.

Capital costs for 2005, 2010, and 2015 are based on proposed engineering, procurement, and construction (EPC) estimates for plants that will be commissioned in 2010, 2015, and 2020. A wet scrubber is included in the EPC costs for new pulverized coal plants. Owners' costs of 20% for coal, nuclear, and combined-cycle gas plants and 10% for simple-cycle gas plants provide an "all-in" cost. These owners' costs are based on national averages and include transmission and interconnection, land, permitting, and other costs. As with wind systems, an additional 20% of the capital costs listed in Table B-13 is applied to coal and nuclear generation technology in New England, representing siting difficulties.

B.4.2 Fuel Prices

Fuel prices for natural gas and coal are derived from reference projections from the AEO (EIA 2007b). These tables provide the prices in each census region, which are then assigned to a NERC subregion in WinDS. Prices in the AEO are projected to 2030. Beyond 2030, WinDS projects that fuel prices will increase at the same national annual average rate as the AEO's 2030 projection.

Figure B-4 illustrates the projected fossil fuel prices in constant $US2005. The 20% Wind Scenario uses the reference AEO fuel price forecast for coal because government agencies and the private sector regularly use that forecast to make planning and investment decisions. The New York Mercantile Exchange futures prices for natural gas for May 2007 through 2012 exceed the AEO's high fuel price forecast over that period. Also, under the current set of technology cost and performance assumptions, the WinDS model tends to select natural gas-fueled technology over coal-fueled technology. To provide a conservative estimate while representing a more traditional mix of conventional generation technology, the AEO high natural gas price forecast has been implemented.

The price of uranium fuel in WinDS is constant at $0.5/MMBtu (Denholm and Short 2006).

Table B-13. Cost and performance characteristics for conventional generation (US$2006)

	Install Date	Capital Cost ($/kW)	Fixed O&M ($/MW/yr)	Variable O&M ($/MWh)	Heat Rate (Btu/kWh)
Gas CT	2005	625	7,700	12.0	11,560
	2010	750	6,600	2.8	8,900
	2015	750	6,600	2.8	8,900
	2020	750	6,600	2.8	8,900
	2030	750	6,600	2.8	8,900
Gas-CC	2005	780	14,400	3.0	6,870
	2010	780	14,400	3.0	6,870
	2015	780	14,400	3.0	6,870
	2020	780	14,400	3.0	6,870
	2030	780	14,400	3.0	6,870
New Coal (SC)	2005	2,120	35,300	1.7	9,470
	2010	2,180	35,300	1.7	9,200
	2015	2,240	35,300	1.7	9,100
	2020	2,240	35,300	1.7	9,000
	2030	2,240	35,300	1.7	9,000
Coal - IGCC	2005	2,750	38,100	3.9	9,000
	2010	2,840	38,100	3.9	9,000
	2015	2,840	38,100	3.9	8,900
	2020	2,840	38,100	3.9	8,800
	2030	2,840	38,100	3.9	8,580
Nuclear	2005	3,260	90,000	0.5	10,400
	2010	3,170	90,000	0.5	10,400
	2015	3,020	90,000	0.5	10,400
	2020	2,940	90,000	0.5	10,400
	2030	2,350	90,000	0.5	10,400

Notes: New nuclear plants may not be constructed before 2010. O&M costs do not include fuel. Heat rate is net heat rate (including internal plant loads).

Source: Black & Veatch 2007

B

Figure B-4. Projected coal and natural gas prices in WinDS to 2030

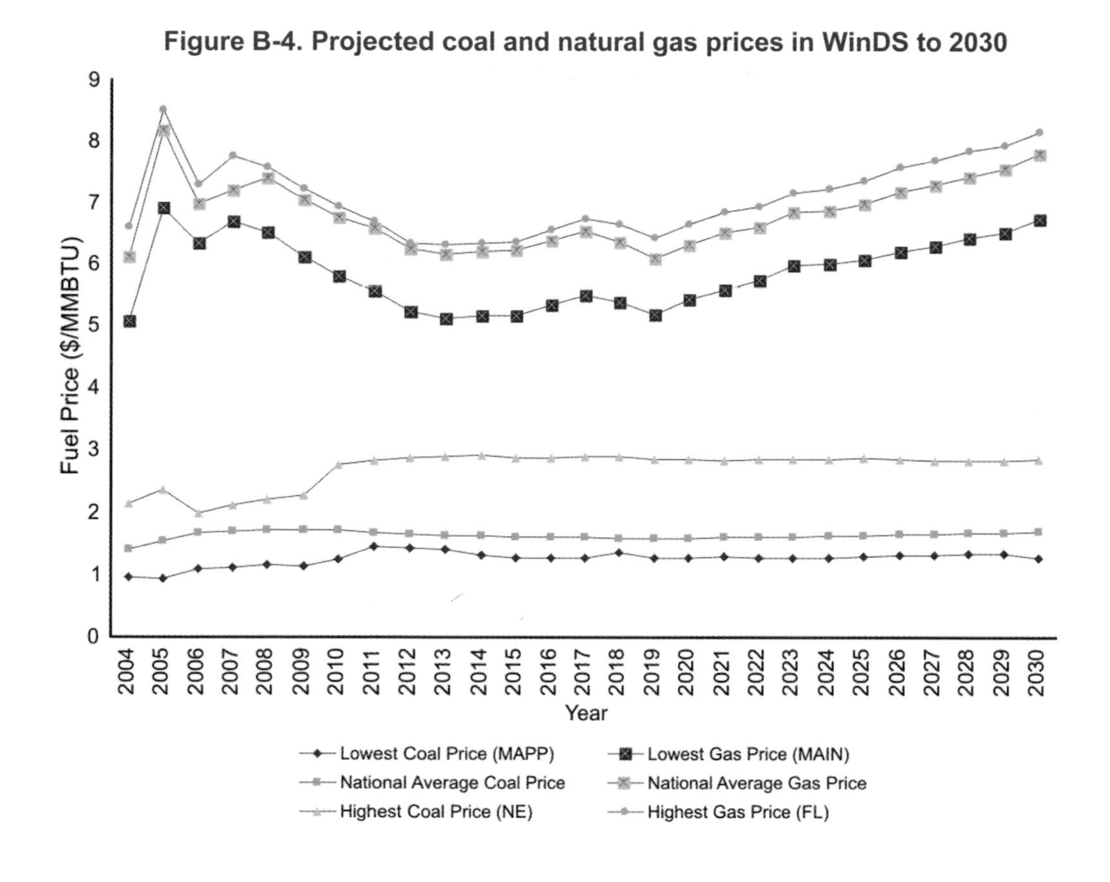

B.5 Transmission

Three types of transmission systems can be used to transport wind power around the country:

- **Existing grid:** It is assumed that 10% of the existing grid can be used for new wind capacity, either by improving the grid or by tapping existing unused capacity. A GIS optimization determines the distance at which a particular wind farm will have to be built to connect to the grid (based on the assumption that the closest wind installation will access the grid first at the least cost). In this way, a supply curve of costs to access the grid is created for each class of wind in each region. Additionally, the model assumes a pancake-type fee may be charged for crossing between balancing areas . The supply curves described earlier are based on this type of transmission and the GIS optimization described here. In the near term, one can expect that most wind will be built and will use the existing grid without needing to build excessive amounts of new transmission lines, but as higher penetration levels are reached, the existing grid will be insufficient.

  Existing transmission capacity is estimated using a database of existing lines (length and voltage) from RDI/Platts (Platts Energy Market Data; see http://www.platts.com). This database is translated into a megawatt capacity as a function of kilovolt (kV) rating and length (Weiss and Spiewak 1998).

- **New lines:** The model has the ability to build straight-line transmission lines between any of the 358 wind regions. The line is built exactly to the size necessary to transmit the desired megawatts and the cost of building that transmission line is accounted for in the model.

  AWEA experts indicate that new transmission line capacity might be constructed for any generation technology for an average cost of $1,600/MW-mile. Based on input from the AWEA expert panel, regional transmission cost variations include an additional 40% in New England and New York; 30% in PJM East (New Jersey and Delaware); 20% in PJM West (Maryland, West Virginia, Pennsylvania, Ohio, parts of Illinois, Indiana, and Virginia); and 20% in California.

  The WinDS model assumes that 50% of the cost of new transmission is borne by the generation technology for which the new transmission is being built (wind or conventional); the other half is borne by the ratepayers within a region (because of the reliability benefits to all users associated with new transmission). This 50–50 allocation, which is common in the industry, was recently adopted for the 15-state Midwest Independent Transmission System Operator (Midwest ISO) region. New wind transmission lines that carry power across the main interconnects are not cost-shared with other technology. In the WinDS model, this sharing of costs is implied by reducing the cost of new transmission associated with a particular capacity by 50%. This means that the relative costs of transmission and capacity capital are in line with the model's assumption. The remaining 50% of transmission costs are integrated into the final cost value outputs from the model, resulting in accurate total transmission costs.

- **In-region transmission:** Within any of the 358 wind regions, the model can build directly from a wind resource location to a load within the same region. A second GIS-generated supply curve is used within the model to assign a cost for this transmission.

A fourth type of transmission, used predominantly by conventional capacity and called general transmission, can be built as well. This is limited because conventional capacity can generally be built in the region where it is needed, thereby obviating the need for new transmission.

WinDS uses a transmission loss rate of 0.236 kW/MW-mile. This value is based on the loss estimates for a typical transmission circuit (Weiss and Spiewak 1998). The assumed typical line is a 200-mile, 230-kV line rated at 170 megavolt amperes (MVA; line characteristics derived from EPRI [1983]).

To emulate large regional planning structures based on that of the Midwest ISO, there is essentially no wheeling fee between balancing areas used in this analysis (although the model has the capability to model such a fee). The wind penetration is limited to 25% energy in each of the three interconnects: Western, Eastern, and ERCOT.

B.6 Treatment of Resource Variability

The variability of wind resources can impact the electrical grid in several ways. One useful way to examine these impacts is to categorize them in terms of time, ranging from multiyear planning issues to small instantaneous fluctuations in output.

At the longest time interval, a utility's capacity expansion plans might call for the construction of more nameplate generation capacity. To meet this need, planners can plan to build conventional dispatchable capacity or wind. The variability of wind output precludes the planners from considering 1 MW of nameplate wind capacity to be the same as 1 MW of nameplate dispatchable capacity. The wind capacity cannot be counted on to be available when electricity demand is at its peak. Actually, conventional capacity cannot be considered 100% available, either. The difference is in the degree of availability. Conventional generators are available 80% to 98% of the time. However, wind energy is available at varying levels that average about 30% to 45% of the time, depending on the quality of the wind site. For planning purposes, this lack of availability can be handled in the same way—a statistical treatment that calculates how much more load can be added to the system for each megawatt of additional nameplate wind or conventional capacity or effective load carrying capability (ELCC).

Wind's ELCC is less than that of conventional capacity because (1) the wind availability is less conventional fuel availability and (2) at any given instant, energy output from a new wind farm can be heavily correlated with the output from existing wind farms. In other words, if the wind is not blowing at one wind site, there is a reasonable chance that it is not blowing at another nearby site. On the other hand, there is essentially no correlation between the outputs of any two conventional generation plants.

Fortunately, there are ways to partly mitigate both the low availability of the wind resource and its correlation between sites. In the past 20 years, the capacity factors of new wind installations have improved considerably. This is attributable to better site exploration and characterization and to improvements in the wind turbines (largely higher towers).

The correlation in wind output between sites can also be reduced. Increasing the distance between sites and the terrain features that separate them reduces the chance that two sites will experience the same wind at the same time. Figure B-5 shows this correlation as a function of distance between sites in an east–west direction and in a north–south direction (Simonsen and Stevens 2004). With its multiple regions, WinDS is able to approximate the distance between sites and, therefore, the correlation between their outputs. WinDS uses the correlation between sites to estimate the variation in wind output from the total set of wind farms supplying power to a particular region.

Between each two-year optimization period and for each demand region, WinDS updates its estimate of the marginal ELCC associated with adding wind of each resource class in each wind supply region to meet demand within a NERC region. This marginal ELCC is a strong function of the wind capacity factor and the distance from the existing wind systems to the new wind site. It is also a weak function of the demand region's LDC and the size and forced outage rates of conventional capacity. This marginal ELCC is assumed to be the capacity value of each megawatt of that

Figure B-5. Distance between wind sites and correlation with power output

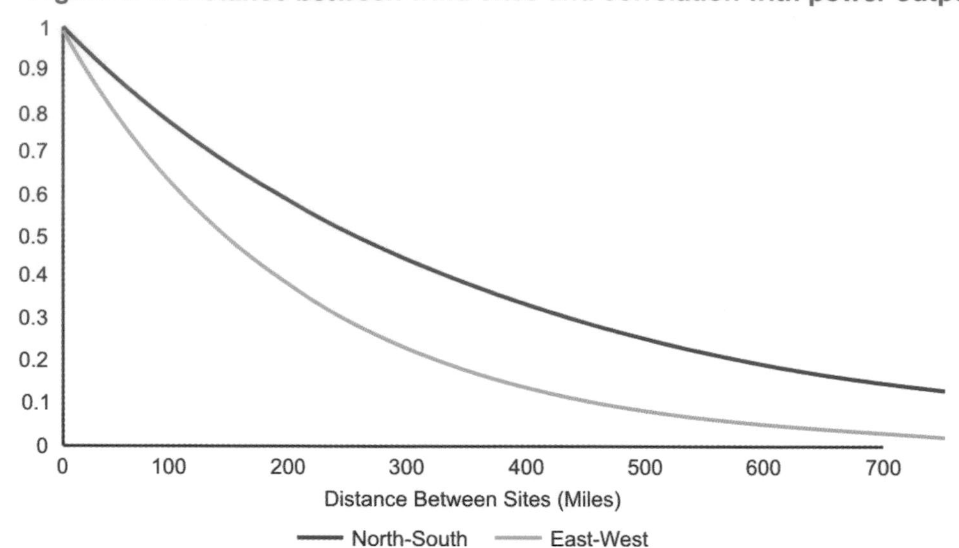

wind class added in the next period in that wind supply region to serve the NERC region's demand.

All other factors being equal, when expanding wind capacity, WinDS will select the next site in a region that is as far from the existing sites as possible to ensure the lowest correlation and the highest ELCC for the next wind site. (From a practical standpoint, all factors are never "equal," and WinDS considers the trade-offs between ELCC and wind site quality, transmission availability and cost, and local siting costs.)

Generally, for the first wind site supplying a demand region, these capacity values (ELCCs) are almost equal to the peak season capacity factor. As the wind penetrates to higher levels, though, the ELCC can decline to almost zero in an individual wind supply region.

The next time frame of major interest is the day ahead. Utilities generally make decisions on which generating units to commit to generation the day before they are actually committed. To comply with these unit-commitment procedures, independent power plant owners can be expected to bid for firm capacity a day ahead. This can be problematic for wind generator owners. For example, if the wind owner bids to provide firm capacity and the wind does not blow as forecast, the owner may have to make up the difference by purchasing power on the real-time market. If the purchased power costs more per kilowatt-hour than the owner is being paid for the day-ahead bid, the owner will lose money.

Not all of today's electric grid systems operate day-ahead and real-time markets. California, for example, allows a monthly balancing of bid and actual wind generation that is much more tolerant of the inaccuracies in forecasting wind a day ahead of time. In all cases, however, the imbalances can be offset with adequate operating reserves. To capture the essence of the unit-commitment issue, WinDS estimates the impact of wind variability on the need for operating reserves (which include quick-start and spinning reserves) that can rapidly respond to changes in wind output. The operating reserves are assumed to be a linear function of the variance in the sum of generation (both wind and conventional) minus load. Because the variability of wind is statistically independent of load variability and forced outages, the total variance can be calculated as the sum of the variance associated with the normal (i.e., no wind) operating reserve and the total variance (over all the wind supply regions) in the wind output over the reconciliation period.

Before each two-year optimization, WinDS calculates the marginal operating reserve additions required by the next unit of wind added in a particular wind supply region from a particular wind class. The resulting value is the difference between the operating reserve required by the total system with the new wind and the operating reserve required by the total system if there were no new wind installations in that region. This value is then used throughout the next two-year linear program optimization as the marginal operating reserve requirement induced by the next megawatt of wind addition in that region of that wind resource class.

In the shortest time interval, regulation reserves must compensate for instantaneous changes in wind output. Regulation reserves are normally provided by automatic generation control of conventional generators whose output can be automatically adjusted to compensate for small voltage changes on the grid. Fortunately, these instantaneous changes in wind output do not all occur at the same time, even from wind turbines within the same wind farm. This lack of correlation over time and the ease with which conventional generators can respond allows this second-order cost to be reasonably ignored.

WinDS assumes that the wind generated energy delivered to a specific demand region in a specific time slice in excess of the total load for that region/time slice will be lost. In addition, WinDS also statistically accounts for surplus wind lost within a time slice because of variations in load and wind within the time slice.

WinDS includes three options for mitigating the impact of resource variability. The first option is to add conventional generators that can provide spinning reserve (e.g., gas-CC) and quick-start capabilities (combustion turbines). The second, and usually least costly, option is to allow the dispersion of new wind installations to reduce the correlation of the outputs from different wind sites. Finally, the model can allow for storage of electricity at the wind site, which is usually the most costly option. The storage option was not available within this analysis and is currently being developed for the model.

B

B.7 Federal and State Energy Policy

The WinDS accounts for all currently enacted federal and state emission standards, renewable portfolio standards (RPS), and tax credits.

B.7.1 Federal Emission Standards

WinDS provides the ability to add a national cap on **CO_2** emissions from electricity production. WinDS can also account for a tax for CO_2 emissions. However, neither a carbon cap nor a tax is implemented in the 20% Wind Scenario.

Emissions of **SO_2** are capped at the national level. WinDS uses a cap that corresponds roughly to the 2005 Clean Air Interstate Rule (CAIR), replacing the previous limits established by the 1990 Clean Air Act Amendments (CAAA). The CAIR rule divides the United States into two regions. WinDS uses the U.S. Environmental Protection Agency's (EPA) estimate of the effective national cap on

SO_2 resulting from the CAIR rule (EPA 2005). Table B-14 shows the SO_2 cap used in WinDS.

Table B-14. National SO_2 emission limit schedule in WinDS

Year	2003	2010	2015	2020	2030
National SO_2 Emissions (Million Tons)	10.6	6.1	5.0	4.3	3.5

(EPA 2005)

WinDS currently allows unrestrained **NO_x** emissions. . The NO_x cap from CAIR can be added, but the net effect on the overall competitiveness of coal is expected to be relatively small (EIA 2003).

WinDS currently allows unrestrained **Mercury** emissions. The Clean Air Mercury Rule (see http://www.epa.gov/camr/index.htm) is a cap and trade regulation, which is expected to be met largely by the CAIR requirements. Control technologies for SO_2 and NO_x that are required for CAIR are expected to capture enough mercury to largely meet the cap goals. As a result, the incremental cost of mercury regulations is very low and is not modeled in WinDS (EIA 2003).

B.7.2 Federal Energy Incentives

Several classes of incentives have been applied to wind systems at the federal level. These incentives generally have the effect of reducing the cost of producing energy from renewable sources. A production tax credit (PTC) offsets the tax liability of companies based on the amount of energy produced. This analysis assumes that the current PTC will be available for wind through 2008 (see Table B-15).

B

Table B-15. Federal renewable energy incentives

Name	Value	Notes and Source
Renewable Energy PTC	\$19/MWh	Applies to wind. No limit to the aggregated amount of incentive. Value is adjusted for inflation to US\$2006. Expires end of 2008.

(U.S. Congress 2005)

B.7.3 State Energy Incentives

Several states also offer production and investment incentives for renewable energy resource development. Table B-16 lists the values used in WinDS. However, in the 20% Wind Scenario these incentives are overwhelmed by the specification of wind energy generation in each year through 2030.

Table B-16. State renewable energy incentives

State	PTC $/ MWh	ITC	Assumed State Corporate Tax Rate
Iowa		5.00%	10.0%
Idaho		5.00%	7.60%
Minnesota		6.50%	9.8%
New Jersey		6.00%	9.0%
New Mexico	10		7.0%
Oklahoma	2.5		6.0%
Utah		4.75%	5.0%
Washington		6.50%	0.0%
Wyoming		4.00%	0.0%

Investment and production tax credit data from IREC 2006
Tax rates from: www.taxadmin.org/fta/rate/corp_inc.html

B.7.4 State Renewable Portfolio Standards

A number of states have developed Renewable Portfolio Standards (RPS), and states can put capacity mandates in place as an alternative or supplement to an RPS (see Table B-17). A capacity mandate requires a utility to install a certain fixed capacity of renewable energy generation. Unless prohibited by law, a state might also meet requirements by importing electricity.

Table B-17. State RPS requirements as of August 2005

State	RPS Start Year[2]	RPS Full Imple-mentation[3]	Penalty in $/MWh	WinDS Assumed RPS Fraction[4]	Legislated RPS Fraction (%)	Load Fraction[5]
Arizona	2001	2025	50	0.0079	1.1	1
California	2003	2017	5	0.034	20	0.63
Colorado	2007	2015	50	0.044	10	0.69
Connecticut	2004	2010	55	0.013	10	0.94
Delaware	2007	2019	25	0.056	10	0.75
Illinois	2004	2013	10	0.062	15	0.92
Massachusetts	2003	2009	50	0.026	4	0.85
Maryland	2006	2019	20	0.045	7.5	0.8
Minnesota	2002	2015	10	0.072	1,125 MW	1
Montana	2008	2015	10	0.075	15	0.9
New Jersey	2005	2008	50	0.029	6.5	1
New Mexico	2006	2011	10	0.026	10	0.53
Nevada	2003	2015	10	0.133	20	0.89
New York	2006	2013	5	0.035	25	0.84
Oklahoma	2005	2016	50	0.05	See Note 6	1
Oregon	2002	2020	5	0.078	See Note 6	1
Pennsylvania	2007	2020	45	0.014	8	0.98
Rhode Island	2007	2019	55	0.069	15	0.99
Texas	2003	2015	50	0.01	5,880 MW	1
Vermont	2005	2012	10	0.05	See Note 6	1
Wisconsin	2001	2011	10	0.006	2.2	0.75

Notes:
1) RPS data as of 8/16/05. Source: IREC 2006.

2) RPS Start Year is the "beginning" of the RPS program. The RPS is ramped linearly to the full implementation year.

3) RPS Full Implementation is the year that the full RPS fraction must be met. WinDS assumes the fraction met is ramped up linearly between the start year and the full implementation year.

4) WinDS Assumed RPS Fraction is the fraction of state demand that must be met by wind by the full implementation year. This value is based on the total state RPS requirement and adjusted to estimate the fraction actually provided by wind since WinDS does not currently include other renewables such as biomass cofiring and certain hydro projects.

5) Load fraction is the fraction of the total state load that must meet the RPS. In certain locations, municipal or cooperative power systems may be exempt from the RPS.

6) Several states have special funds set aside to promote renewables. The net increase in wind due to these funds was estimated and applied as an effective RPS.

B.8 Electricity Sector Direct Cost Calculation

The objective of the electricity sector direct cost calculation is to determine the difference in system-wide costs where 20% wind penetration is required compared to the case where no new wind generation is installed after 2006. The goal was to estimate the cost per kilowatt-hour of wind produced and the cost per kilowatt-hour of the total load met. The resulting numbers for both scenarios are reported in Appendix A.

To gather necessary costs from the WinDS model, it was programmed to calculate costs incurred in each year of the simulation from 2008 through 2030 for both cases (with and without wind). These costs are then broken into subgroups, including wind capital costs; conventional energy capital costs; wind and conventional transmission build costs (including the full transmission cost, not just the portion shared by each generator); and conventional fuel costs.

Because the impacts of reduced fuel demand and wind turbines installed in the years immediately preceding 2030 are not evident until after 2030, the cost impacts beyond 2030 are estimated. To arrive at the estimate, the model assumes that wind generation would linearly decay from 2030 to 2050 and that the conventional fuel and O&M savings would also linearly decay to 0 from 2030 to 2050. This is a conservative approach because it assumes that the wind farms are retired linearly.

Finally, all costs (including the approximated costs after 2030) are discounted back to 2006. The WinDS model is run with an 8.5% real weighted cost of capital to represent a typical utility perspective. In evaluating a policy such as an RPS, a social discount rate of 7% should be used in accordance with Office of Management and Budget guidelines (OMB 1992). This lower rate effectively places higher (higher than a utility's 8.5% discount rate) value on benefits and costs encountered further in the future. The total cost difference then becomes the difference in the present value of the two cost streams. To find the cost per kilowatt-hour (levelized cost) of wind produced, the total cost difference is levelized to satisfy the following formula:

$\sum$ wind generation$_t$ * LC /$(1+d)^t$ = PV of costs in 20% case – PV of costs in no wind case

As a second result, to find the cost per kilowatt-hour of total generation, replace wind generation with total generation in the preceding formula. The complete equation to calculate the present value of costs used in the preceding equation is as follows:

$$PV_{Costs} = a + b + c$$

$$a = \sum_{t=2006}^{2030} ((CapCostNewCapacity_t + CapCostNewTransmission_t + O\&MCost_t + FuelCost_t) / (1 + d)^{(t-2006)})$$

$$b = \sum_{t=2031}^{2050} (WindO\&MCostsCapBuiltBy2030_t / (1 + d)^{(t-2006)})$$

$$c = \sum_{t=2031}^{2050} (((ConvO\&M_{2030} + Fuel_{Cost,2030}) FractionNotRetiredWind) / (1 + d^{(t-2006)}))$$

where

FractionNotRetiredWind = Fraction of wind generation remaining from wind capacity installed prior to 2031 in the 20% wind case

B.9 References & Suggested Further Reading

Black & Veatch. 2007. *20 % Wind Energy Penetration in the United States: A Technical Analysis of the Energy Resource*. Walnut Creek, CA.

Denholm, P., and W. Short. 2006. *Documentation of WinDS Base Case.* Version AEO 2006 (1). Golden, CO: National Renewable Energy Laboratory (NREL). http://www.nrel.gov/analysis/winds/pdfs/winds_data.pdf.

EERE (U.S. DOE Office of Energy Efficiency and Renewable Energy). Wind and Hydropower Technologies Program: Wind Powering America Web site. http://www.eere.energy.gov/windandhydro/windpoweringamerica/.

EIA (Energy Information Administration). 2000. Cross Reference of States to Federal Regions, NERC Regions, and Census Divisions. Washington, DC: EIA. http://www.eia.doe.gov/cneaf/electricity/ipp/html1/tb5p01.html.

EIA. 2002. *Upgrading Transmission Capacity for Wholesale Electric Power Trade.* Washington, DC: EIA. http://www.eia.doe.gov/cneaf/pubs_html/feat_trans_capacity/w_sale.html.

EIA. 2003. Analysis of S. 485, the Clear Skies Act of 2003, and S. 843, the Clean Air Planning Act of 2003. SR/OIAF2003-03(2003). Washington, DC: EIA.

EIA. 2004a. *The Electricity Market Module of the National Energy Modeling System; Model Documentation Report.* DOE/EIA-M068(2004). Washington, DC: EIA.

EIA. 2004b. *Analysis of Senate Amendment 2028, the Climate Stewardship Act of 2003.* SR/OIAF/2004-06. Washington, DC: EIA.

B

EIA. 2006a. *Assumptions to the Annual Energy Outlook 2006 with Projections to 2030.* . Washington, DC: EIA. http://www.eia.doe.gov/oiaf/archive/aeo06/assumption/index.html

EIA. 2006b. *Supplemental Tables to the Annual Energy Outlook 2006*. Washington, DC: EIA. http://www.eia.doe.gov/oiaf/archive/aeo06/supplement/index.html.

EIA. 2006c. *Energy and Economic Impacts of H.R.5049, the Keep America Competitive Global Warming Policy Act.* SR/OIAF/2006-03.Washington, DC: EIA.

EIA. 2007a. *Annual Energy Outlook 2007 with Projections to 2030.* Washington, DC: EIA. Report No. DOE/EIA-0383. http://www.eia.doe.gov/oiaf/archive/aeo07/index.html

EIA. 2007b. *Supplemental Tables to the Annual Energy Outlook.* Tables 60 through 72. Washington, DC: EIA. http://www.eia.doe.gov/oiaf/aeo/supplement/index.html.

EIA. 2007c. *Assumptions to the Annual Energy Outlook 2007 with Projections to 2030.* Washington, DC: EIA. http://www.eia.doe.gov/oiaf/aeo/assumption/index.html

EIA. 2007d. *Energy Market and Economic Impacts of a Proposal to Reduce Greenhouse Gas Intensity with a Cap and Trade System.* SR/OIAF/2007-01. Washington, DC: EIA. http://www.epa.gov/cleanenergy/egrid/index.htm.

Elliott, D.L., and M.N. Schwartz. 1993. *Wind Energy Potential in the United States.* PNL-SA-23109. NTIS No. DE94001667. Richland, WA: Pacific Northwest Laboratory (PNL).

EPA (U.S. Environmental Protection Agency). 1996. "Compilation of Air Pollutant Emission Factors, AP-42." In *Volume I: Stationary Point and Area Sources.* Fifth edition. Washington, DC: EPA. http://www.epa.gov/ttn/chief/ap42/.

EPA. 2005a. *Clean Air Interstate Rule, Charts and Tables.* Washington, DC: EPA. http://www.epa.gov/cair/charts_files/cair_emissions_costs.pdf.

EPA. 2005b. eGRID Emissions & Generation Resource Integrated Database. Washington, DC: EPA Office of Atmospheric Programs.

EPRI (Electric Power Research Institute). 1983. *Transmission Line Reference Book, 345-kV and Above*. Second edition. Palo Alto, CA: EPRI.

IREC (Interstate Renewable Energy Council). 2006. Database of State Incentives for Renewable & Efficiency (DSIRE). http://www.dsireusa.org/.

Kalnay, E., M. Kanamitsu, R. Kistler, W. Collins, D. Deaven, L. Gandin, M. Iredell, et al. 1996. "The NCEP/NCAR 40-Year Reanalysis Project." *Bulletin of the American Meteorological Society*. 77, 437–471. http://ams.allenpress.com/perlserv/?request=res-loc&uri=urn%3Aap%3Apdf%3Adoi%3A10.1175%2F1520-0477%281996%29077%3C0437%3ATNYRP%3E2.0.CO%3B2.

McDonald, A., and L. Schrattenholzer. 2001. "Learning Rates for Energy Technologies." *Energy Policy* 29, 255–261.

Musial, W., and S. Butterfield. 2004. *Future for Offshore Wind Energy in the United States*. NREL/CP-500-36313. Golden, CO: NREL.

B

NREL. 2006. National Wind Technology Center Website, "About the Program," http://www.nrel.gov/wind/uppermidwestanalysis.html.

NREL. 2006. *Projected Benefits of Federal Energy Efficiency and Renewable Energy Programs-FY2007 Budget Request.* NREL/TP-320-39684. Golden, CO: NREL. http://www1.eere.energy.gov/ba/pdfs/39684_00.pdf.

OMB (Office of Management and Budget). 1992. *Guidelines and Discount Rates for Benefit-Cost Analysis of Federal Programs.* Circular A-94.Washington, DC: OMB. http://www.whitehouse.gov/OMB/circulars/index.html.

PA Consulting Group. 2004. "Reserve Margin Data." *Energy Observer* 2, July.

PNL. 1987. *Wind Energy Resource Atlas of the United States.* DOE/CH 10093-4. Richland, WA: PNL.

Sabeff, L., R. George, D. Heimiller, and A. Milbrandt. 2004. *Regional Data and GIS Representation: Methods, Approaches & Issues. Presentation to Scoping Workshop for GIS Regionalization for EERE Models.* http://www.nrel.gov/analysis/workshops/pdfs/brady_gis_workshop.pdf.

Simonsen, T., and B. Stevens. 2004. *Regional Wind Energy Analysis for the Central United States.* Grand Forks, ND: Energy and Environmental Research Center. http://www.undeerc.org/wind/literature/Regional_Wind.pdf.

U.S. Congress. 2005. *Domenici-Barton Energy Policy Act of 2005.* Washington, DC: 109th Congress.

Weiss, L., and S. Spiewak. 1998. *The Wheeling and Transmission Manual.* Lilburn,GA: The Fairmont Press Inc.

Wiser, R., and M. Bolinger. 2007. *Annual Report on U.S. Wind Power Installation, Cost, and Performance Trends: 2006.* DOE/GO-102007-2433. Golden, CO: NREL. http://www.osti.gov/bridge.

Notes: Many of the assumptions about conventional generation and fuel prices are drawn from the EIA's National Energy Modeling System. This information is published in the AEO, which consists of three documents: the main AEO (which focuses on results); the supplemental tables (which contain additional details on results at the regional level); and the assumptions (which presents input details). Several sources for emissions data are available from the EPA, including the AP-42 series of documents. Detailed emissions estimates for different combustion technologies and emissions controls can be found in the AP-42 series. The eGRID database estimates emissions rates from existing plants, based on measured fuel use and continuous emissions monitoring system data measurement.

B

B

Appendix C. Wind-Related Jobs and Economic Development

This appendix details the economic model used to project the employment and economic development impacts of the 20% Wind Scenario described in Appendix A. Ramping up wind capacity and electricity output from wind would displace jobs and economic activity elsewhere. However, identifying such transfers accurately would be very difficult. Therefore, the impacts cited here do not constitute impacts to the U.S. economy overall but are specific to the wind industry and related industries. The impacts were calculated using the Jobs and Economic Development Impacts (JEDI) model, based in part on data from the Wind Deployment System (WinDS) model (developed by the National Renewable Energy Laboratory [NREL]). Appendix A summarizes the WinDS modeled scenario, and specific assumptions are described in Appendix B. Cost and performance projections for this analysis were supplied by Black & Veatch (Black & Veatch 2007) and are detailed in Appendix B.

The 20% Wind Scenario was constructed by specifying annual wind energy generation for every year from 2007 to 2030. The specifications were based on a trajectory proposed in an NREL study (Laxson, Hand, and Blair 2006). The NREL study forced the WinDS model to reach the 20% level for wind-generated electricity by 2030. The investigators evaluated aggressive near-term growth rates followed by sustainable levels of wind capacity installations that would maintain electricity generation levels at 20% and accommodate the repowering of aging wind installations beyond 2030. The 20% wind by 2030 trajectory was implemented in WinDS by calculating the percentage of annual energy production from wind at an increase of approximately 1% per year. Figure C-1 illustrates the energy generation trajectory proposed by the NREL study with the corresponding annual wind capacity installations that the WinDS model projects will meet these energy-generation percentages.

The combined cost, technology, and operational assumptions in the WinDS model show that an annual installation rate of about 16 gigawatts per year (GW/year) reached by 2018 could result in generation capacity capable of supplying 20% of the nation's electricity demand by 2030. This annual installation rate is affected by the quality of wind resources selected for development as well as future wind turbine performance. The declining annual installed capacity after 2024 is an artifact of the prescribed energy generation from the NREL study, which did not consider technology improvement and wind resource variation. The NREL study provides an upper level of about 20 GW/year, because turbine performance is unchanged over time and only one wind resource power class was assumed. Based on the wind resource data and the projected wind technology improvements presented in this report, sustaining a level of annual installations at approximately 16 GW/year beyond 2030 would accommodate the repowering of aging wind turbine equipment along with increased electricity demand, so that the nation's energy demand would

Figure C-1. Prescribed annual wind technology generation as a percentage of national electricity demand from Laxson, Hand, and Blair (2006) and corresponding annual wind capacity installation for 20% Wind Scenario from WinDS model.

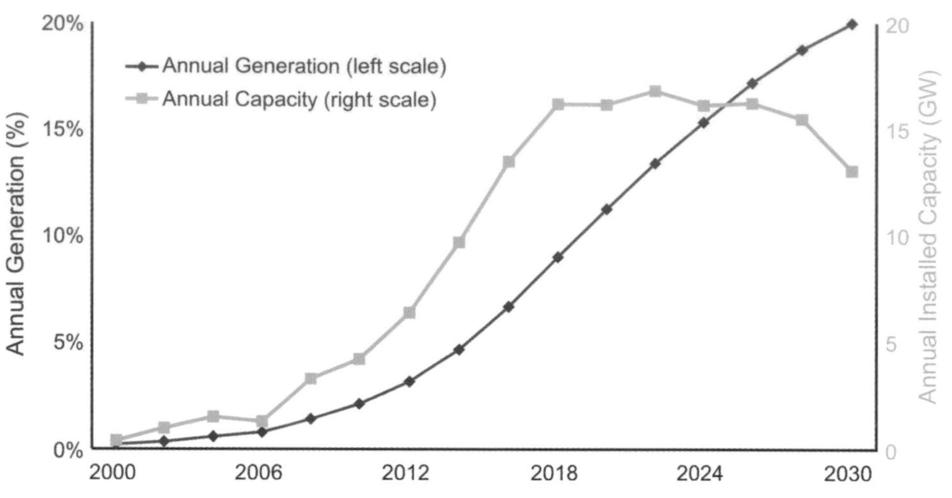

continue to be met by 20% wind. This installation level could maintain energy production of 20% of the nation's demand. Additionally, this scenario shows that this level of wind development could accommodate the repowering of aging wind turbine equipment. Specific policy incentives necessary for this growth, such as a production tax credit (PTC) or carbon regulation policy, are not modeled.

To obtain 20% of U.S. electricity from wind by 2030, changes in the wind power and electricity industries would need to be made. These changes, which are discussed in the body of this report, include advances in domestic manufacturing of wind turbine components; training, labor, and materials for installation of wind farms and operations and maintenance (O&M) functions; and improvements in wind technology and electric power system infrastructure. This appendix covers the output from the JEDI model, which shows the potential employment impacts from this scenario along with other impacts to the United States associated with new wind installations.

C

C.1 The JEDI Model

C.1.1 Model Description

The JEDI model was developed in 2002 for NREL to demonstrate the state and local economic development impacts associated with developing wind power plants in the United States. These impacts include employment numbers created in the wind power sector, and the increase in overall economic activity associated with the construction and operating phases of new wind power. The JEDI spreadsheet-based model for wind is free and available to the public. It can be downloaded from the Wind Powering America website: www.windpoweringamerica.gov. Documentation is listed on the same site. For questions, please contact Marshall Goldberg at mrgassociates@earthlink.net or Suzanne Tegen at suzanne_tegen@nrel.gov.

JEDI was initially designed to estimate economic impacts to state economies. Subsequent enhancements made the model capable of performing county, regional, and national analyses as well. This particular analysis focuses primarily on economic impacts for the United States as a whole, although some state and regional results are presented.

To calculate economic impacts, the model relies on investment and expenditure data from the 20% Wind Scenario for the period between 2007 and 2030. The model also uses industry multipliers that trace supply linkages in the economy. For example, the analysis shows how wind turbine purchases benefit not only turbine manufacturers, but also the fabricated metal industries and other businesses that supply inputs (goods and services) to those manufacturers.

The model evaluates three separate impacts for each expenditure: direct, indirect, and induced.

- **Direct impacts** are the on-site or immediate effects created by spending money for a new wind project. In the JEDI model, the construction phase includes the on-site jobs of the contractors and crews hired to construct the plant as well as their managers and staffs. Direct impacts also include jobs at the manufacturing plants that build the turbines as well as the jobs at the factories that produce the towers and blades.[18]
- **Indirect impacts** refer to the increase in economic activity that occurs, for example, when a contractor, vendor, or manufacturer receives payment for goods or services and in turn is able to pay others who support their business. This includes the banker who finances the contractor and the accountant who keeps the contractor's books, as well as the steel mills, electrical part manufacturers, and suppliers of other necessary materials and services.
- **Induced impacts** are the changes in wealth that result from spending by people directly and indirectly employed by the project. For example, when plant workers and other local workers receive income from expenditures related to the plant, they in turn purchase food, clothing, and other goods and services from local business.

The sum of these three impacts is the total impact from the turbine's construction. Figure C-2 illustrates this ripple effect, from direct impacts to induced impacts. This figure excludes the impacts on other energy sectors as wind power displaces other sources of energy.

JEDI relies on U.S.-specific multipliers and personal expenditure patterns. These multipliers—for patterns of employment, wage and salary income, output (economic activity), and personal spending (expenditure)—are adapted from the IMPLAN Professional Software model (Minnesota IMPLAN Group, Inc., Stillwater, Minnesota; see http://www.implan.com). The IMPLAN® model is based on U.S. industry and census data. Spending from new investments (e.g., purchases of equipment and services) to construct and operate wind plants is matched with the appropriate multipliers for each industry sector (e.g., construction, electrical

[18] When an impact analysis is conducted in this manner, the definitions of *direct* and *indirect* are changed somewhat. Typically, the change in final demand to an industry (in this instance the wind industry) is seen as the direct effect. In the JEDI model, the direct effect includes what are usually called first-round indirect effects (e.g., demand to manufacturers and other goods and service suppliers). The JEDI indirect effects are all subsequent rounds of the industry indirect effects.

Figure C-2. Wind's economic ripple effect

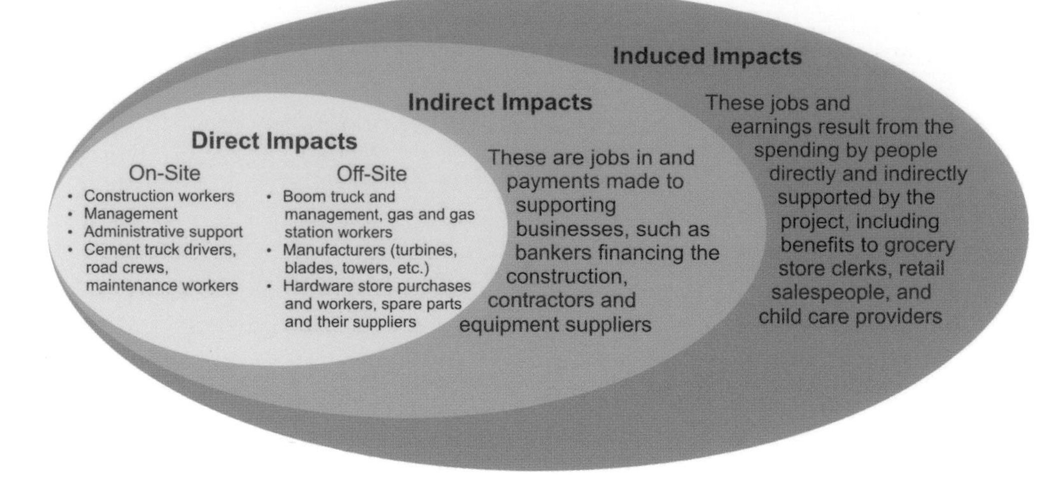

equipment, machinery, professional services, and others) affected by the change in expenditure.

Outputs from the JEDI model are reported for two distinct phases: the construction phase and the annual operations phase. The construction period outputs represent the entire construction period (typically one year for a utility-scale wind project, although this can vary depending on the size of the project). The outputs for the operating period represent the jobs and economic impacts created for one year of operation.

C.1.2 Caveats

Before noting the specific economic impacts from the 20% Wind Scenario, it is important to underscore several caveats about the JEDI model.

First, the model is considered static. As such, it relies on inter-industry relationships and personal consumption patterns at the time of the analysis. The model does not account for feedback through demand, increases, or reductions that could result from price changes. Similarly, the model does not account for feedback from inflationary pressures or potential constraints on local labor and money supplies. In addition, the model assumes that adequate local resources and production and service capabilities are available to meet the level of local demand identified in the model's assumptions. For new power plants, the model does not automatically take into account improvements in industry productivity over time, changes during construction, or changes in O&M processes (e.g., production recipe for labor, materials, and service cost ratios). To adjust for advancements in technology or changes in wages and salaries, the model is run with new cost assumptions (e.g., once with a construction cost of $1,650/kW and again with a construction cost – excluding construction financing - of $1,610/kW).

Second, the intent of using the JEDI model is to construct a reasonable profile of investments (e.g., wind power plant construction and operating costs) to demonstrate the economic impacts that will likely result during the construction and operating periods. Given the potential for future changes in wind power plant costs beyond those identified, and potential changes in industry and personal consumption patterns in the economy noted earlier, the analysis is not intended to provide a

C

precise forecast, but rather an estimate of overall economic impacts in the wind energy sector from specific scenarios.

Third, because the analysis and results are specific to developing new land-based and offshore wind power plants only, this is considered a gross analysis. The results do not reflect the net impacts of construction or operation of other types of electricity-generating power plants or replacement of existing power generation resources to meet growing needs.

Fourth, the analysis assumes that the output from the wind power plants and the specific terms of the power purchase agreements generate sufficient revenues to accommodate the equity and debt repayment and annual operating expenditures.

And finally, the analysis period is 2007 through 2030; additional impacts beyond these years are not considered.

C.2 Wind Scenario Inputs

To assess the economic development from the addition of 293 GW of wind technology in the United States, the authors relied on inputs from the WinDS model. The detailed cost and performance projections can be found in Appendix B of this report.

Table C-1 summarizes the wind data assumptions used in the JEDI model. The cost data are allocated into expenditure categories. Each category includes the portion of the expenditure that goes to the local area, which in this case is the entire United States.

Table C-1. JEDI wind modeling assumptions

Category	Land-Based	Shallow Offshore	Total
Period of Analysis	2007-2030	2007-2030	
Nameplate Capacity	239.5 GW	53.9 GW	293.4 GW
Number of Turbines	79,130	17,976	97,106
Turbine Size	1500–5000 kW	3000 kW	
Technology Cost[1] per kW			
2007	$1650	$2400	
2010	$1650	$2300	
2015	$1610	$2200	
2020	$1570	$2150	
2025	$1530	$2130	
2030	$1480	$2100	
O&M Costs			
Fixed[2]	$11.50/kW	$15.00/kW	
Variable[3]			
2004	$7.00/MWh	$21.00/MWh	
2010	$5.50/MWh	$18.00/MWh	
2015	$5.00/MWh	$16.00/MWh	
2020	$4.60/MWh	$14.00/MWh	
2025	$4.50/MWh	$13.00/MWh	

C

Category	Land-Based	Shallow Offshore	Total
2030	$4.40/MWh	$11.00/MWh	
U.S. Spending			
Labor	100%	100%	
Materials and Services	100%	100%	
Equipment (Manufacturing Transition)[4]			
Major Components			
Blades	50% in 2007 to 80% in 2030		
Towers	26% in 2007 to 50% in 2030		
Machine Heads	20% in 2007 to 42% in 2030		
Sub-Components	10% in 2007 to 30% in 2030		

Notes: 1. All dollar values are 2006 dollars. Technology costs exclude construction financing costs and regional cost variations that result from increased population density, elevation, or other considerations that are included in the WinDS model. Thus, the cumulative investment costs presented in this study are lower than those presented in Appendix A. 2. Fixed costs include land lease cost. 3. Variable costs include property taxes. 4. Refers to U.S. manufacturing/assembly for turbines, blades and towers. For purposes of this modeling, the transition (percentage of U.S. manufacturing/assembly) is assumed to occur at an average annual rate over the 24-year period.

As explained earlier, the JEDI model uses project expenditures—or spending—for salaries, services, and materials to calculate the total economic impacts. Table C-2 summarizes the expenditure data used in the analysis.

Table C-2. Wind plant expenditure data summary (in millions)

Category	Onshore	Offshore	All Wind
Total Cumulative Construction Cost (2007-2030)	$379,343	$115,790	$495,133
Domestic Spending	$200,192	$94,690	$294,882
Total Annual Operational Expenses in 2030 (300 GW)	$63,618	$20,765	$84,383
Direct O&M Costs	$4,394	$2,861	$7,255
Other Annual Costs	$59,224	$17,904	$77,128
Property Taxes	$1,533	$345	$1,877
Land Lease	$639	$144	$783

Notes: All dollar values are 2006 dollars. All dollars represent millions of dollars. Though some of the money spent during construction leaves the country, all O&M spending is domestic.

C.3 Findings

As Table C-3 indicates, developing 293 GW of new land-based and offshore wind technologies from 2007 to 2030 could have significant economic impacts for the entire United States. Cumulative economic activity from the construction phase alone will reach more than $944 billion for direct, indirect, and induced activity in the nation. This level of economic activity stimulates an annual average of more than 250,000 workers required for employment in the wind power and related sectors from 2007 forward. Of these average annual positions, the wind industry supports 70,000 full-time workers in construction-related sectors, including more than 47,000 full-time workers directly in construction and 22,000 workers in manufacturing. As noted earlier, this estimate does not take into account the offsetting effects on employment in other energy sectors.

Table C-3. U.S. construction-related economic impacts from 20% wind

Average Annual Impacts	Jobs	Earnings	Output	
Direct Impacts	72,946	$5,221	$12,217	
Construction Sector Only	47,020	$3,547		
Manufacturing Sector Only	22,346	$1,446		
Other Industry Sectors	3,580	$228		
Indirect Impacts	66,035	$3,008	$11,377	
Induced Impacts	119,774	$4,483	$15,749	
Total Impacts (Direct, Indirect, Induced)	258,755	$12,712	$39,343	
Total Construction Impacts 2007-2030	**Jobs**	**Earnings**	**Output**	**NPV of Output**
Direct Impacts	1,750,706	$125,305	$293,197	$111,153
Construction Sector Only	1,128,479	$85,129		
Manufacturing Sector Only	536,305	$34,706		
Other Industry Sectors	85,922	$5,471		
Indirect Impacts	1,584,842	$72,197	$273,057	$103,541
Induced Impacts	2,874,582	$107,591	$377,984	$143,367
Total Impacts (Direct, Indirect, Induced)	6,210,129	$305,093	$944,238	$358,061

Note: All dollar values are millions of 2006 dollars. Average annual Jobs are full-time equivalent for each year of the construction period. Cumulative jobs are total full-time equivalent for the 24-year construction period from 2007 through 2030. The NPV column shows the net present value of the output column with a discount rate of 7%, per guidance from the Office of Management and Budget.

Under this scenario, the wind industry would produce 305 GW/year. By 2020, the economic activity generated from annual operations of the wind turbines would exceed $27 billion/year. The number of wind plant workers alone would grow to more than 28,000/year, and total wind-related employment would exceed 215,000 workers (see Table C-4).

Table C-4. U.S. operations-related economic impacts from 20% wind

Operation of 300 GW in 2030	Jobs	Earnings	Output	
Direct Impacts	76,667	$3,643	$8,356	
Plant Workers Only	28,557	$1,617		
Nonplant Workers	48,110	$2,026		
Indirect Impacts	37,785	$1,624	$5,642	
Induced Impacts	102,126	$3,822	$13,429	
Total Impacts (Direct, Indirect, Induced)	216,578	$9,090	$27,427	

C

Total Operation Impacts 2007-2030	Jobs	Earnings	Output	NPV of Output
Direct Impacts	1,163,297	$55,907	$122,463	$26,072
Property Tax			$1,877	$760
Land Lease			$783	$317
Other Direct Impacts			$119,804	$24,996
Plant Workers Only	482,578	$27,458		
Nonplant Workers	680,719	$28,449		
Indirect Impacts	561,107	$24,118	$84,008	$17,674
Induced Impacts	1,591,623	$59,572	$209,286	$42,569
Total Impacts (Direct, Indirect, Induced)	3,316,027	$139,596	$415,757	$86,315

Note: All dollar values are millions of 2006 dollars. Operation jobs in 2030 are full-time equivalent for operation of the 305 GW fleet existing in 2030. Cumulative jobs are total full-time equivalent for the 24-year construction period from 2007 through 2030. The NPV column shows the net present value of the output column with a discount rate of 7%, per guidance from the Office of Management and Budget.

Figure C-3 shows the economic impacts from direct, indirect, and induced impacts .

Figure C-3. Annual direct, indirect and induced economic impacts from 20% scenario

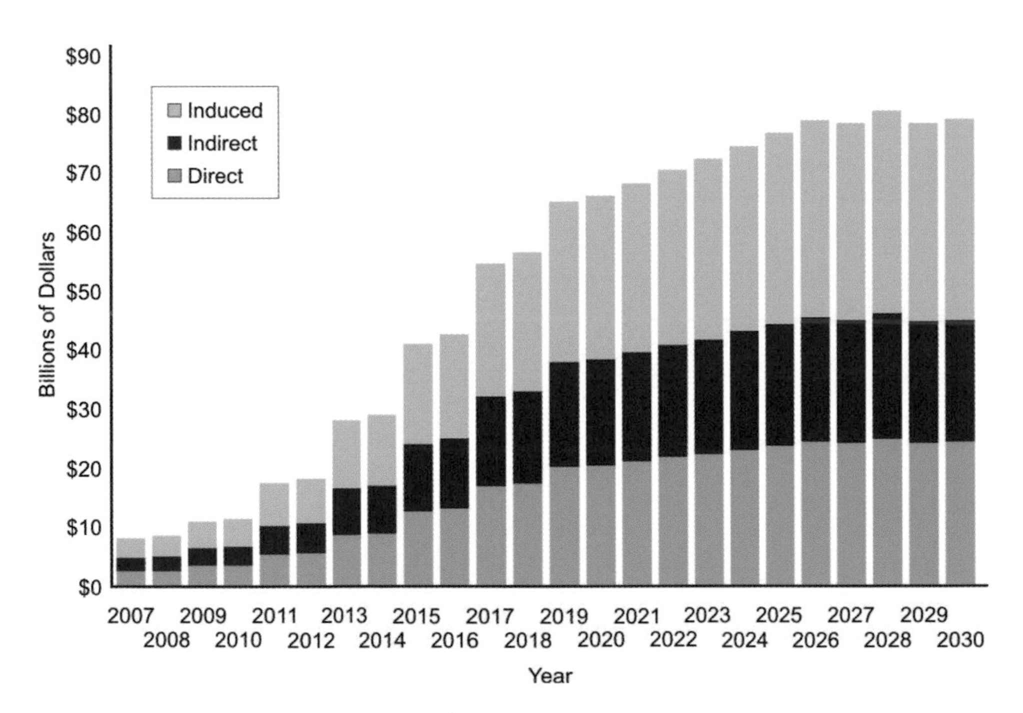

Figure C-4 displays the total economic impacts on a relative basis. The impacts of both the construction and the operation phases are included for the entire period from 2007 through 2030.

The 20% Wind Scenario shows the U.S. wind industry growing from its current 3 GW/year in 2007 to a sustained 16 GW/year by around 2018. In the following sections, employment impacts in the wind industry are divided into three major industry sectors: manufacturing, construction, and operations. Each sector is

Figure C-4. Total economic impacts of 20% wind energy by 2030 on a relative basis

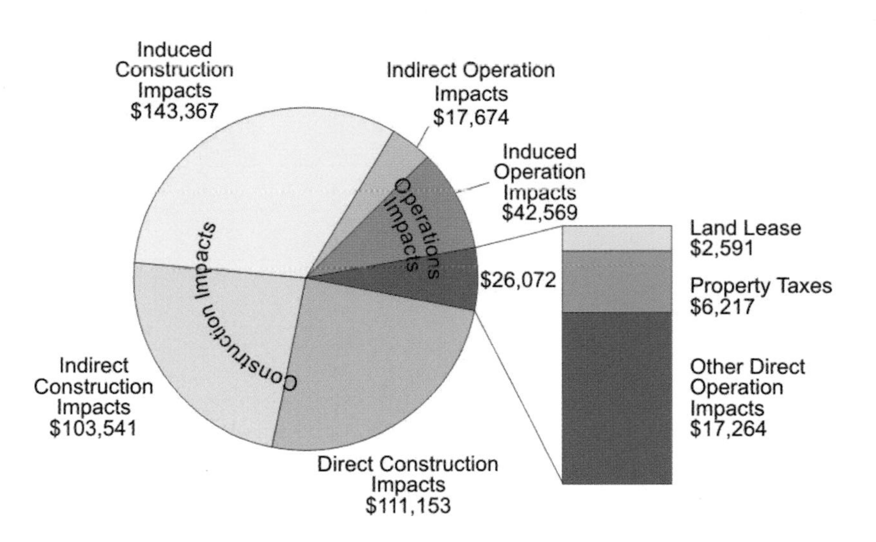

described during the year of its maximum employment supported by the wind industry.

The JEDI model estimates the number of jobs supported by one project throughout the economy, as well as the total economic output from the project. Results from the JEDI model do not include macroeconomic effects. Instead, the model focuses on jobs and impacts supported by specific wind projects. In other words, the employment estimates from the JEDI model look only at gross economic impacts from this 20% Wind Scenario.

C.4 Manufacturing Sector

The 20% Wind Scenario includes the prospect of significantly expanding wind power manufacturing capabilities in the United States. In 2026, this level of wind development supports more than 32,000 U.S. manufacturing full-time workers, including land-based and offshore wind projects. These employment impacts are directly related to producing the major components and subcomponents for the turbines, towers, and blades installed in the United States. Although the level of domestic wind installations declines after 2021 in the scenario modeled, the manufacturing and construction industries have the potential to maintain a high level of employment and expand further to meet increasing global demand.

To estimate the potential location for manufacturing jobs, data from a non-governmental organization, Renewable Energy Policy Project (REPP), report were used (Sterzinger and Svrcek 2004). The REPP report identified existing U.S. companies with the technical potential to enter the wind turbine market. The map in Figure C-5 was created using the percentages of manufacturing capability in each state and JEDI's manufacturing jobs output. Again, these potential manufacturing jobs from the REPP report are based on technical potential existing in 2004, without assuming increased productivity or expansion over time. The data also assumes that existing facilities that manufacture components similar to wind turbine components are modified. Most of the manufacturing jobs in this scenario are located in the Great Lakes region, where manufacturing jobs are currently being lost. Even states

C

Figure C-5. Potential manufacturing jobs created by 2030

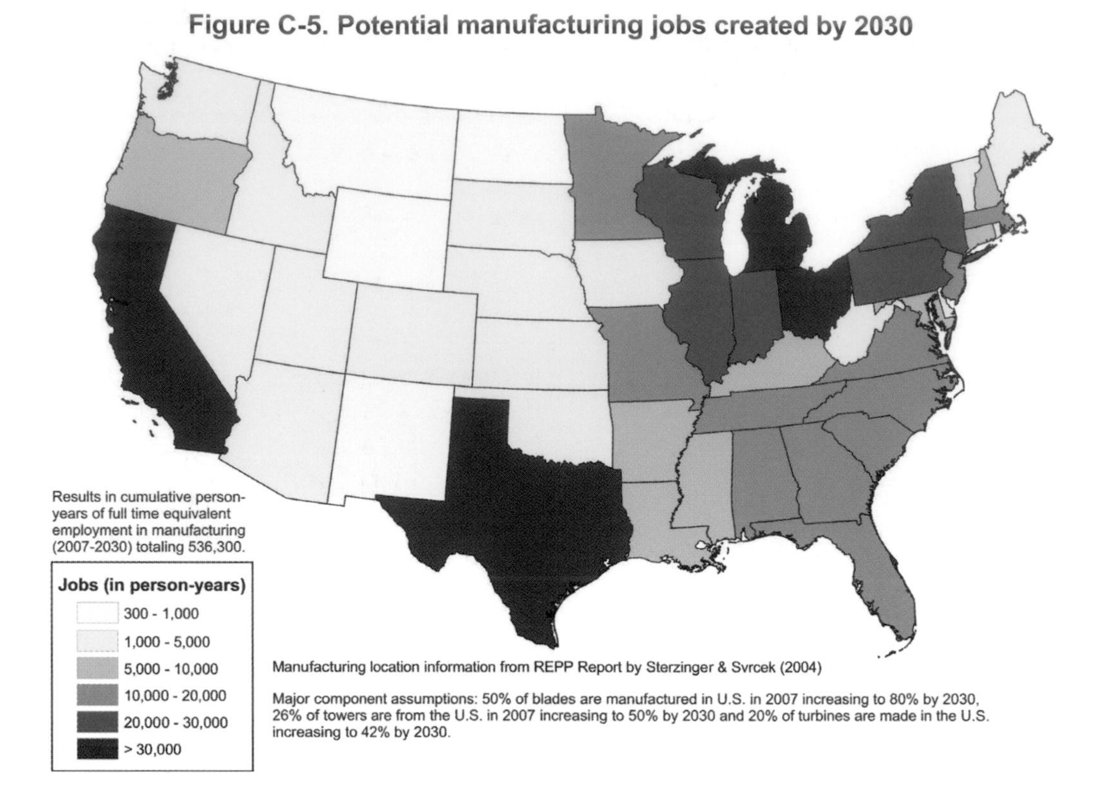

without a significant wind resource can be impacted economically from new manufacturing jobs (e.g., southeastern US).

C.5 Construction Sector

The year 2021 represents the height of the wind plant construction period, with 16.7 GW of wind having been brought online. In that year, more than 65,000 construction industry workers are assumed to be employed and $54.5 billion is generated in the U.S. economy from direct, indirect, and induced construction spending.

To reach the 20% Wind Scenario, today's wind power industry would have to grow from 9,000 annual construction jobs in 2007 to 65,000 new annual construction jobs in 2021. Construction jobs could be dispersed throughout the United States. Assuming the 16 GW/year capacity can be maintained into the future, including the replacement of outdated wind plants, the industry could maintain 20% electricity from wind as demand grows. In this scenario, the construction sector would experience the largest increase in jobs, followed by the operations sector, and then by the manufacturing sector. Figure C-6 shows the direct employment impact on the construction sector, the manufacturing sector and the operations sector (plant workers only).

Figure C-7 shows employment impacts during the same years, but adds the indirect and induced jobs. The bottom three bars (manufacturing, construction, and operations—including plant workers and other direct jobs) are direct jobs only. This chart depicts the large impact from the indirect and induced job categories, compared to the initial direct expenditures in the direct categories.

Figure C-6. Direct manufacturing, construction, and operations jobs supported by the 20% Wind Scenario

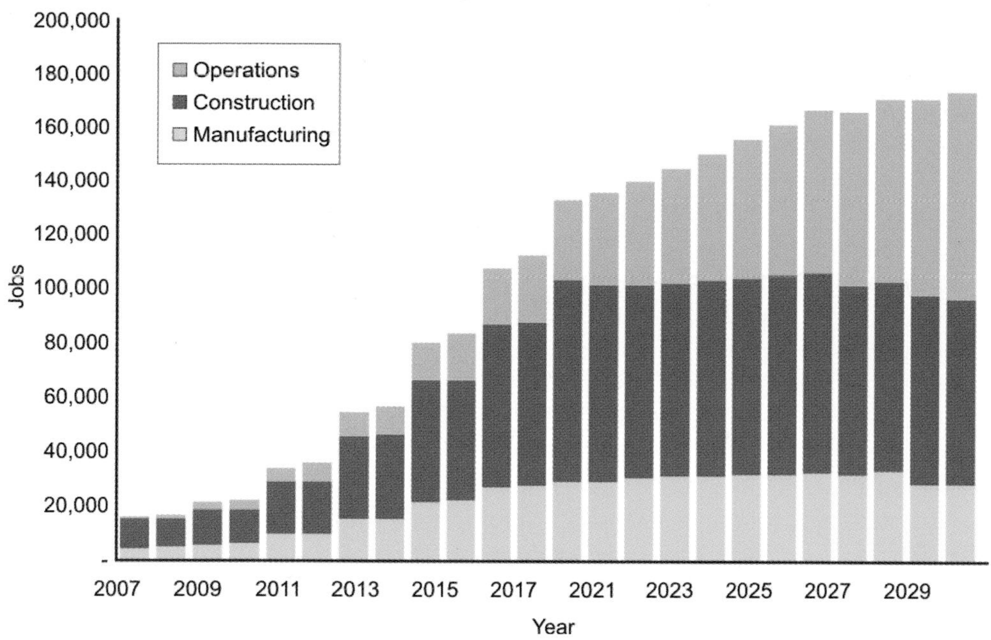

Figure C-7. Jobs per year from direct, indirect, and induced categories

In the last ten years of the scenario, the wind industry could support 500,000 jobs, including over 150,000 direct jobs.

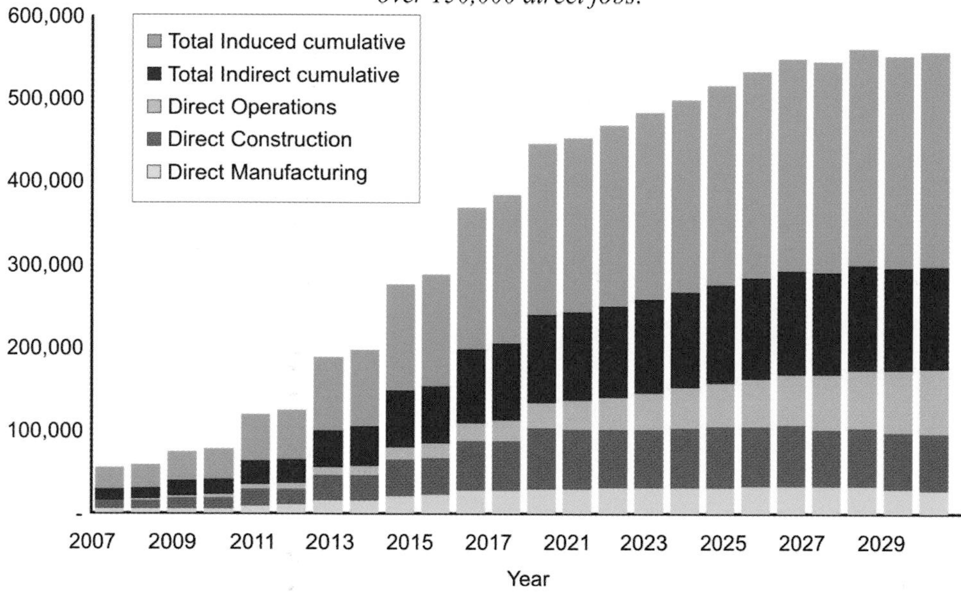

C.6 Operations Sector

JEDI predicts that in 2030, employment of more than 215,000 total operations workers (direct, indirect, and induced) will exist to maintain 293 GW of wind capacity. This includes more than 28,000 direct O&M jobs and 48,000 other direct jobs related to operating a wind plant (e.g., utility services and subcontractors). JEDI predicts that in 2030, land-based and offshore wind project operations will have a total economic impact of $27 billion. Operations employment would be dispersed across the country and is likely to be near wind installations. Rural Americans, in particular, could realize significant positive impacts from this scenario in the form of landowner payments and property taxes. Counties use property taxes to improve roads and schools, along with other vital infrastructure. More than $8.8 billion is estimated in property taxes and land lease payments between 2007 and 2030, which could be an important boost for rural communities.

Figure C-8 shows the results of JEDI analysis, performed on a state-by-state basis, in the form of impacts to each North American Electric Reliability Corporation (NERC) region. The individual state impacts were summed to calculate the NERC region impacts. These total impacts are lower than those from the JEDI analysis for the entire country because any job or dollar flowing out of state is considered monetary leakage (in the U.S. analysis, the model considers the whole country to be "local").

Figure C-8 shows jobs in job-years, which are FTE jobs counted in each year in which they exist. For example, if a maintenance worker holds one job for 20 years, this is shown as 20 job-years. For this figure, jobs during construction are assumed to last for one year. Jobs during the operations period are assumed to last for 20 years. Economic impacts are direct, indirect, and induced. Because it represents impacts from 305 GW of new wind starting in 2004 and ending in 2030, Figure C-8 shows three additional years when compared to other results.

Figure C-8. Jobs and economic impacts by NERC region

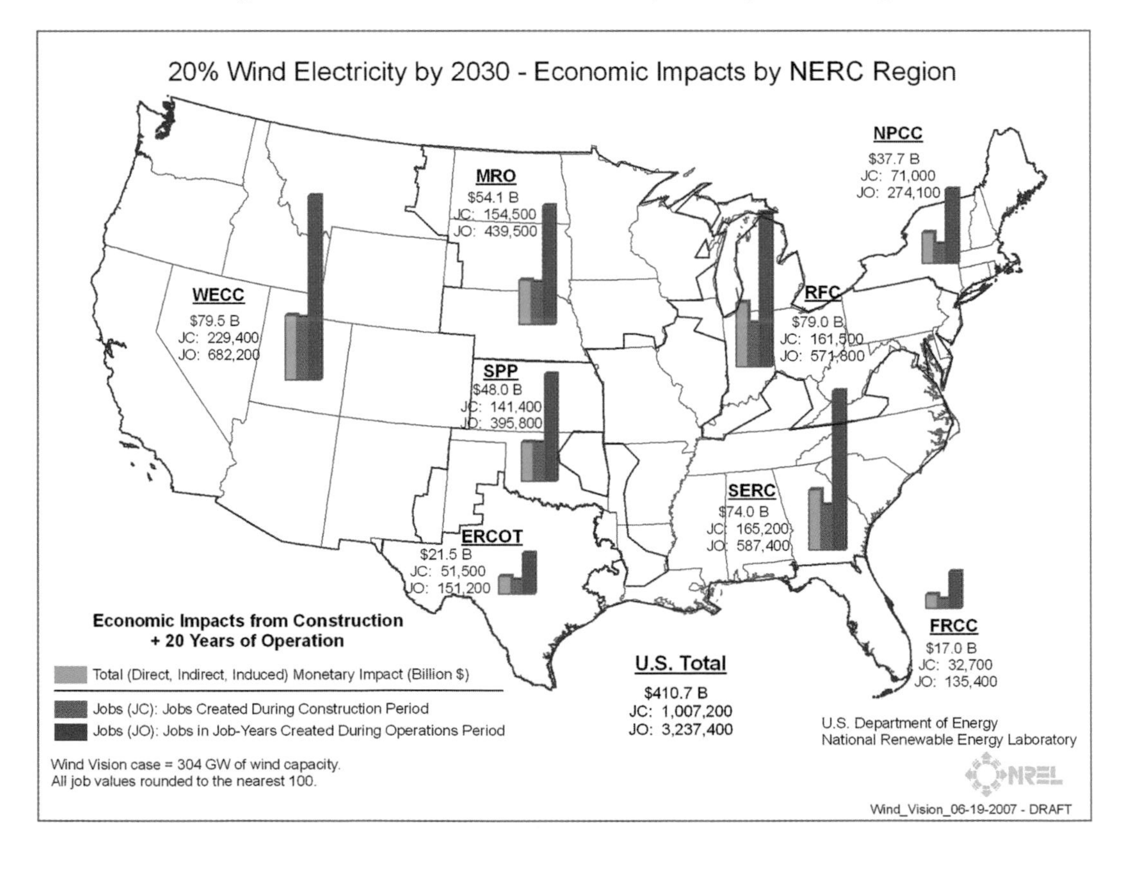

C.7 Conclusion

As a nation, the United States has made much progress recently in developing its wind resources. However, advancements in wind technologies and the projected increasing demand for electricity, will provide significant opportunities to further develop this domestic renewable resource. Actions toward this goal, as identified in the 20% Wind Scenario, offer residents and businesses in the rural and urban United States potential for economic development opportunities and potential for employment.

The United States is a prime location for developing wind resources and new wind manufacturing facilities. At the same time, relocating or expanding existing industries can give businesses opportunities to meet many of the material needs associated with wind technology manufacturing, installation, and facility operation.

In many areas of the country, renewable resources provide an opportunity to boost the local economy significantly. Wind plants offer employment during construction and continue to support permanent jobs during operation. Today, tax revenues from wind plants help to fund local schools, hospitals, and government services.

Based on the scenario presented in this report, a new and expanding wind manufacturing industry can meet 20% of our domestic electricity needs through 2030.

C.8 References

Laxson, A., M. Hand, and N. Blair. 2006. *High Wind Penetration Impact on U.S. Wind Manufacturing Capacity and Critical Resources.* NREL/TP-500-40482. Golden, CO: National Renewable Energy Laboratory.

MIG IMPLAN. "IMPLAN Professional Software." Stillwater, MN: Minnesota IMPLAN Group, Inc. (MIG) www.implan.com/software.html.

Black & Veatch. 2007 *20 % Wind Energy Penetration in the United States: A Technical Analysis of the Energy Resource.* Walnut Creek, CA

Sterzinger, G., and M. Svrcek. 2004. *Wind Turbine Development: Location of Manufacturing Activity.* Washington, DC: Renewable Energy Policy Project (REPP).

Appendix D. Lead Authors, Reviewers and Other Contributors

The U.S. Department of Energy would like to acknowledge the authors and reviewers listed below. This technical report is the culmination of contributions from more than 90 individuals and more than 50 organizations since June 2006. Their contributions and support were important throughout the development of this report. The final version of this document was prepared by the U.S. Department of Energy. Overall report reviewers included the U.S. Department of Energy, National Renewable Energy Laboratory, American Wind Energy Association, and other selected National Laboratory staff.

Report Lead Editors and Coordinators
Steve Lindenberg, U.S. Department of Energy, Brian Smith and Kathy O'Dell, National Renewable Energy Laboratory (NREL), Ed DeMeo, Renewable Energy Consulting Services*, (Team co-manager) and Bonnie Ram, Energetics Incorporated* (Team co-manager)

Report Production, Editing, and Graphic Images

Donna Heimiller	NREL, WinDS maps and graphics
Cliff Scher	Energetics Incorporated, Document Version Control
Russell Raymond	Energetics Incorporated, Document Version Control
Wendy Wallace	Energetics Incorporated, Document Version Control
Julie Chappell	Energetics Incorporated, Graphics Lead
Tommy Finamore	Energetics Incorporated, Cover Graphic Design
Susan Kaczmarek	Energetics Incorporated, Document Layout Coordinator
GE Energy	Cover graphic photographs

Members of the following advisory group supplied strategic guidance:

Rashid Abdul	Mitsubishi Power Systems
Stan Calvert	U.S. Department of Energy
Edgar DeMeo*	Renewable Energy Consulting Services, Inc.
Robert Gates	Clipper Windpower
Robert Gramlich	American Wind Energy Association
Thomas O. Gray	American Wind Energy Association
Steven Lindenberg	U.S. Department of Energy
James Lyons	GE Global Research
Brian McNiff	McNiff Light Industries
Bentham Paulos	Energy Foundation
Bonnie Ram*	Energetics Incorporated
Janet Sawin	Worldwatch Institute
Brian Smith	National Renewable Energy Laboratory
J. Charles Smith	Utility Wind Integration Group
Randall Swisher	American Wind Energy Association
Robert Thresher	National Renewable Energy Laboratory
James Walker	enXco

* Support provided under subcontract to the National Renewable Energy Laboratory

D

Chapter 1. Executive Summary and Overview of the 20% Wind Scenario

Elizabeth Salerno	American Wind Energy Association
Robert Gramlich	American Wind Energy Association
Alison Silverstein	Consultant
Paget Donnelly	Energetics Incorporated
Edgar DeMeo	Renewable Energy Consulting Services, Inc.
Larry Flowers	National Renewable Energy Laboratory
Thomas O. Gray	American Wind Energy Association
Maureen Hand	National Renewable Energy Laboratory
Bonnie Ram	Energetics Incorporated
Brian Smith	National Renewable Energy Laboratory

Chapter 2. Wind Turbine Technology

Michael Robinson*	National Renewable Energy Laboratory
Paul Veers	Sandia National Laboratories
Sandy Butterfield	National Renewable Energy Laboratory
Jim Greene	National Renewable Energy Laboratory
Walter Musial	National Renewable Energy Laboratory
Robert Thresher	National Renewable Energy Laboratory
Edgar DeMeo	Renewable Energy Consulting Services, Inc.
Robert Gramlich	American Wind Energy Association
Robert Poore	Global Energy Concepts, LLC
Scott Schreck	National Renewable Energy Laboratory
Alison Silverstein	Consultant
Brian Smith	National Renewable Energy Laboratory
James Walker	*enXco*
Lawrence Willey	GE Energy
Jose Zayas	Sandia National Laboratories
Rashid Abdul	Mitsubishi
Jim Ahlgrimm	U.S. Department of Energy
James Lyons	GE Global Research
Amir Mikhail	Clipper Windpower

Chapter 3. Manufacturing, Material and Resources

Lawrence Willey*	GE Energy
Corneliu Barbu	GE Energy (formerly)
Maureen Hand	National Renewable Energy Laboratory
Edgar DeMeo	Renewable Energy Consulting Services, Inc.
Kate Gordon	Apollo Alliance
Steve Lockard	TPI Composites
Brian O'Hanlon	U.S. Department of Commerce
Elizabeth Salerno	American Wind Energy Association
Brian Siu	Apollo Alliance
Brian Smith	National Renewable Energy Laboratory
Paul Veers	Sandia National Laboratories
James Walker	enXco

* Lead authors and advisors for each chapter are shown in **bold.** The final versions of the chapters are the sole responsibility of the U.S. Department of Energy. Task force members are underlined, and Task Force chairpersons are identified with an asterisk. Reviewers are shown in *italics*. Reviewers did not help to draft the chapters they reviewed. Their participation is not meant to imply that they or their respective organizations either agree or disagree with the findings of the effort.

Stephen Connors	Massachusetts Institute of Technology
Brian McNiff	McNiff Light Industries

Chapter 4. Transmission and Integration into the U.S. Electric System

J. Charles Smith*	Utility Wind Integration Group
Robert Gramlich	American Wind Energy Association
Mark Ahlstrom	WindLogics
Jeff Anthony	American Wind Energy Association
Jack Cadogan	U.S. DOE Retired
James Caldwell	Los Angeles Department of Water and Power
Henri Daher	National Grid USA
Edgar DeMeo	Renewable Energy Consulting Services, Inc.
Ken Donohoo	Electric Reliability Council of Texas
Abraham Ellis	Public Service Company of New Mexico
Douglas Faulkner	Puget Sound Energy
Robert Fullerton	Western Area Power Administration
Stephen Gehl	Electric Power Research Institute
Jay Godfrey	American Electric Power
John Holt	National Rural Electric Cooperative Association
Karen Hyde	Xcel Energy
Mike Jacobs	American Wind Energy Association
Brendan Kirby	Oak Ridge National Laboratory
Ronald L. Lehr	American Wind Energy Association
Charles Linderman	Edison Electric Institute
Michael Milligan	National Renewable Energy Laboratory
Dale Osborn	Midwest Independent System Operator
Philip Overholt	U.S. Department of Energy
Brian Parsons	National Renewable Energy Laboratory
Richard Piwko	GE Energy
Steve Ponder	Sierra Pacific Resources
Craig Quist	PacifiCorp
Kristine Schmidt	Xcel Energy
Matthew Schuerger	Energy Systems Consulting Services, LLC
Alison Silverstein	Consultant
Beth Soholt	Wind on the Wires
John Stough	AEP
Robert Thomas	Cornell University
Gary Thompson	Nebraska Public Power District
Robert Zavadil	EnerNex
Ellen Lutz	U.S. Department of Energy (formerly)

Chapter 5. Wind Power Siting and Environmental Effects

Laurie Jodziewicz*	American Wind Energy Association
Bonnie Ram	Energetics Inc.
James Walker	enXco
Wayne Walker	Horizon Wind Energy
Abby Arnold	Resolve
John Coequyt	Greenpeace
Edgar DeMeo	Renewable Energy Consulting Services, Inc.
Nathanael Greene	Natural Resources Defense Council

* Lead authors and advisors for each chapter are shown in **bold.** Task force members are underlined and Task Force chairpersons are identified with an asterisk. Reviewers are shown in *italics*.

Alan Nogee	Union of Concerned Scientists
Janet Sawin	Worldwatch Institute
Alison Silverstein	Consultant
Tom Weis	enXco Consultant
Katherine Kennedy	Natural Resources Defense Council (formerly)
Jim Lindsay	Florida Power & Light Company
Laura Miner	U.S. Department of Energy (formerly)
Robert Thresher	National Renewable Energy Laboratory

Chapter 6. Wind Power Markets

The contributions of this Task Force (Markets and Stakeholders) spanned a broad spectrum of issues, and are reflected in many of the chapters in this report.

Larry Flowers*	National Renewable Energy Laboratory
Ronald L. Lehr	American Wind Energy Association
David Olsen	Center for Energy Efficiency and Renewable Technologies
Brent Alderfer	Community Energy
Jeff Anthony	American Wind Energy Association
Lori Bird	National Renewable Energy Laboratory
Lisa Daniels	Windustry
Trudy Forsyth	National Renewable Energy Laboratory
Robert Gough	Intertribal Council on Utility Policy
Steven Lindenberg	U.S. Department of Energy
Walter Musial	National Renewable Energy Laboratory
Kevin Rackstraw	Clipper Windpower
Roby Robichaud	U.S. Department of Energy
Susan Sloan	American Wind Energy Association
Tom Wind	Wind Utility Consulting
Bob Anderson	Bob Anderson Consulting
Ruth Baranowski	National Renewable Energy Laboratory
Edgar DeMeo	Renewable Energy Consulting Services, Inc.
Robert Fullerton	Western Area Power Administration
Robert Gramlich	American Wind Energy Association
Karen Hyde	Xcel Energy
Bonnie Ram	Energetics Incorporated
Kristine Schmidt	Xcel Energy
Michael Skelley	Horizon Wind Energy
Brian Smith	National Renewable Energy Laboratory
Dennis Lin	U.S. Department of Energy
Roby Roberts	Goldman Sachs
Wayne Walker	Horizon Wind Energy

D

* Lead authors and advisors for each chapter are shown in **bold.** Task force members are underlined and Task Force chairpersons are identified with an asterisk. Reviewers are shown in *italics*.

Appendices A, B, and C and Supporting Analysis Task Force

Maureen Hand[*] (A, B, C)	National Renewable Energy Laboratory
Nate Blair (A, B)	National Renewable Energy Laboratory
Suzanne Tegen (C)	National Renewable Energy Laboratory
Mark Bolinger (A)	Lawrence Berkeley National Laboratory
Dennis Elliott (B)	National Renewable Energy Laboratory
Ray George (B)	National Renewable Energy Laboratory
Marshall Goldberg (C)	MRG Associates
Donna Heimiller (B)	National Renewable Energy Laboratory
Tracy Hern (A)	Western Resource Advocates
Bart Miller (A)	Western Resource Advocates
Ric O'Connell (A, B)	Black & Veatch
Marc Schwartz (B)	National Renewable Energy Laboratory
Ryan Wiser (A)	Lawrence Berkeley National Laboratory
Jeff Anthony	American Wind Energy Association
Steven Clemmer	Union of Concerned Scientists
Edgar DeMeo	Renewable Energy Consulting Services, Inc.
Robert Gramlich	American Wind Energy Association
Christopher Namovicz	U.S. DOE Energy Information Administration
Elizabeth Salerno	American Wind Energy Association
Alison Silverstein	Consultant
Brian Smith	National Renewable Energy Laboratory
Ian Baring-Gould	National Renewable Energy Laboratory
Jack Cadogan	U.S. DOE Retired
Eric Gebhardt	GE Energy
Gary Jordan	GE Energy
Brian Parsons	National Renewable Energy Laboratory
Ryan Pletka	Black & Veatch
Walter Short	National Renewable Energy Laboratory
Martin Tabbita	GE Energy
Hanson Wood	enXco
Michael DeAngelis	Sacramento Municipal Utility District
Alejandro Moreno	U.S. Department of Energy

The **Communications and Outreach Task Force** advised on outreach strategy and facilitated engagement of key stakeholders. Members of this task force include:

Mary McCann-Gates[*]	Clipper Windpower
Jill Pollyniak[*]	Clipper Windpower
Thomas O. Gray	American Wind Energy Association
Susan Williams Sloan	American Wind Energy Association
Peggy Welsh	Energetics Incorporated

D

Workshops and Outreach

Two strategic workshops took place during the course of this work. At the first of these, held August 17–18, 2006, attendees developed the initial statement of the 20% Wind Scenario and defined work plans. At the second, held November 9–10, 2006, participants shared and discussed preliminary results and obtained input from a group of invited individuals from key stakeholder sectors. Previously, these

[*] Lead authors and advisors for each chapter are shown in **bold.** Task force members are underlined and Task Force chairpersons are identified with an asterisk. Reviewers are shown in *italics*.

individuals had been external to the effort. Many of the authors, reviewers, and task force members listed in this appendix attended one or both of these workshops.

The invited participants at the November workshop brought along important feedback and perspectives from their respective sectors that have helped to shape this report. Some also reviewed sections of the report. Their participation is not meant to imply that they or their respective organizations either agree or disagree with the findings of the effort. These participants are listed below:

Aaron Brickman	U.S. Department of Commerce
Jennifer DeCesaro	Clean Energy Group
Michael Fry	American Bird Conservancy
Matt Gadow	DMI Industries
Stephen Gehl	Electric Power Research Institute
David Hamilton	Sierra Club
John Holt	National Rural Electric Cooperative Association
Robert Hornung	Canadian Wind Energy Association
Karen Hyde	Xcel Energy
Ed Ing	Law Office of Edwin T. C. Ing
Debra Jacobson	DJ Consulting
Miles Keogh	National Association of Regulatory Utility Commissioners
Charles Linderman	Edison Electric Institute
Steve Lockard	TPI Composites
Craig Mataczynski	Renewable Energy Systems Americas
Christopher Namovicz	U.S. Department of Energy, Energy Information Administration
Alan Nogee	Union of Concerned Scientists
Jim Presswood	Natural Resources Defense Council
Kristine Schmidt	Xcel Energy
Linda Silverman	U.S. Department of Energy
Brian Siu	Apollo Project
Kate Watson	Horizon Wind Energy

On November 28, 2006, a topical outreach workshop was held with representatives from nongovernmental organizations concerned about wildlife conservation and the environment. Participants discussed the early findings of the Environment and Siting Task Force and offered insights into issues important to their organizations. Workshop attendees are listed below. Their participation is not meant to imply that they or their respective organizations either agree or disagree with the findings of the effort.

Matthew Banks	World Wildlife Fund
Laura Bies	The Wildlife Society
Brent Blackwelder	Friends of the Earth
John Coequyt	Greenpeace
Amy Delach	Defenders of Wildlife
Tom Franklin	Izaak Walton League
Michael Fry	American Bird Conservancy
Robert Gramlich	American Wind Energy Association
Tony Iallonardo	National Audubon Society
Laurie Jodziewicz	American Wind Energy Association
Katie Kalinowski	Resolve/National Wind Coordinating Collaborative
Katherine Kennedy	Natural Resources Defense Council

D

Betsy Loyless	National Audubon Society
Laura Miner	U.S. Department of Energy
Amber Pairis	Association of Fish and Wildlife Agencies
Cliff Scher	Energetics Incorporated
Kate Smolski	Greenpeace
Robert Thresher	National Renewable Energy Laboratory
James Walker	enXco
Wayne Walker	Horizon Wind Energy
Tim Warman	National Wildlife Federation
Tom Weis	enXco
Peggy Welsh	Energetics Incorporated
Marchant Wentworth	Union of Concerned Scientists

D

D

Appendix E. Glossary

Area control error (ACE): The instantaneous difference between net actual and scheduled interchange, taking into account the effects of frequency deviations.

Balancing area (balancing authority area): The collection of generation, transmission, and loads within the metered boundaries of the balancing authority. The balancing authority maintains load-resource balance within this area.

Before-and-after control impact (BACI): A schematic method used to trace environmental effects from substantial anthropogenic changes to the environment. The overall aim of the method is to estimate the state of the environment before and after any change and the specific objectives is to compare changes at reference sites (or control sites) with the actual area of impact.

Bus: An electrical conductor that serves as a common connection for two or more electrical circuits.

Bus-bar: The point at which power is available for transmission.

Cap and trade: An established policy tool that creates a marketplace for emissions. Under a cap and trade program, the government regulates the aggregate amount of a type of emissions by setting a ceiling or cap. Participants in the program receive allocated allowances that represent a certain amount of pollutant and must purchase allowances from other businesses to emit more than their given allotment.

Capability: The maximum load that a generating unit, generating station, or other electrical apparatus can carry under specified conditions for a given period of time without exceeding approved limits of temperature and stress.

Capacity: The amount of electrical power delivered or required for which manufacturers rate a generator, turbine, transformer, transmission circuit, station, or system.

Capacity factor (CF): A measure of the productivity of a power plant, calculated as the amount of energy that the power plant produces over a set time period, divided by the amount of energy that would have been produced if the plant had been running at full capacity during that same time interval. Most wind power plants operate at a capacity factor of 25% to 40%.

Capacity penetration: The ratio of the nameplate rating of the wind plant capacity to the peak load. For example, if a 300-megawatt (MW) wind plant is operating in a zone with a 1,000 MW peak load, the capacity penetration is 30%. The capacity penetration is related to the energy penetration by the ratio of the system load factor to the wind plant capacity factor. For example, say that the system load factor is 60% and the wind plant capacity factor is 40%. In this case, and with an energy penetration of 20%, the capacity penetration would be 20% × 0.6/0.4, or 30%.

E

Capital costs: The total investment cost for a power plant, including auxiliary costs.

Carbon dioxide (CO_2): A colorless, odorless, noncombustible gas present in the atmosphere. It is formed by the combustion of carbon and carbon compounds (such as fossil fuels and biomass); by respiration, which is a slow form of combustion in animals and plants; and by the gradual oxidation of organic matter in the soil. CO_2 is a greenhouse gas that contributes to global climate change.

Carbon monoxide (CO): A colorless, odorless, but poisonous combustible gas. Carbon monoxide is produced during the incomplete combustion of carbon and carbon compounds, such as the fossil fuels coal and petroleum.

Circuit: An interconnected system of devices through which electrical current can flow in a closed loop.

Competitive Renewable Energy Zones (CREZ): A mechanism of the renewable portfolio standard in Texas designed to ensure that the electricity grid is extended to prime wind energy areas. The designation of these areas directs the Electric Reliability Council of Texas to develop plans for transmission lines to these areas that will connect them with the grid. See also "Electric Reliability Council of Texas" and "renewable portfolio standard."

Conductor: The material through which electricity is transmitted, such as an electrical wire.

Conventional fuel: Coal, oil, and natural gas (fossil fuels); also nuclear fuel.

Cycle: In AC electricity, the current flows in one direction from zero to a maximum voltage, then back down to zero, then to a maximum voltage in the opposite direction. This comprises one cycle. The number of complete cycles per second determines the frequency of the current. The standard frequency for AC electricity in the United States is 60 cycles.

Dispatch: The physical inclusion of a generator's output onto the transmission grid by an authorized scheduling utility.

Distribution: The process of distributing electricity. Distribution usually refers to the series of power poles, wires, and transformers that run between a high-voltage transmission substation and a customer's point of connection.

Effective load-carrying capability (ELCC): The amount of additional load that can be served at the target reliability level by adding a given amount of generation. For example, if adding 100 MW of wind could meet an increase of 20 MW of system load at the target reliability level, the turbine would have an ELCC of 20 MW, or a capacity value of 20% of its nameplate value.

Electricity generation: The process of producing electricity by transforming other forms or sources of energy into electrical energy. Electricity is measured in kilowatt-hours.

Electric Reliability Council of Texas (ERCOT): One of the 10 regional reliability councils of the North American Electric Reliability Council. ERCOT is a membership-based 501(c)(6) nonprofit corporation, governed by a board of directors and subject to oversight by the Public Utility Commission of Texas and the Texas Legislature. ERCOT manages the flow of electric power to approximately 20 million customers in Texas, representing 85% of the state's electric load and 75% of the Texas land area. See also "North American Electric Reliability Council."

Energy: The capacity for work. Energy can be converted into different forms, but the total amount of energy remains the same.

Energy penetration: The ratio of the amount of energy delivered from one type of resource to the total energy delivered. For example, if 200 megawatt-hours (MWh) of wind energy supplies 1,000 MWh of energy consumed, wind's energy penetration is 20%.

Externality: A consequence that accompanies an economic transaction, where that consequence affects others beyond the immediate economic actors and cannot be limited to those actors.

Feed-in law: A legal obligation on utilities to purchase electricity from renewable sources. Feed-in laws can also dictate the price that renewable facilities receive for their electricity.

Frequency: The number of cycles through which an alternating current passes per second, measured in hertz.

Gearbox: A system of gears in a protective casing used to increase or decrease shaft rotational speed.

Generator: A device for converting mechanical energy to electrical energy.

Gigawatt (GW): A unit of power, which is instantaneous capability, equal to one million kilowatts.

Gigawatt-hour (GWh): A unit or measure of electricity supply or consumption of one million kilowatts over a period of one hour.

Global warming: A term used to describe the increase in average global temperatures caused by the greenhouse effect.

Green power: A popular term for energy produced from renewable energy resources.

Greenhouse effect: The heating effect that results when long-wave radiation from the sun is trapped by greenhouse gases produced by natural and human activities.

Greenhouse gases (GHGs): Gases such as water vapor, CO_2, methane, and low-level ozone that are transparent to solar radiation, but opaque to long-wave radiation. These gases contribute to the greenhouse effect.

Grid: A common term that refers to an electricity transmission and distribution system. See also "power grid" and "utility grid."

Grid codes: Regulations that govern the performance characteristics of different aspects of the power system, including the behavior of wind plants during steady-state and dynamic conditions. These fundamentally technical documents contain the rules governing the operations, maintenance, and development of the transmission system and the coordination of the actions of all users of the transmission system.

Heat rate: A measure of the thermal efficiency of a generating station. Commonly stated as British thermal units (Btu) per kilowatt-hour. *Note:* Heat rates can be expressed as either gross or net heat rates, depending whether the electricity output is gross or net generation. Heat rates are typically expressed as net heat rates.

Instantaneous penetration: The ratio of the wind plant output to load at a specific point in time, or over a short period of time.

Investment tax credit (ITC): A tax credit that can be applied for the purchase of equipment such as renewable energy systems.

Kilowatt (kW): A standard unit of electrical power, which is instantaneous capability equal to 1,000 watts.

Kilowatt-hour (kWh): A unit or measure of electricity supply or consumption of 1,000 watts over a period of one hour.

Leading edge: The surface part of a wind turbine blade that first comes into contact with the wind.

Lift: The force that pulls a wind turbine blade.

Load (electricity): The amount of electrical power delivered or required at any specific point or points on a system. The requirement originates at the consumer's energy-consuming equipment.

Load factor: The ratio of the average load to peak load during a specified time interval.

Load following: A utility's practice in which more generation is added to available energy supplies to meet moment-to-moment demand in the utility's distribution system, or in which generating facilities are kept informed of load requirements. The goal of the practice is to ensure that generators are producing neither too little nor too much energy to supply the utility's customers.

Megawatt (MW): The standard measure of electricity power plant generating capacity. One megawatt is equal to 1,000 kilowatts or 1 million watts.

Megawatt-hour (MWh): A unit or energy or work equal to1,000 kilowatt-hours or 1 million watt-hours.

Met tower: A meteorological tower erected to verify the wind resource found within a certain area of land.

E

Modified Accelerated Cost Recovery System (MACRS): A U.S. federal system through which businesses can recover investments in certain property through depreciation deductions over an abbreviated asset lifetime. For solar, wind, and geothermal property placed in service after 1986, the current MACRS property class is five years. With the passage of the Energy Policy Act of 2005, fuel cells, microturbines, and solar hybrid lighting technologies became classified as five-year property as well.

Nacelle: The cover for the gearbox, drivetrain, and generator of a wind turbine.

Nameplate rating: The maximum continuous output or consumption in MW of an item of equipment as specified by the manufacturer.

Nondispatchable: The timing and level of power plant output generally cannot be closely controlled by the power system operator. Other factors beyond human control, such as weather variations, play a strong role in determining plant output.

Nitrogen oxides (NO_x): The products of all combustion processes formed by the combination of nitrogen and oxygen. NO_x and sulfur dioxide (SO_2) are the two primary causes of acid rain.

Power: The rate of production or consumption of energy.

Power grid: A common term that refers to an electricity transmission and distribution system. See also "utility grid."

Power marketers: Business entities engaged in buying and selling electricity. Power marketers do not usually own generating or transmission facilities, but take ownership of the electricity and are involved in interstate trade. These entities file with the Federal Energy Regulatory Commission (FERC) for status as a power marketer.

Power Purchase Agreement (PPA): A long-term agreement to buy power from a company that produces electricity.

Power quality: Stability of frequency and voltage and lack of electrical noise on the power grid.

Public Utility Commission: A governing body that regulates the rates and services of a utility.

Public Utility Regulatory Policies Act (PURPA) of 1978: As part of the National Energy Act, PURPA contains measures designed to encourage the conservation of energy, more efficient use of resources, and equitable rates. These measures included suggested retail rate reforms and new incentives for production of electricity by cogenerators and users of renewable resources.

Production tax credit (PTC): A U.S. federal, per-kilowatt-hour tax credit for electricity generated by qualified energy resources. Originally enacted as part of the Energy Policy Act of 1992, the credit expired at the end of 2001, was extended in March 2002, expired at the end of 2003, was renewed on October 4, 2004 and was then extended through December 31, 2008.

Radioactive waste: Materials remaining after producing electricity from nuclear fuel. Radioactive waste can damage or destroy living organisms if it is not stored safely.

Ramp rate: The rate at which load on a power plant is increased or decreased. The rate of change in output from a power plant.

Renewable energy: Energy derived from resources that are regenerative or that cannot be depleted. Types of renewable energy resources include wind, solar, biomass, geothermal, and moving water.

Regional Greenhouse Gas Initiative (RGGI): An agreement among 10 northeastern and mid-Atlantic states to reduce CO_2 emissions. Through the initiative, the states will develop a regional strategy to control GHGs. Fundamental to the agreement is the implementation of a multistate cap and trade program to induce a market-based emissions controlling mechanism.

Renewable energy credit (REC) or certificate: A mechanism created by a state statute or regulatory action to make it easier to track and trade renewable energy. A single REC represents a tradable credit for each unit of energy produced from qualified renewable energy facilities, thus separating the renewable energy's environmental attributes from its value as a commodity unit of energy. Under a REC regime, each qualified renewable energy producer has two income streams—one from the sale of the energy produced, and one from the sale of the RECs. The RECs can be sold and traded and their owners can legally claim to have purchased renewable energy.

Renewable portfolio standard (RPS): Under such a standard, a certain percentage of a utility's overall or new generating capacity or energy sales must be derived from renewable resources (e.g., 1% of electric sales must be from renewable energy in the year 200x). An RPS most commonly refers to electricity sales measured in megawatt-hours, as opposed to electrical capacity measured in megawatts.

Restructuring: The process of changing the structure of the electric power industry from a regulated guaranteed monopoly to an open competition among power suppliers.

Rotor: The blades and other rotating components of a wind turbine.

Solar energy: Electromagnetic energy transmitted from the sun (solar radiation).

Sulfur dioxide (SO_2): A colorless gas released as a by-product of combusted fossil fuels containing sulfur. The two primary sources of acid rain are SO_2 and NO_x.

Trade wind: The consistent system of prevailing winds occupying most of the tropics. Trade winds, which constitute the major component of the general circulation of the atmosphere, blow northeasterly in the northern hemisphere and southeasterly in the southern hemisphere. The trades, as they are sometimes called, are the most persistent wind system on Earth.

Turbine: A term used for a wind energy conversion device that produces electricity. See also "wind turbine."

Turbulence: A swirling motion of the atmosphere that interrupts the flow of wind.

Utility grid: A common term that refers to an electricity transmission and distribution system. See also "power grid."

Variable-speed wind turbines: Turbines in which the rotor speed increases and decreases with changing wind speeds. Sophisticated power control systems are required on variable-speed turbines to ensure that their power maintains a constant frequency compatible with the grid.

Volt (V): A unit of electrical force.

Voltage: The amount of electromotive force, measured in volts, between two points.

Watt (W): A unit of power.

Watt-hour (Wh): A unit of electricity consumption of one watt over the period of one hour.

Wind: Moving air. The wind's movement is caused by the sun's heat, the earth, and the oceans, which force air to rise and fall in cycles.

Wind energy: Energy generated by using a wind turbine to convert the mechanical energy of the wind into electrical energy. See also "wind power."

Wind generator: A wind energy conversion system designed to produce electricity.

Wind power: Power generated by using a wind turbine to convert the mechanical power of the wind into electrical power. See also "wind energy."

Wind power density: A useful way to evaluate the wind resource available at a potential site. The wind power density, measured in watts per square meter, indicates the amount of energy available at the site for conversion by a wind turbine.

Wind power class: A scale for classifying wind power density. There are seven wind power classes, ranging from 1 (lowest wind power density) to 7 (highest wind power density). In general, sites with a wind power class rating of 4 or higher are now preferred for large-scale wind plants.

Wind power plant: A group of wind turbines interconnected to a common utility system.

Wind resource assessment: The process of characterizing the wind resource and its energy potential for a specific site or geographical area.

Wind speed: The rate of flow of wind when it blows undisturbed by obstacles.

Wind speed profile: A profile of how the wind speed changes at different heights above the surface of the ground or water.

Wind turbine: A term used for a device that converts wind energy to electricity.

Wind turbine rated capacity: The amount of power a wind turbine can produce at its rated wind speed.

Windmill: A wind energy conversion system that is used primarily to grind grain. Windmill is commonly used to refer to all types of wind energy conversion systems.

Engineering

DE RE METALLICA, Georgius Agricola. The famous Hoover translation of greatest treatise on technological chemistry, engineering, geology, mining of early modern times (1556). All 289 original woodcuts. 638pp. $6^3/_4$ x 11. 0-486-60006 8

FUNDAMENTALS OF ASTRODYNAMICS, Roger Bate et al. Modern approach developed by U.S. Air Force Academy. Designed as a first course. Problems, exercises. Numerous illustrations. 455pp. $5^3/_8$ x $8^1/_2$. 0-486-60061-0

DYNAMICS OF FLUIDS IN POROUS MEDIA, Jacob Bear. For advanced students of ground water hydrology, soil mechanics and physics, drainage and irrigation engineering and more. 335 illustrations. Exercises, with answers. 784pp. $6^1/_8$ x $9^1/_4$. 0-486-65675-6

THEORY OF VISCOELASTICITY (SECOND EDITION), Richard M. Christensen. Complete consistent description of the linear theory of the viscoelastic behavior of materials. Problem-solving techniques discussed. 1982 edition. 29 figures. xiv+364pp. $6^1/_8$ x $9^1/_4$. 0-486-42880-X

MECHANICS, J. P. Den Hartog. A classic introductory text or refresher. Hundreds of applications and design problems illuminate fundamentals of trusses, loaded beams and cables, etc. 334 answered problems. 462pp. $5^3/_8$ x $8^1/_2$. 0-486-60754-2

MECHANICAL VIBRATIONS, J. P. Den Hartog. Classic textbook offers lucid explanations and illustrative models, applying theories of vibrations to a variety of practical industrial engineering problems. Numerous figures. 233 problems, solutions. Appendix. Index. Preface. 436pp. $5^3/_8$ x $8^1/_2$. 0-486-64785-4

STRENGTH OF MATERIALS, J. P. Den Hartog. Full, clear treatment of basic material (tension, torsion, bending, etc.) plus advanced material on engineering methods, applications. 350 answered problems. 323pp. $5^3/_8$ x $8^1/_2$. 0-486-60755-0

A HISTORY OF MECHANICS, René Dugas. Monumental study of mechanical principles from antiquity to quantum mechanics. Contributions of ancient Greeks, Galileo, Leonardo, Kepler, Lagrange, many others. 671pp. $5^3/_8$ x $8^1/_2$. 0-486-65632-2

STABILITY THEORY AND ITS APPLICATIONS TO STRUCTURAL MECHANICS, Clive L. Dym. Self-contained text focuses on Koiter postbuckling analyses, with mathematical notions of stability of motion. Basing minimum energy principles for static stability upon dynamic concepts of stability of motion, it develops asymptotic buckling and postbuckling analyses from potential energy considerations, with applications to columns, plates, and arches. 1974 ed. 208pp. $5^3/_8$ x $8^1/_2$. 0-486-42541-X

BASIC ELECTRICITY, U.S. Bureau of Naval Personnel. Originally a training course; best nontechnical coverage. Topics include batteries, circuits, conductors, AC and DC, inductance and capacitance, generators, motors, transformers, amplifiers, etc. Many questions with answers. 349 illustrations. 1969 edition. 448pp. $6^1/_2$ x $9^1/_4$. 0-486-20973-3

ROCKETS, Robert Goddard. Two of the most significant publications in the history of rocketry and jet propulsion: "A Method of Reaching Extreme Altitudes" (1919) and "Liquid Propellant Rocket Development" (1936). 128pp. $5\frac{3}{8} \times 8\frac{1}{2}$. 0-486-42537-1

STATISTICAL MECHANICS: PRINCIPLES AND APPLICATIONS, Terrell L. Hill. Standard text covers fundamentals of statistical mechanics, applications to fluctuation theory, imperfect gases, distribution functions, more. 448pp. $5\frac{3}{8} \times 8\frac{1}{2}$. 0-486-65390-0

ENGINEERING AND TECHNOLOGY 1650–1750: ILLUSTRATIONS AND TEXTS FROM ORIGINAL SOURCES, Martin Jensen. Highly readable text with more than 200 contemporary drawings and detailed engravings of engineering projects dealing with surveying, leveling, materials, hand tools, lifting equipment, transport and erection, piling, bailing, water supply, hydraulic engineering, and more. Among the specific projects outlined-transporting a 50-ton stone to the Louvre, erecting an obelisk, building timber locks, and dredging canals. 207pp. $8\frac{3}{8} \times 11\frac{1}{4}$. 0-486-42232-1

THE VARIATIONAL PRINCIPLES OF MECHANICS, Cornelius Lanczos. Graduate level coverage of calculus of variations, equations of motion, relativistic mechanics, more. First inexpensive paperbound edition of classic treatise. Index. Bibliography. 418pp. $5\frac{3}{8} \times 8\frac{1}{2}$. 0-486-65067-7

PROTECTION OF ELECTRONIC CIRCUITS FROM OVERVOLTAGES, Ronald B. Standler. Five-part treatment presents practical rules and strategies for circuits designed to protect electronic systems from damage by transient overvoltages. 1989 ed. xxiv+434pp. $6\frac{1}{8} \times 9\frac{1}{4}$. 0-486-42552-5

ROTARY WING AERODYNAMICS, W. Z. Stepniewski. Clear, concise text covers aerodynamic phenomena of the rotor and offers guidelines for helicopter performance evaluation. Originally prepared for NASA. 537 figures. 640pp. $6\frac{1}{8} \times 9\frac{1}{4}$. 0-486-64647-5

INTRODUCTION TO SPACE DYNAMICS, William Tyrrell Thomson. Comprehensive, classic introduction to space-flight engineering for advanced undergraduate and graduate students. Includes vector algebra, kinematics, transformation of coordinates. Bibliography. Index. 352pp. $5\frac{3}{8} \times 8\frac{1}{2}$. 0-486-65113-4

HISTORY OF STRENGTH OF MATERIALS, Stephen P. Timoshenko. Excellent historical survey of the strength of materials with many references to the theories of elasticity and structure. 245 figures. 452pp. $5\frac{3}{8} \times 8\frac{1}{2}$. 0-486-61187-6

ANALYTICAL FRACTURE MECHANICS, David J. Unger. Self-contained text supplements standard fracture mechanics texts by focusing on analytical methods for determining crack-tip stress and strain fields. 336pp. $6\frac{1}{8} \times 9\frac{1}{4}$. 0-486-41737-9

STATISTICAL MECHANICS OF ELASTICITY, J. H. Weiner. Advanced, self-contained treatment illustrates general principles and elastic behavior of solids. Part 1, based on classical mechanics, studies thermoelastic behavior of crystalline and polymeric solids. Part 2, based on quantum mechanics, focuses on interatomic force laws, behavior of solids, and thermally activated processes. For students of physics and chemistry and for polymer physicists. 1983 ed. 96 figures. 496pp. $5\frac{3}{8} \times 8\frac{1}{2}$. 0-486-42260-7

Physics

OPTICAL RESONANCE AND TWO-LEVEL ATOMS, L. Allen and J. H. Eberly. Clear, comprehensive introduction to basic principles behind all quantum optical resonance phenomena. 53 illustrations. Preface. Index. 256pp. 5⅜ x 8½.
0-486-65533-4

QUANTUM THEORY, David Bohm. This advanced undergraduate-level text presents the quantum theory in terms of qualitative and imaginative concepts, followed by specific applications worked out in mathematical detail. Preface. Index. 655pp. 5⅜ x 8½. 0-486-65969-0

ATOMIC PHYSICS (8th EDITION), Max Born. Nobel laureate's lucid treatment of kinetic theory of gases, elementary particles, nuclear atom, wave-corpuscles, atomic structure and spectral lines, much more. Over 40 appendices, bibliography. 495pp. 5⅜ x 8½. 0-486-65984-4

A SOPHISTICATE'S PRIMER OF RELATIVITY, P. W. Bridgman. Geared toward readers already acquainted with special relativity, this book transcends the view of theory as a working tool to answer natural questions: What is a frame of reference? What is a "law of nature"? What is the role of the "observer"? Extensive treatment, written in terms accessible to those without a scientific background. 1983 ed. xlviii+172pp. 5⅜ x 8½. 0-486-42549-5

AN INTRODUCTION TO HAMILTONIAN OPTICS, H. A. Buchdahl. Detailed account of the Hamiltonian treatment of aberration theory in geometrical optics. Many classes of optical systems defined in terms of the symmetries they possess. Problems with detailed solutions. 1970 edition. xv + 360pp. 5⅜ x 8½. 0-486-67597-1

PRIMER OF QUANTUM MECHANICS, Marvin Chester. Introductory text examines the classical quantum bead on a track: its state and representations; operator eigenvalues; harmonic oscillator and bound bead in a symmetric force field; and bead in a spherical shell. Other topics include spin, matrices, and the structure of quantum mechanics; the simplest atom; indistinguishable particles; and stationary-state perturbation theory. 1992 ed. xiv+314pp. 6⅛ x 9¼. 0-486-42878-8

LECTURES ON QUANTUM MECHANICS, Paul A. M. Dirac. Four concise, brilliant lectures on mathematical methods in quantum mechanics from Nobel Prize-winning quantum pioneer build on idea of visualizing quantum theory through the use of classical mechanics. 96pp. 5⅜ x 8½. 0-486-41713-1

THIRTY YEARS THAT SHOOK PHYSICS: THE STORY OF QUANTUM THEORY, George Gamow. Lucid, accessible introduction to influential theory of energy and matter. Careful explanations of Dirac's anti-particles, Bohr's model of the atom, much more. 12 plates. Numerous drawings. 240pp. 5⅜ x 8½. 0-486-24895-X

ELECTRONIC STRUCTURE AND THE PROPERTIES OF SOLIDS: THE PHYSICS OF THE CHEMICAL BOND, Walter A. Harrison. Innovative text offers basic understanding of the electronic structure of covalent and ionic solids, simple metals, transition metals and their compounds. Problems. 1980 edition. 582pp. 6⅛ x 9¼. 0-486-66021-4

CATALOG OF DOVER BOOKS

A TREATISE ON ELECTRICITY AND MAGNETISM, James Clerk Maxwell. Important foundation work of modern physics. Brings to final form Maxwell's theory of electromagnetism and rigorously derives his general equations of field theory. 1,084pp. $5\frac{3}{8}$ x $8\frac{1}{2}$. Two-vol. set. Vol. I: 0-486-60636-8 Vol. II: 0-486-60637-6

MATHEMATICS FOR PHYSICISTS, Philippe Dennery and Andre Krzywicki. Superb text provides math needed to understand today's more advanced topics in physics and engineering. Theory of functions of a complex variable, linear vector spaces, much more. Problems. 1967 edition. 400pp. $6\frac{1}{2}$ x $9\frac{1}{4}$. 0-486-69193-4

INTRODUCTION TO QUANTUM MECHANICS WITH APPLICATIONS TO CHEMISTRY, Linus Pauling & E. Bright Wilson, Jr. Classic undergraduate text by Nobel Prize winner applies quantum mechanics to chemical and physical problems. Numerous tables and figures enhance the text. Chapter bibliographies. Appendices. Index. 468pp. $5\frac{3}{8}$ x $8\frac{1}{2}$. 0-486-64871-0

METHODS OF THERMODYNAMICS, Howard Reiss. Outstanding text focuses on physical technique of thermodynamics, typical problem areas of understanding, and significance and use of thermodynamic potential. 1965 edition. 238pp. $5\frac{3}{8}$ x $8\frac{1}{2}$. 0-486-69445-3

THE ELECTROMAGNETIC FIELD, Albert Shadowitz. Comprehensive undergraduate text covers basics of electric and magnetic fields, builds up to electromagnetic theory. Also related topics, including relativity. Over 900 problems. 768pp. $5\frac{5}{8}$ x $8\frac{1}{4}$. 0-486-65660-8

GREAT EXPERIMENTS IN PHYSICS: FIRSTHAND ACCOUNTS FROM GALILEO TO EINSTEIN, Morris H. Shamos (ed.). 25 crucial discoveries: Newton's laws of motion, Chadwick's study of the neutron, Hertz on electromagnetic waves, more. Original accounts clearly annotated. 370pp. $5\frac{3}{8}$ x $8\frac{1}{2}$. 0-486-25346-5

EINSTEIN'S LEGACY, Julian Schwinger. A Nobel Laureate relates fascinating story of Einstein and development of relativity theory in well-illustrated, nontechnical volume. Subjects include meaning of time, paradoxes of space travel, gravity and its effect on light, non-Euclidean geometry and curving of space-time, impact of radio astronomy and space-age discoveries, and more. 189 b/w illustrations. xiv+250pp. $8\frac{3}{8}$ x $9\frac{1}{4}$. 0-486-41974-6

THE VARIATIONAL PRINCIPLES OF MECHANICS, Cornelius Lanczos. Philosophic, less formalistic approach to analytical mechanics offers model of clear, scholarly exposition at graduate level with coverage of basics, calculus of variations, principle of virtual work, equations of motion, more. 418pp. $5\frac{3}{8}$ x $8\frac{1}{2}$. 0-486-65067-7

Paperbound unless otherwise indicated. Available at your book dealer, online at **www.doverpublications.com**, or by writing to Dept. GI, Dover Publications, Inc., 31 East 2nd Street, Mineola, NY 11501. For current price information or for free catalogues (please indicate field of interest), write to Dover Publications or log on to **www.doverpublications.com** and see every Dover book in print. Dover publishes more than 400 books each year on science, elementary and advanced mathematics, biology, music, art, literary history, social sciences, and other areas.